Solving Problems with NMR Spectroscopy

Solving Problems with NMR Spectroscopy

Second Edition

Atta-ur-Rahman
International Center for Chemical and Biological Sciences
(H. E. J. Research Institute of Chemistry and Dr. Panjwani Center
for Molecular Medicine and Drug Research), University of Karachi,
Karachi-75270, Pakistan

Muhammad Iqbal Choudhary
International Center for Chemical and Biological Sciences
(H. E. J. Research Institute of Chemistry and Dr. Panjwani Center
for Molecular Medicine and Drug Research), University of Karachi,
Karachi-75270, Pakistan

Atia-tul-Wahab
Dr. Panjwani Center for Molecular Medicine and Drug Research
(International Center for Chemical and Biological Sciences),
University of Karachi, Karachi-75270, Pakistan

AMSTERDAM • BOSTON • HEIDELBERG • LONDON
NEW YORK • OXFORD • PARIS • SAN DIEGO
SAN FRANCISCO • SINGAPORE • SYDNEY • TOKYO
Academic Press is an Imprint of Elsevier

Academic Press is an imprint of Elsevier
125, London Wall, EC2Y 5AS, UK
525 B Street, Suite 1800, San Diego, CA 92101-4495, USA
225 Wyman Street, Waltham, MA 02451, USA
The Boulevard, Langford Lane, Kidlington, Oxford OX5 1GB, UK

Notices
Knowledge and best practice in this field are constantly changing. As new research and experience broaden our understanding, changes in research methods, professional practices, or medical treatment may become necessary.

Practitioners and researchers must always rely on their own experience and knowledge in evaluating and using any information, methods, compounds, or experiments described herein. In using such information or methods they should be mindful of their own safety and the safety of others, including parties for whom they have a professional responsibility.

To the fullest extent of the law, neither the Publisher nor the authors, contributors, or editors, assume any liability for any injury and/or damage to persons or property as a matter of products liability, negligence or otherwise, or from any use or operation of any methods, products, instructions, or ideas contained in the material herein.

British Library Cataloguing-in-Publication Data
A catalogue record for this book is available from the British Library

Library of Congress Cataloging-in-Publication Data
A catalog record for this book is available from the Library of Congress

ISBN: 978-0-12-411589-7

For information on all Academic Press publications
visit our website at http://store.elsevier.com/

Working together
to grow libraries in
developing countries

www.elsevier.com • www.bookaid.org

Table of Contents

Foreword

The second edition of the book *Solving Problems with NMR Spectroscopy* is aimed to strengthen the understanding of how an NMR spectrometer functions. This revised version of the book takes the same problem-solving approach as the highly praised first edition, published in 1996. The book focuses on describing the basic principles of NMR spectroscopy and explains in detail the functioning of an NMR spectrometer. The optimum use of this powerful technique is introduced step by step, and common problems encountered by the practitioners and users of NMR spectroscopy are described in an easy-to-understand manner. The real strength of the book is its highly practical approach in describing both the concepts and applications of NMR spectroscopy.

The second edition introduces a number of new topics, including developments in NMR hardware, such as cryogenically cooled probes, new probeheads, high-field magnets, and DNP–NMR, as well as innovative pulse sequences, such as DOSY, concatenated NMR techniques, and PANSY. Particularly interesting is a new chapter on sensitivity issues in NMR spectroscopy and their currently available applications, which have driven most of the developments in this field. Another chapter on recent developments in NMR spectroscopy updates the readers about the changing landscape in this field. Over 180 penetrating problems and their well-described solutions help to reinforce and test the understanding of the readers about various aspects of modern NMR spectroscopy. Many of these problems focus on developing the interpretation skills of the readers in various types of NMR spectra toward structure determination. The use of color printing and improved figures enhance the readability of the text.

The revised edition of *Solving Problems with NMR Spectroscopy* by Atta-ur-Rahman, M. Iqbal Choudhary, and Atia-tul-Wahab is certainly a very useful addition to the NMR literature. I am confident that the book will receive wide appreciation both from students as well as professionals.

Professor Dr. Richard R. Ernst
Nobel Prize in Chemistry, 1991
Zurich, 2015

Chapter 1

The Basics of Modern NMR Spectroscopy

1.1 WHAT IS NMR?

Nuclear magnetic resonance (NMR) spectroscopy is the study of molecules by recording the interaction of radiofrequency (Rf) electromagnetic radiations with the nuclei of molecules placed in a strong magnetic field. Zeeman first observed the strange behavior of certain nuclei when subjected to a strong magnetic field at the end of the nineteenth century, but practical use of the so-called "Zeeman effect" was made only in the 1950s when NMR spectrometers became commercially available.

Like all other spectroscopic techniques, NMR spectroscopy involves the interaction of the material being examined with electromagnetic radiation. Why do we use the word "electromagnetic radiation"? This is so because each ray of light (or any other type of electromagnetic radiation) can be considered to be a sine wave that is made up of two mutually perpendicular sine waves that are exactly in phase with each other, i.e., their maxima and minima occur at exactly the same point of line. One of these two sine waves represents an oscillatory *electric field*, while the second wave (that oscillates in a plane perpendicular to the first wave) represents an oscillating magnetic field – hence the term "*electromagnetic*" radiation.

Cosmic rays, which have a very high frequency (and a short wavelength), fall at the highest energy end of the known electromagnetic spectrum and involve frequencies greater than 3×10^{20} Hz. Radiofrequency (Rf) radiation, which is

Solving Problems with NMR Spectroscopy. http://dx.doi.org/10.1016/B978-0-12-411589-7.00001-2

TABLE 1.1 The Electromagnetic Spectrum

Radiation	Wavelength (nm) λ	Frequency (Hz) v	Energy (kJ mol^{-1})
Cosmic rays	$<10^{-3}$	$>3 \times 10^{20}$	$>1.2 \times 10^8$
Gamma rays	10^{-1} to 10^{-3}	3×10^{18} to 3×10^{20}	1.2×10^6 to 1.2×10^8
X-rays	10 to 10^{-1}	3×10^{16} to 3×10^{18}	1.2×10^4 to 1.2×10^6
Far ultraviolet rays	200 to 10	1.5×10^{15} to 3×10^{16}	6×10^2 to 1.2×10^4
Ultraviolet rays	380 to 200	8×10^{14} to 1.5×10^{15}	3.2×10^2 to 6×10^2
Visible light	780 to 380	4×10^{14} to 8×10^{14}	1.6×10^2 to 3.2×10^2
Infrared rays	3×10^4 to 780	10^{13} to 4×10^{14}	4 to 1.6×10^2
Far infrared rays	3×10^5 to 3×10^4	10^{12} to 10^{13}	0.4 to 4
Microwaves	3×10^7 to 3×10^5	10^{10} to 10^{12}	4×10^{-3} to 0.4
Radiofrequency (Rf) waves	10^{11} to 3×10^7	10^6 to 10^{10}	4×10^{-7} to 4×10^{-3}

the type of radiation that concerns us in NMR spectroscopy, occurs at the other (the lowest energy) end of the electromagnetic spectrum and involves energies of the order of 100 MHz (1 MHz = 10^6 Hz). Gamma rays, X-rays, ultraviolet rays, visible light, infrared rays, microwaves and radiofrequency waves all fall between these two extremes. The various types of radiations and the corresponding ranges of wavelength, frequency, and energy are presented in Table 1.1.

Electromagnetic radiation also exhibits behavior characteristic of particles, in addition to its wave-like character. Each quantum of radiation is called a *photon*, and each photon possesses a discrete amount of energy, which is directly proportional to the frequency of the electromagnetic radiation. The strength of a chemical bond is typically around 400 kJ mol^{-1}, so that only radiations above the visible region will be capable of breaking bonds. But infrared, microwaves, and radio-frequency radiations will not be able to do so.

Let us now consider how electromagnetic radiation can interact with a particle of matter. Quantum mechanics (the field of physics dealing with energy at the atomic level) stipulates that in order for a particle to absorb a photon of electromagnetic radiation, the particle must first exhibit a uniform periodic motion with a frequency that exactly matches the frequency of the absorbed radiation. When these two frequencies exactly match, the electromagnetic fields can "*constructively*" interfere with the oscillations of the particle. The system is then said to be "*in resonance*" and absorption of *Rf* energy can take place. Nuclear magnetic resonance involves the immersion of nuclei in a magnetic field, and then matching the frequency at which they are precessing with electromagnetic radiation of exactly the same frequency so that energy absorption can occur.

1.1.1 The Birth of a Signal

Certain nuclei, such as ^{1}H, ^{2}H, ^{13}C, ^{15}N, and ^{19}F, possess a spin angular momentum and hence a corresponding magnetic moment μ, given by

$$\mu = \frac{\gamma h [I(I+1)]^{1/2}}{2\pi} \tag{1.1}$$

where h is Planck's constant and γ is the magnetogyric ratio (also called gyromagnetic ratio). When such nuclei are placed in a magnetic field B_0, applied along the z-axis, they can adopt one of $2I + 1$ quantized orientations, where I is the spin quantum number of the nucleus (Fig. 1.1). Each of these orientations corresponds to a certain energy level:

$$E = -\mu_z B_0 = -\frac{m_1 \gamma h B_0}{2\pi} \tag{1.2}$$

where m_1 is the magnetic quantum number of the nucleus and μ_z is the magnetic moment. In the lowest energy orientation, the magnetic moment of the nucleus is most closely aligned with the external magnetic field (B_0), while in the highest energy orientation it is least closely aligned with the external field. Organic chemists are most frequently concerned with ^{1}H and ^{13}C nuclei, both of which have a spin quantum number (I) of 1/2, and only two quantized orientations are therefore allowed, in which the nuclei are either aligned parallel to the applied field (lower energy orientation) or antiparallel to it (higher energy orientation). The nuclei with only two quantized orientations are called dipolar nuclei.

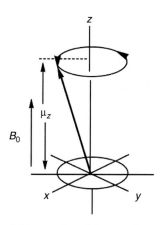

FIGURE 1.1 Representation of the precession of the magnetic moment about the axis of the applied magnetic field, B_0. The magnitude μ_z, of the vector corresponds to the Boltzmann excess in the lower energy (α) state.

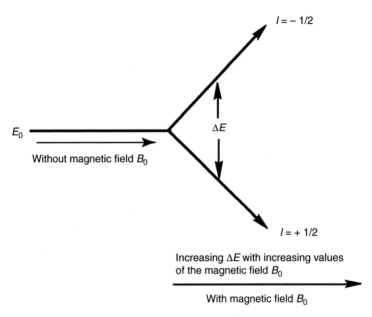

$I = -1/2$

E_0

ΔE

Without magnetic field B_0

$I = +1/2$

Increasing ΔE with increasing values
of the magnetic field B_0

With magnetic field B_0

FIGURE 1.2 The energy difference between the two energy states ΔE increases with increasing value of the applied magnetic field B_0, with a corresponding increase in sensitivity.

Transitions from the lower energy level to the higher energy level can occur by absorption of radiofrequency radiation of the correct frequency. The energy difference ΔE between these energy levels is proportional to the external magnetic field (Fig. 1.2), as defined by the equation $\Delta E = \gamma h B_0/2\pi$. In frequency terms, this energy difference corresponds to

$$v_0 = \frac{\gamma B_0}{2\pi} \tag{1.3}$$

Before being placed in a magnetic field, the nucleus is spinning on its axis, which is stationary. The external magnetic field (like that generated by the NMR magnet) causes the spinning nucleus to exhibit a characteristic wobbling motion (precession) often compared to the movement of a gyroscopic top before it topples, when the two ends of its axis no longer remain stationary but trace circular paths in opposite directions (Fig. 1.3). If a radiofrequency field is now applied in a direction perpendicular to the external magnetic field and at a frequency that exactly matches the precessional frequency ("Larmor" frequency) of the nucleus, absorption of energy will occur and the nucleus will suddenly "flip" from its lower energy orientation (in which its magnetic moment was processing in a direction aligned with the external magnetic field) to the higher energy orientation, in which it is aligned in the opposite direction. It can then relax back to the lower energy state through *spin-lattice relaxation* (T_1) by transfer of energy to the assembly of surrounding molecules ("lattice"), or by

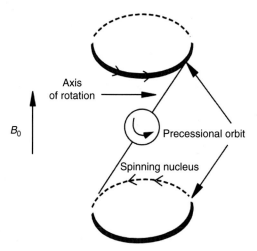

Axis
of rotation

B_0

Precessional orbit

Spinning nucleus

FIGURE 1.3 Precessional or Larmor motion of an NMR active nucleus in magnetic field B_0. Every nucleus has an inherently different range of precession frequencies, depending on its magnetogyric ratio (γ).

spin-spin relaxation (T_2), involving transfer of energy to a neighboring nucleus. The change in the impedance of the oscillator coils caused by the relaxation is measured by the detector as a signal in the form of a decaying beat pattern, known as a *free induction decay* (FID) (Fig. 1.4), which is stored in the computer memory and converted by a mathematical operation known as Fourier transformation to the conventional NMR spectrum.

Thus, excitations caused by absorption of radiofrequency energy cause nuclei to migrate to a higher energy level, while relaxations cause them to flip back to the lower energy level, and an equilibrium state is soon established. Interestingly, the interaction with the radiofrequency causes certain nuclei to excite to the higher energy state(s) and others to fall back to the lower energy state(s). This relaxation process is termed as *induced relaxation* (see the Glossary section), which is different from spontaneous relaxation recorded as an FID (Section 2.1.3). It is the net *Rf* absorption due to the difference in the populations in the two states which leads to the NMR signal.

In nuclei with positive magnetogyric ratios, such as ^{1}H or ^{13}C, the lower energy state will correspond to the +1/2 state, and the higher energy state to the −1/2 state, but in nuclei with negative magnetogyric states, for example, ^{29}Si or ^{15}N, the opposite will be true.

Magnetogyric ratio (γ) is not a "magic number." It is a measurable quantity for any charged particle (in case of NMR it is a rotating nucleus). Equation 1.4 is used for the measurement of magnetogyric ratio (γ):

$$\gamma = \frac{q}{2m}$$

(1.4)

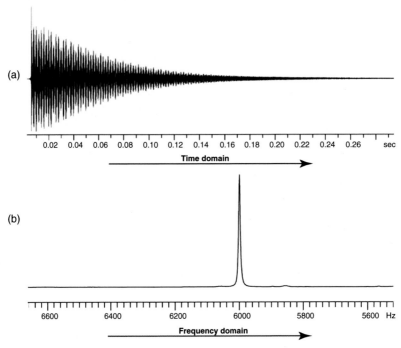

FIGURE 1.4 (a) Free induction decay (FID) in the time domain. (b) Fourier transformation of the time domain signal yields the conventional frequency domain spectrum.

where q is the charge and m is the mass of the charged particle. The magnetogyric ratios of some important nuclei are given in Table 1.2 (Harris, 1989).

If the populations of the upper and lower energy states were equal, then no energy difference between the two states of the nucleus (in its parallel and antiparallel orientations) would exist and no NMR signal would be observed. However, at equilibrium there is a slight excess ("Boltzmann excess") of nuclei

TABLE 1.2 Magnetogyric Ratios of Some Important NMR-Active Nuclei

Nucleus	Magnetogyric Ratio, γ (10^7 Rad T^{-1} s^{-1})	Larmor Frequency, ν_0 (MHz) for $B_0 = 7.0461$ T
^1H	26.7520	300.000
^{13}C	6.7283	75.435
^{15}N	−2.712	30.410
^{19}F	25.181	282.282
^{29}Si	−5.3188	59.601
^{31}P	10.841	121.442

(a)   (b)

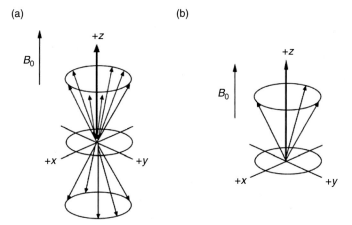

FIGURE 1.5   (a) Vector representation displaying a greater number of spins aligned with the magnetic field B_0. (b) Excess spin population (Boltzmann distribution excess) aligned with B_0 results in a bulk magnetization vector in the $+z$ direction.

in the lower energy (α) state as compared to the upper energy (β) state, and it is this difference in the populations of the two levels that is responsible for the NMR signal (Fig. 1.5). The ratio of the populations between the two states is given by the Boltzmann equation:

$$\frac{N_\beta}{N_\alpha} = \exp\left(\frac{-\Delta E}{kT}\right) \tag{1.5}$$

where N_α is the population of the lower energy state, N_β is the population of the upper energy state k is the Boltzmann constant and T is the temperature.

On a 100 MHz instrument, if there are a million nuclei in the lower energy level, there will be 999,987 in the upper energy level, yielding only a tiny excess of 13 nuclei in the lower energy state. It is this tiny excess that is detected by the NMR spectrometer as a signal. Since the signal intensity is dependent on the population difference between nuclei in the upper and lower energy states, and since the population difference depends on the strength of the applied magnetic field (B_0), the signal intensities will be significantly higher on instruments with more powerful magnets. Nuclear Overhauser enhancement (Section 6.2), polarization transfer (Section 4.2), or most recently dynamic nuclear polarization (DNP) techniques (Section 3.6.1) can also be employed to enhance the population of the ground state over that of the upper higher energy state to obtain a more intense signal.

Problem 1.1

Why are nuclei with odd atomic mass or number generally NMR active?

Problem 1.2

Is it correct that practically every element in the periodic table can be analyzed by NMR spectroscopy?

Problem 1.3

What is a nuclear spin?

Problem 1.4

What is meant by the relaxation time?

Problem 1.5

What is induced relaxation and how does it contribute in understanding the population difference between lower (α) and upper (β) energy states?

Problem 1.6

What will happen if the radiofrequency pulse is applied for an unusually long time?

Problem 1.7

From the discussion in Section 1.1, can you summarize the factors affecting the population difference between the lower energy state (N_α) and the upper energy state (N_β). How is the population difference related to the NMR signal strength?

Problem 1.8

As mentioned in the text, there is only a slight excess of nuclei in the ground state (about 13 protons in a million protons at 100 MHz). Would you expect that in the case of a ^{13}C-NMR experiment, the same population difference will prevail?

Problem 1.9

Explain what is meant by the Larmor frequency and what is its importance in an NMR experiment?

Problem 1.10

What are the factors on which Larmor frequency depends, and what does that means in terms of selective detection of one type of nucleus in the presence of another type (e.g., ^{1}H in the presence of ^{13}C or *vice versa* on a 9.4 T or 400 MHz NMR spectrometer).

Problem 1.11

How magnetogyric ratios (γ) of ^{1}H, ^{13}C, ^{15}N, and ^{2}H (deuterium) relate with their Larmor frequencies (v)?

What is "magnetogyric ratio" of a nucleus and how does it affect (1) the energy difference between two states, and (2) the sensitivity of the nuclear species to the NMR experiment?

1.2 INSTRUMENTATION

NMR spectrometers have improved significantly, particularly in the last few decades, with the development of very stable superconducting magnets and of computers that allow measurements over long time periods under homogeneous field conditions. Repetitive scanning and signal accumulation allow NMR spectra to be obtained with very small sample quantities.

There were two types of NMR spectrometers in the 1990s—continuous wave (CW) and pulsed Fourier transform (FT). The latter have now largely replaced the CW instruments. In the CW instruments, the oscillator frequency was kept constant while the magnetic field was changed gradually. The value of the magnetic field at which a match (*in-resonance* condition) is reached between the oscillator frequency and the frequency of nuclear precession depends on the shielding effects that the protons experience (in the case of ^{1}H-NMR). Different protons will therefore sequentially undergo transitions between their respective lower and upper energy levels at different values of the changing applied magnetic field as and when the oscillator frequency matches exactly their respective Larmor frequencies during the scan, and corresponding absorption signals will be observed. One limitation of this procedure was that at any given moment, only protons resonating at a particular chemical shift can be subjected to excitation at the appropriate value of the magnetic field, and it is therefore necessary to *sequentially* excite the protons that have differing precessional frequencies in a given molecule. A given set of protons will therefore be scanned for only a small fraction of the total scan time, with other protons or base line noise being scanned for the rest of the time.

Fortunately, an alternative method of excitation was developed. This involves the application of a short but intense radiofrequency pulse extending over the entire bandwidth of frequencies in which the nuclei to be observed resonate, so that all the nuclei falling within the region are excited simultaneously. As a result, the total scan time is made independent of the sweep width *W*. The relaxations that occur immediately after this excitation process are measured as exponentially decaying waves (FID) and are converted to NMR spectra by Fourier transformation. Such instruments, called pulse Fourier transform (PFT) NMR spectrometers (Fig. 1.6), have now replaced the earlier CW instruments. The NMR measurements on the earlier CW instruments were in the *frequency domain,* involving the measurement of the signal amplitude as a function of frequency. The sample in such experiments was subjected to a *weak* field, and the energy *absorbed* was measured. In pulse NMR, the sample is subjected to a *strong* burst of radiofrequency energy; when the pulse is switched off, the

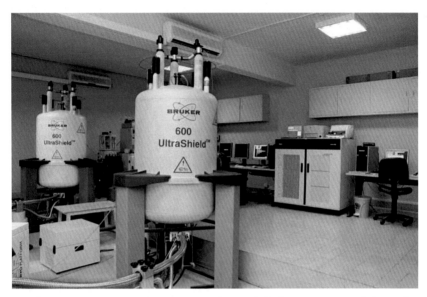

FIGURE 1.6 A 600 MHz NMR spectrometer. The console is the computer-controlled recording and measuring system; the superconducting magnet is in front.

energy *emitted* by the relaxing nuclei is measured. Thus, the CW NMR experiment may be considered as providing an absorption spectrum, while the pulse NMR experiment affords an "emission" spectrum (nonradiative release of energy, see Section 2.1.3).

Researchers need to be aware of some basic features of NMR spectrometers, briefly presented here.

1.2.1 The Magnet

The heart of the NMR spectrometer is the magnet. Modern high-field NMR spectrometers have oscillators with frequencies of up to 1000 MHz (1 GHz) or more (Yanagisawa et al., 2014; Bascunan et al., 2011; Haase et al., 2005). The solenoid in these magnets is made of a niobium alloy (NbTi or Nb_3Sn) wire. When dipped in liquid helium ($-269°C$), the resistance to the flow of electrons becomes almost zero. So, once charged, the *"superconducting"* magnets become permanently magnetized and can exhibit a magnetic field without consuming electricity. The liquid helium is housed in an inner container, with liquid nitrogen in an outer container to minimize the loss of helium by evaporation. A large balloon can be connected to the magnet to collect the evaporated helium gas, for subsequent liquefaction and recycling. In places where liquid helium is not readily available, it is advisable to order special magnet Dewars along with the instrument, with long helium hold times. Fitted with such special Dewars, 500 MHz instruments need to be refilled only about once a year. In early 2000, ultrashielded and ultrashielded + ultrastabilized (US^2) NMR magnets were developed with long-term stable magnetic field, compact designs, and low helium evaporation/consumption. This new technology also provides the added advantage of reducing the stray fields and protection against external electromagnetic field disturbances. Superconducting magnets are very stable, allowing measurements to be made over long periods with little or no variation of the magnetic field (B_0). With the advancements in superconducting magnets and refrigeration/insulation technologies, ultrahigh field NMR spectrometers have become available which are especially suited for the analysis of insensitive nuclei (^{13}C, ^{15}N, ^{31}P, ^{17}O) in structural biology research.

More recently, earthfield NMR (EF-NMR) (270 μT) has been introduced for commercial applications. EF-NMR uses the globally available, homogenous magnetic field of the earth for detection (Ross et al., 2012; Melton and Pollak, 1971). This generally requires a large amount of sample, and only a limited number of experiments can be performed (Katz et al., 2012; Liao et al., 2010; Halse et al., 2009). This technology can make low field NMR available in countries and regions where availability of liquid helium is still an issue, although the experiments are of very limited use (Section 9.1.2). Similarly ultralow field and low field pulse NMR (LFP-NMR) spectrometers are being developed for various applications.

Similarly PFT-NMR spectrometers with permanent magnets (45–90 MHz) have returned back to the marketplace as robust machine for routine identification of known compounds or medium scale synthetic chemistry work (Sections 9.1.1, and 9.1.2).

Problem 1.13

Which of the following conditions will yield better NMR results?
1. More sample with measurement on a lower MHz NMR spectrometer.
2. Less sample with the use of a higher MHz NMR spectrometer.

Problem 1.14

Describe the effect of the magnet's power B_0 on the separation of the nuclei in the frequency spectrum. Do changes in magnetic power B_0 also affect the coupling constant?

Problem 1.15

Do I get a higher resolution if I record the spectrum on a higher field instrument? In other words, will the resolution be better on a 600 MHz instrument as compared to a 300 MHz instrument?

1.2.2 The Probe

The probe, situated between the field gradient coils in the bore of the magnet, consists of a cylindrical metal tube that transmits the pulses to the sample and receives the resulting NMR signals. The glass tube containing the sample solution is lowered gently onto a cushion of air from the top of the magnet into the upper regions of the probe. The probe, which is inserted into the magnet from the bottom of the cryostat, is normally kept at room temperature, as is the sample tube. The sample is spun on its axis in a stream of air to minimize the effects of any magnetic field inhomogeneities. The gradient coils are also kept at room temperature. For recording ^{1}H-NMR and ^{13}C-NMR spectra a dual ^{1}H/^{13}C probe is recommended, which, although having a somewhat (10–20%) lower sensitivity than the dedicated ^{1}H probe, has the advantage of avoiding frequent changing of the probe, retuning, and reshimming. If other nuclei (e.g., ^{15}N, ^{19}F, ^{31}P) are to be studied, then broad-band multinuclear probes can be used, although the sensitivity of such probes is lower than that of "dedicated" probes. Special inverse probes were introduced to conduct inverse NMR experiments (Section 3.3.2.2). Solid-state NMR probes, more properly known as magic angle spinning (MAS) probes, are also readily available for special purposes (Section 9.2).

We also need to choose the probe diameter to accommodate 3, 5, 10, or 15 mm sample tubes. In wide-bore magnets, the probes can be several centimeters in diameter, allowing insertion of larger sample tubes (and even small animals, such as cockroaches and mice). Normally, the 5 mm probe is used, unless sample solubility is a critical limitation, when it may become necessary to use a larger quantity of sample solution to obtain a sufficiently strong signal. The usual limitation is that of sample quantity rather than sample solubility, and it is often desirable to be able to record good spectra with very small sample quantities. In such situations, we should use the smallest diameter probe possible that affords stronger signals than larger diameter probes with the same amount of sample. Microprobes of diameter 1, 1.5, and 2.5 mm with special sample tubes are particularly useful in such cases, and special NMR tubes are used with it. If, however, the amount of sample available is not a limiting factor, then it may be preferable to use a larger diameter probe to obtain good shim values and to subject as much sample as possible to the NMR experiment so as to obtain a good spectrum in the shortest possible measuring time. Such a situation may arise, for instance, in INADEQUATE spectra (Section 7.7) in which ^{13}C–^{13}C couplings are being observed, and it may be necessary to scan for days to obtain an acceptable spectrum.

The significant improvements in sensitivity achieved during the last 5 years have been largely due to the development of magnets with higher magnetic field, improved probe design, and radiofrequency circuits. Since the probe needs to be located very close to the sample, it must be made of a material with a low magnetic susceptibility; otherwise, it would cause distortions of the static magnetic field B_0, thereby adversely affecting line shape and resolution. Much research has, therefore, been undertaken by NMR spectrometer manufacturers to develop materials that have low magnetic susceptibilities suitable for use in probes. The probe must also have a high field (B_1) homogeneity; i.e., it must be able to receive and transmit radiofrequency signals from and to different regions of the sample solution in a uniform manner. Besides these room temperature probes, new cryogenically cooled probe technology has also been introduced in the last decade.

In a cryogenically cooled probe, the *Rf* coils and preamplifiers are cooled to 10 K ($-263.15°C$) by a uniform injection of gaseous helium, while the sample tube remains at room temperature. This leads to a substantial increase in sensitivity as the signal-to-noise (*S/N*) ratio increases with the reduction of electronic noise at very low temperature. Cryogenically cooled probes are now available in several configurations, such as dual ($^{13}C/^1H$), inverse, triple resonance or triple resonance inverse ($^{15}N/^{13}C/^1H$), and magic angle spinning (MAS) solid-state probes, and in various sizes. Introduction of cryogenically cooled probe technology is one of the most important milestones in NMR spectroscopy due to the tremendous boost in sensitivity achieved. About four-fold sensitivity enhancement achievable by cryogenically cooled probe technology has made it possible to study very small quantities of samples as well as nuclei with low natural abundance (Section 3.3.2.1). Recently, liquid nitrogen cooled cryogenic probes were introduced which should further popularize the use of cryogenically cooled probe technology (Kovacs et al., 2005) (Section 9.1.4).

Typical probe assemblies, room temperature and cryogenically cooled, are shown in Fig. 1.7, while Fig. 1.8 shows the difference in the *S/N* ratio in the ^{1}H-NMR spectra of oxandrolone (0.5 mg) recorded using a cryogenically cooled probe and a room temperature probe, respectively.

Problem 1.16

Which types of NMR spectrometers would give the best sensitivity for recording carbon spectra?

Problem 1.17

Recommend the most suitable probe for each of the following laboratories:
1. A laboratory involved in biochemical work or in analytical studies on natural products.
2. A laboratory involved in the synthesis of phosphorus compounds and organometallic complexes.
3. A laboratory where large-scale synthesis of organic compounds is carried out.
4. A laboratory where various nitrogenous compounds are prepared and studied.
5. A laboratory where structures of labeled proteins of clinical importance in liquid state are deduced.

Problem 1.18

What properties should an "ideal" NMR probe have?

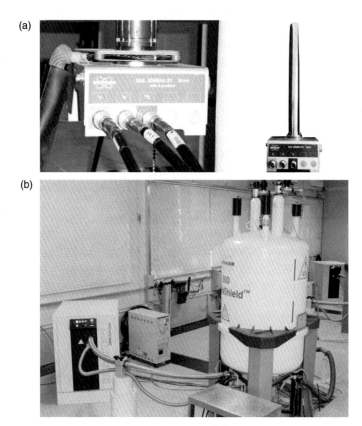

FIGURE 1.7 (a) A typical probe assembly. (b) Cryogenically cooled probe system (probe assembly not shown).

Problem 1.19

Why should one invest in acquiring a cryogenically cooled probe? What are the advantages of cryogenically cooled probes over room temperature probes?

1.2.3 Probe Tuning

Inside the probe is a wire coil that surrounds the sample tube. This wire transmits the radiofrequency pulses to the sample and then receives the NMR signals back from the sample. The probe circuit is tuned to effectively transfer the *Rf* to the sample and sensitively detect the precessing magnetization by matching the resonant frequency of the circuit to the precessional frequency of the nuclei. It is vital that the impedance of the wire be identical to those of the transmitter and receiver to properly perform the dual function of a pulse transmitter and a signal

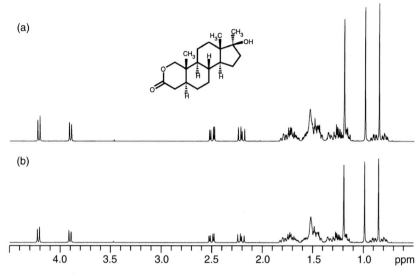

FIGURE 1.8 ^{1}H-NMR of oxandrolone (0.5 mg dissolved in 0.6 mL of CD$_3$OD) was recorded on a (a) 500 MHz NMR spectrometer equipped with cryogenically cooled probe, and (b) 500 MHz NMR spectrometer equipped with room temperature probe using eight scans.

receiver. In addition, the impedance of the coil must be matched with the impedance of the spectrometer electronics (Bendet-Taicher et al., 2014). The probe is tuned and matched by adjusting the two capacitors present inside the probe resonant circuit by a long screw driver near the coil (Fig. 1.9). Adjusting one of the capacitors changes the resonant frequency of the circuit, and this adjustment is carried out so that the circuit resonant frequency precisely matches the precessional frequency of the observed nucleus. The other capacitor controls the impedance of the circuit, and it is adjusted to match the probe impedance (Poeschko et al., 2014).

Normally, it is necessary to adjust these capacitors when the solvent is changed. The two capacitors are adjusted in conjunction with one another, since adjustment of one tends to affect the other and an optimum combination of settings is required. This process is facilitated by employing a directional coupler that is inserted between the probe and the transmitter output (Fig. 1.10). The power of the pulse transmitter reflected from the probe is measured by the directional coupler, and the probe is tuned so that the reflected power is kept to a minimum to obtain the best performance (Poeschko et al., 2014; Daugaard et al., 1981).

Problem 1.20

How does probe tuning affect the quality of the NMR spectrum?

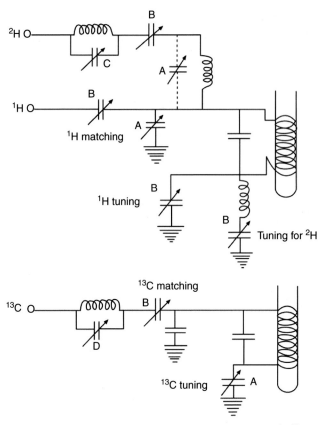

FIGURE 1.9   A schematic representation of a typical resonant circuit for a dual ^{1}H/^{13}C probe. The capacitors A, B, C, and D perform various functions, such as symmetrization and matching of resonance.

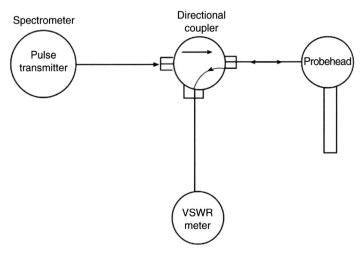

FIGURE 1.10   Use of directional coupler for probe tuning.

1.2.4 Shimming

Modern superconducting magnets have a set of superconducting gradient coils that are adjusted during installation of magnet (and never adjusted by the user). There is, however, another set of printed coils at room temperature that are wrapped around the magnet cylinder and these need to be adjusted from time to time. The weak magnetic fields produced by these coils can be adjusted to simplify any errors in the static field, a process known as *"shimming."* The shim assembly contains many different coils, which have their respective fields aligned with the x-, y-, and z-axes. The NMR probe lies in between the shim assembly, with the sample tube being located in the center of the z-gradient coil. The static field in superconducting magnets lies along the z-axis (in the older iron magnets it was aligned horizontally). The proper adjustment of the vertical z- and z^2-gradients is important, particularly since most of the field inhomogeneities along the x- and y-axes are eliminated by the rapid spinning of the sample tube along the z-axis. It is, therefore, necessary to correct the x- and y-gradients only to the third order (x, x^2, x^3, y, y^2, y^3), while the z-gradients need to be corrected to the fourth or fifth order, particularly on high-field instruments (Pearson, 1991; Chmurny and Hoult, 1990).

The axial shims, i.e., z, z^2, etc. that alter the field on the z-axis only, are corrected while spinning the sample at 15–25 Hz. The radial shims, i.e., ZXY that affect the x and y coordinates should be shimmed without spinning the sample.

Since it is z- and z^2-gradients that have to be adjusted most frequently, the operator had to become proficient in the rapid and optimum adjustment of these gradients each time the sample was changed on the older instruments. This is now done automatically on the more modern instruments. The adjustments afford maximum lock levels, which in turn lead to higher resolution and improved line shape. The intensity of the lock signal (Section 1.2.5) displayed on the lock-level meter or on some other gradient device indicates the field homogeneity, and it is therefore used to monitor the shimming process. In theory, the field generated by each shim coil is independent of the other, but practically there are considerable interactions and the shims must be adjusted interactively.

One feature of the shimming process is the interdependability of the gradients; i.e., changing one set of gradients alters others, so that an already optimized gradient will need to be readjusted if other gradients have been subsequently altered. Good shimming therefore requires patience and perseverance, since there are several gradients to be adjusted and they affect each other. Shimming of the various gradients is therefore not done randomly, since certain gradients affect other gradients to deferring extents. The NMR operator soon recognizes these pairs or small groups of interdependent gradients that need to be adjusted together. The adjustment of x- and y-gradients corresponds to first-order shimming, changes in xy-, xz-, yz-, and $x^2 - y^2$-gradients represent second-order shimming, while optimization of xz^2- and yz^2-gradients is called third-order shimming. It is normally not necessary to alter the xy-, xz-, yz-, xy-, or $x^2 - y^2$-gradients.

Adjustment of the z-gradients affects the line widths, with changes in z-, z^3-, and z^5-gradients altering the symmetrical line broadening and adjustments of z^2 and z^4-gradients causing unsymmetrical line broadening. Changes in the lower order gradients, for example, z or z^2, cause more significant effects than changes in the higher order gradients (z^3, z^4, and z^5). The height and shape of the spinning side bands is affected by changing the horizontal x- and y-gradients, adjustments to these gradients normally being carried out without spinning the sample tube, since field inhomogeneity effects in the horizontal (xy) plane are suppressed by spinning the sample tube. A recommended stepwise procedure for shimming is as follows:

1. First optimize the z-gradient to maximum lock level. Note the maximum value obtained.
2. Then adjust the z^2-gradient, and note carefully the direction in which the z^2-gradient is changed.
3. Again adjust the z-gradient for maximum lock level.
4. Check if the strength of the lock level obtained is greater than that obtained in step 1. If not, then readjust z^2, changing the setting in a direction opposite to that in step 2.
5. Readjust the z-gradient for maximum lock level, and check if the lock level obtained is greater than that in steps 1 and 3.
6. Repeat the preceding adjustments till an optimum setting of z/z^2-gradients is achieved, adjusting the z^2-gradient in small steps in the direction so that maximum lock level is obtained after subsequent adjustment of the z-gradient.
7. If x-, y^2-, or z^2-gradients require adjustment, then follow this by readjustment of the x- and y-gradients, making groups of three (x^2, x, y; y^2, x, y; z^2, x, y). This should be followed by readjustment of the z-gradient.

The main shim interactions are presented in Table 1.3. Note that since adjustments are made for maximum lock signal corresponding to the area of the single solvent line in the deuterium spectrum, a high lock signal will correspond to a high intensity of the NMR lines but will not represent improvement in the *line shape*. The duration and shape of the FID is a better indication of the line shape. Shimming should therefore create an exponential decay of the FID over a long time to produce correct line shapes.

The duration for which an FID is acquired also controls the resolution obtainable in the spectrum. Suppose we have two signals, 500.0 and 500.2 Hz away from the tetramethylsilane (TMS) signal. To observe these two signals separately, we must be able to see the 0.2-Hz difference between them. This would be possible only if these FID oscillations were collected for long enough so that this difference became apparent. If the FID was collected for

TABLE 1.3 Main Shimming Interactions

Gradient Adjusted*	Main Interactions	Subsidiary Interactions
z	–	–
z^2	z	–
x	y	z
y	x	z
xz	x	z
yz	y	z
xy	x, y	–
z^3	z	z^2
z^4	z^2	z, z^3
z^5	z, z^3	$z^2 - z^4$
$x^2 - y^2$	xy	x, y
xz^2	xz	x, z
yz^2	yz	y, z
zxy	xy	x, y, z
$z(x^2 - y^2)$	$x^2 - y^2$	x, y, z
x^3	x	–
y^3	y	–

Alteration in any gradient in the first column will affect the gradients in the second column markedly, while those in the third column will be less affected.

only a second, then 500 oscillations (Hz) would be observed in this time, which would not allow a 0.2-Hz difference to be seen. To obtain a resolution of signals separated by n Hz, we therefore need to collect data for $0.6/n$ seconds. Bear in mind, however, that if the intrinsic nature of the nuclei is such that the signal decays rapidly, i.e., if a particular nucleus has a short T_2^* (Section 4.1.3), then the signals will be broad irrespective of the duration for which the data are collected. As already stated, FIDs that decay over a long time produce sharp lines, whereas fast-decaying FIDs yield broad lines. Thus, to obtain sharp lines, we should optimize the shimming process so that the signal decays slowly.

For longer experiments an automatic shimming system (AUTOSHIM) must be turned on at the start of the experiment (in Bruker NMR spectrometers). This will keep the lock level to the same position.

FIDs have to be accumulated and stored in the computer memory, often over long periods, to obtain an acceptable S/N ratio. During this time there may be small drifts in the magnetic field due to a slight electrical resistance in the magnet solenoid, variations in room temperature, and other outside influences,

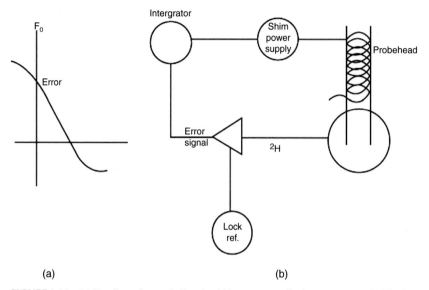

(a)   (b)

FIGURE 1.11   (a) The dispersion mode line should have zero amplitude at resonance. (b) The deuterium lock keeps a constant ratio between the static magnetic field and the radiofrequency. This is achieved by a lock feedback loop, which keeps the frequency of the deuterium signal of the solvent unchanged throughout the experiment.

such as the presence of nearby metal objects. It is therefore desirable to lock the signal onto a standard reference to compensate for these small changes.

The deuterium line of the deuterated solvent is used for this purpose, and the intensity of this lock signal is employed to monitor the shimming process. The deuterium lock prevents any change in the static field or radiofrequency by maintaining a constant ratio between the two. This is achieved *via* a lock feedback loop (Fig. 1.11), which keeps a constant frequency of the deuterium signal. The deuterium line has a dispersion-mode shape; i.e., its amplitude is zero at resonance (at its center), but it is positive and negative on either side (Fig. 1.12). If the receiver reference phase is adjusted correctly, then the signal will be exactly on resonance. If, however, the field drifts in either direction, the detector will experience a positive or negative signal (Fig. 1.13), which will be fed to a coil lying coaxially with the main magnet solenoid. The coil will generate a field that will be added or subtracted from the main field to compensate for the effect of the field drift. The deuterium lock therefore comprises a simple deuterium spectrometer, operating in parallel to the nucleus being observed (Fig. 1.11b).

Problem 1.21

How would you expect the NMR spectrum to be affected if the instrument is poorly shimmed?

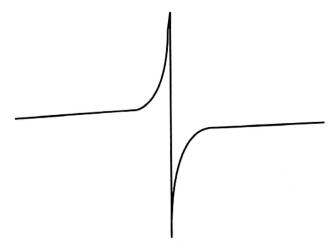

FIGURE 1.12 The dispersion-mode line shape showing the zero amplitude at the center of the peak but nonzero amplitude on each side.

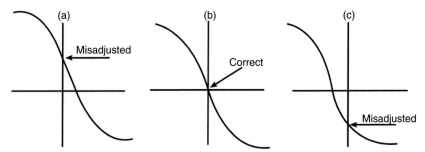

FIGURE 1.13 (a), (c) The reference phase of the receiver is not correctly adjusted. (b) Zero amplitude is achieved by accurate receiver reference phase setting.

Problem 1.22

What is the difference between "shimming" and "tuning" an NMR spectrometer. How are these carried out?

Problem 1.23

How do I tune the probe?

Problem 1.24

Why I am unable to shim?

Problem 1.25

Why it is difficult to shim high field magnets, let us say 900 MHz or 1 GHz (1000 MHz) NMR spectrometers, than their low field versions, i.e., 300 or 400 MHz NMR spectrometers?

1.2.5   Deuterium Lock

Two other parameters that need to be considered in the operation of the lock channel, besides adjusting the receiver reference phase as just described, are (1) *Rf* power, and (2) the gain of the lock signal. If too much *Rf* power is applied to the deuterium nuclei, a state of saturation will result, since they will not be able to dissipate the energy as quickly *via* relaxation processes, producing line broadening and variation of the signal amplitude. It is desirable to achieve the highest transmitter power level that is just below the saturation limit to obtain a good lock signal amplitude. The gain of the lock signal should also be optimized, since too high a lock gain will result in over-amplification of the lock signal, thereby causing excessive noise.

It is important to ensure that the sample is properly locked. For solvents with more than one peak in the proton/deuterium spectrum (ethanol, toluene, and THF are common solvents that have multiple peaks), it is necessary to make sure that the peak being used to lock corresponds to the peak that the software is processed to select. The chemical shift of the solvent must therefore be defined in the software. It is possible to use a mixture of two deuterated solvents, but it is best to avoid doing so unless one is trying to reproduce results reported in the literature obtained with mixed solvents. This is because mixing two solvents will result in a perturbation of the chemical shifts of both solvents and it may cause problems in autolocking and gradient shimming. One should also keep in mind the temperature dependence of the chemical shifts of some solvents. One needs to be particularly careful when using D_2O as a solvent as its chemical shift varies by 0.011 ppm per degree as the temperature is increased. 4,4-Dimethyl-4-silapentane-1-sulfonic acid (DSS) can also be used as a solvent but its shift is also pH dependent. Be careful not to be deceived by a peak at 0.08 ppm as the TMS peak—it is caused by silicone grease impurities in the sample.

1.2.6   Referencing NMR Spectra

There are three methods that are in use for referencing NMR spectra:

1. An internal standard is added to the solution.
2. An external standard is used which is typically a neat liquid.
3. The chemical shift of the lock solvent is used as a reference so that the solvent itself serves as the internal standard. This is now the usual method.

The main problems associated with the use of internal standards are: (1) their chemical shifts change at different dilutions, and (2) removing the standard could become a problem if one wants to recover the sample after recording the NMR spectrum. TMS was often added as an internal standard, but in spite of its relatively inert nature, its chemical shift can vary by more than 0.6 ppm in different organic solvents relative to neat TMS (e.g., benzene +0.30 ppm, chloroform −0.14 ppm).

A neat liquid (TMS) can also be used as a standard in a concentric insert in the NMR tube. This allows the NMR measurement to be carried with two different (unmixed) solvents, one of which is in a separate insert while the other is used for dissolving the solvent. Both Varian, Bruker and Jeol spectrometers use the known absolute deuterium lock frequency, instead of the ^{1}H TMS frequency, as the basis for all multinuclear chemical shift calibrations. This allows one to establish chemical shifts without the need to have TMS in the sample tube. If TMS is present, then TMS can be used to set the absolute zero, but if it is not present, then the deuterium frequency can be used for the lock. Nitrogen chemical shifts can be reported relative to neat liquid ammonia, neat nitromethane, and tetramethylammonium iodide in DMSO-d_6. Biological NMR spectroscopists tend to report nitrogen shifts on the ammonia scale, while organic chemists normally use the nitromethane scale. The two scales differ by 380.2 ppm. Similarly, bio-NMR spectroscopists tend to report ^{13}C shifts relative to DSS (4,4-dimethyl-4-silapentane-1-sulfonic acid) in D_2O which is offset from TMS scale (1% in CDCl$_3$) by a few hundredths of a ppm.

Problem 1.26

What physical changes would you expect in the shape of the NMR signal if the deuterium lock is not applied during data acquisition?

Problem 1.27

I am facing difficulties in locking on. What do I do?

1.2.7 NMR Sample Tubes

In order to measure the NMR spectrum, the sample must first be dissolved in about 0.5 mL of an appropriate solvent. This is a deuterated solvent, such as CDCl$_3$ or one that has no protons such as CCl$_4$, as otherwise the NMR spectrum will be masked by a large solvent peak. About 5–10 mg of the solid sample (or 10 μL of liquid sample) is normally used though NMR spectra of much smaller sample amounts can also be measured.

NMR tubes are typically made of borosilicate glass. Two important specifications of NMR tubes are concentricity and camber. Concentricity refers to the variation in the radial centers, when measured at the inner and outer walls, while camber refers to the "*straightness*" of the NMR tube. If values of either of these are poor, then the homogeneity of the sample will be reduced, resulting in poor quality NMR spectra. If the NMR tube has poor camber, it can wobble when spun, resulting in the appearance of spinning side bands.

The sample is placed in the NMR sample tube which is usually a cylindrical glass tube with a length of 17 cm and a diameter of 5 cm. The reference

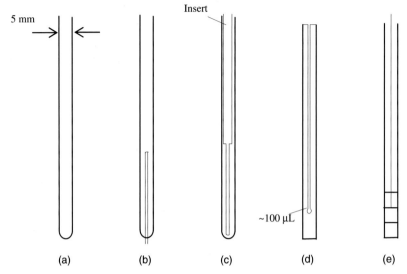

FIGURE 1.14   NMR sample tubes: (a) for measurement with an internal standard; (b) and (c) with capillaries for measurements with an external standard; (d) for microsamples; and (e) Shigemi tubes.

compound, usually tetramethylsilane (TMS), is then added. As TMS has a low boiling point (26°C) and a high vapor pressure, it may be difficult to handle, so deuterated solvents containing 5% TMS are available which may be used. Alternatively, the deuterated solvent itself can be used as a reference, and its frequency difference with respect to TMS inserted into the software. If only a very small amount of sample is available, special NMR tubes with a smaller diameter (1–3 mm) or a Shigemi tube may be used (Section 3.4). These are shown in Fig. 1.14.

Problem 1.28

What should be the volume of solvent that I should use when recording an NMR spectrum?

Problem 1.29

Deuterium (^{2}H) is the most common lock signal. ^{1}H, ^{7}Li, and ^{19}F have also been employed for this purpose. Can we use a nucleus as lock signal while recording a spectrum of the same nuclei?

Problem 1.30

Why is the population difference between the two orientations (α and β for dipole nuclei) very small, whereas in other spectrophotometric techniques (such as UV, IR, Raman, etc.) the entire population is either in the ground energy state or in the excited energy state (after absorption of relevant electromagnetic radiation)?

SOLUTIONS TO PROBLEMS

1.1 Nuclei with odd atomic mass or number, i.e., with odd number of neutrons or protons create an inherent unsymmetry of nuclear spin. Subatomic particles (electron, neutron, and proton) exist in spin pairs. Their odd number leaves at least one of them as an unpaired species, creating an unsymmetry in the nuclear motion. The resulting nuclear spin makes such nuclei NMR active. Examples include ^{13}C, ^{15}N, ^{31}P, ^{19}F, ^{17}O, ^{1}H, etc.

1.2 Practically, every element in the periodic table has one or more isotopes with odd number of neutrons, which makes them NMR active. For example, ^{17}O and ^{13}C isotopes of oxygen and carbon, respectively, are NMR active, whereas their main isotopes are NMR inactive. It is therefore possible to analyze almost every element in the periodic table by NMR spectroscopy.

1.3 All nuclei carry a charge because of the presence of protons. In some nuclei, this charge may spin (such as in the ^{13}C nucleus) while in other nuclei it may not spin and they will not be observed in the NMR spectrum (as in the ^{12}C nucleus). In those nuclei in which the charge is spinning, the nuclei behave like tiny bar magnets, and they can become aligned either in the direction parallel to the applied field (lower energy or α orientation), or in the opposite direction (aligned against it (higher energy or β orientation).

Nuclei such as ^{1}H, ^{13}C, and ^{31}P and have a uniform spherical charge distribution and they are said to have a spin of 1/2. Nuclei in which the charge distribution is nonspherical (quadrupolar nuclei) have spin number I of 1, 3/2, or higher (in steps of 1/2). As stated above, spin 1/2 nuclei have two orientations (with or against the field), spin 1 nuclei will have three orientations, spin 3/2 nuclei will have four orientations, and so on. Thus, deuterium which has spin 1 will exist in three different orientations. A carbon spectrum of $CDCl_3$ therefore shows a 1:1:1 triplet of carbon because of its coupling with attached deuterium.

1.4 After the application of a pulse, the nucleus takes a certain time to relax to its equilibrium position at an exponential rate. The relaxation time is the "time constant" of the exponential curve. For a nucleus to relax to 95% of its equilibrium value, it will take five time constants.

Relaxation can occur by two different processes. In one process (T_1 or spin lattice relaxation), the excited nuclear spin relaxes by giving up its energy to the surrounding lattice resulting in the magnetization along the z-axis relaxing back to its equilibrium position. In the second relaxation process (T_2 relaxation), the magnetization in the x–y plane spreads out by the excited spins exchanging energy with each other, till the net magnetization in the x–y plane (transverse magnetization) becomes zero.

1.5 Applied radiofrequency (*Rf*), when in-phase with the varying range of Larmor frequencies of a particular nucleus, can induce relaxation by accepting energy from in-phase nuclei and make them fall to lower energy states. As a result, absorption of applied radiofrequency leads to excitation of nuclei in the lower energy state(s) to the upper energy state(s). The release of radiofrequency from certain in-phase nuclei in upper energy state(s) helps them to relax to lower energy state(s). It is the net difference in absorption and release of energy due to population difference (Boltzmann distribution excess) which gives birth to an NMR signal. This is unlike other spectroscopic techniques (such as UV and IR spectrophotometry), where entire populations in the lower energy state are excited to the upper energy state through the absorption of specific electromagnetic radiation.

1.6 By absorption of continuous energy from the radiofrequency source, transitions of the nuclei occur to the higher energy state, β; then by relaxation processes, they revert to the lower energy state, α, and an equilibrium is established with a slight Boltzmann excess in the lower α state. This slight excess of population in the lower state will be eliminated on continuous irradiation, and a state of *saturation* will be reached, since the nuclei will not be able to dissipate the extra energy *via* relaxation processes.

1.7 The population difference between two energy levels α and β is directly proportional to the energy difference ΔE.

$$\frac{N_\beta}{N_\alpha} = \exp\left(\frac{-\Delta E}{kT}\right)$$

The energy difference ΔE depends on the strength of the external magnetic field B_0 and the magnetogyric ratio γ. A stronger applied magnetic field B_0 will cause a correspondingly larger separation (ΔE) between the two energy levels and result in a larger difference in the populations of the two levels. Similarly, the energy difference ΔE also depends on the magnetogyric ratio of the nuclear species under observation:

$$\Delta E = \frac{h\gamma B_0}{2\pi}$$

Since the magnetogyric ratio determines the sensitivity of a nuclear species to the external magnetic field, it has a profound effect on the strength of the NMR signals. For instance, ^1H nuclei will have a Larmor frequency of 300 MHz at an external magnetic field of 7.046 T, while ^{13}C nuclei will resonate with a Larmor frequency of only 75.435 MHz in the same magnetic field, since the γ for ^{13}C is about a quarter of the γ for ^1H. The signal strength is determined by γ^3, so a ^{13}C signal will be about 64 times weaker than an ^1H signal [$(1/4)^3 = 1/64$]. In practice, a ^{13}C

signal is over 6000 times weaker, because it occurs in only 1.1% natural abundance.

1.8 No. Since the magnetogyric ratio of ^{13}C is roughly one-fourth that of ^{1}H, the population difference between the two states (α and β) of ^{13}C nuclei will therefore be about 64 times [$(1/4)^3 = 1/64$] less than that of the ^{1}H nuclei (^{13}C enriched samples).

1.9 The frequency with which a nucleus precesses around the applied magnetic field is called its Larmor precessional frequency. The Larmor frequency depends on the strength of the applied magnetic field B_0 and on the magnetogyric ratio of the nucleus. When the radiofrequency B_1 is applied in a direction perpendicular to the external magnetic field B_0, and when the value of B_1 matches exactly the Larmor frequency, absorption of energy occurs causing the nucleus to "flip" to a higher energy orientation. This represents the process of excitation. Through a relaxation process, the nuclei can then relax back to the lower energy level. The energy released during the relaxation process is recorded as an FID (free induction decay), which is then converted into NMR signals through a mathematical operation called Fourier transformation.

1.10 The Larmor or precession frequency of a nucleus depends on the following two factors:
   1. magnetic flux density (B_0),
   2. magnetogyric ratio (γ).

   As the magnetogyric ratios (γ) of nuclei are different, they all have specific Larmor frequencies which can be selectively studied without the interference of the other nuclei. For example, though both ^{13}C and ^{1}H in an NMR sample tube experience the same (B_0), but the magnetogyric ratio of ^{13}C is ¼ of the ^{1}H. As a result, while ^{1}H resonates with the Larmor frequency of 400 MHz, the ^{13}C will resonate with the Larmor frequency of 100 MHz. This allows selective detection of one type of nucleus possible, without the interference of the other. The same phenomenon applies to selective detection of ^{17}O, ^{15}N, ^{19}F, ^{31}P, etc.

1.11 Magnetogyric ratio (γ) is related to the Larmor frequency through following equation:

$$v = \frac{\gamma}{2\pi} B_0$$

where B_0 is applied magnetic field and γ magnetogyric ratio. The magnetogyric ratios of ^{1}H, ^{13}C, ^{15}N, and ^{2}H are 26.7520, 6.7283, -2.712, and 4.1065 (10^7 Rad T^{-1} s^{-1}), respectively. At a given magnetic field (B_0), let say in a 9.4 T (400 MHz), the ^{1}H nuclei will resonate with the highest Larmor frequency, i.e., 400 MHz, followed by ^{13}C ($\sim$100 MHz), ^{2}H ($\sim$61.5 MHz), and ^{15}N ($\sim$$-40.5$ MHz).

1.12 The magnetogyric ratio of a nuclear spin represents its response toward the external magnetic field. Nuclear species with larger magnetogyric ratios have larger differences in energy levels ΔE in comparison to nuclear species of smaller magnetogyric ratios, when placed under an external magnetic field of the same strength B_0:

$$\Delta E = \frac{h\gamma B_0}{2}$$

where h is Plank's constant, γ is magnetogyric ratio, and B_0 is applied magnetic field.

1.13 A higher MHz NMR spectrometer (high field NMR spectrometers) is always a better choice, since the sensitivity of the experiment is proportional to the frequency of the measurement. Moreover, with highly concentrated solutions, the presence of some solid particles can cause an increase in T_1 (FID will be short), and line broadening of the NMR signals will result. Therefore, an optimum concentration (say, 25–50 millimolar solution) is recommended. Of course, ^{1}H-NMR spectra can be readily measured at much lower concentrations, though higher concentrations are necessary for recording ^{13}C-NMR spectra.

1.14 The magnetic field-strength B_0 has a direct relationship with the Larmor frequencies of nuclei: the stronger the magnetic field, the greater the difference, in Hertz (not in ppm), between magnetically nonequivalent nuclei having differing chemical shifts. Moreover, the population excess of the lower energy level over the upper energy level increases with increasing magnetic field, B_0, leading to a corresponding increase in the sensitivity of the NMR experiment. The magnitude of the coupling constants, however, remains unaffected by the magnetic field strength.

1.15 No, this is a common misconception. You do NOT necessarily improve resolution by going to a higher field instrument. In fact, the resolution on the higher field instrument may be poorer than that on the lower field instrument. By going to higher fields, you will achieve a greater *dispersion* of signals, so that nuclei with different chemicals shifts will spread farther apart, leading to simplification of multiplets by reduced second-order perturbations. If the multiplets are overlapping in the low field instrument, then they will separate at higher fields, making interpretation easier. Some nuclei, such as ^{31}P, may show worse resolution at higher fields because of an intrinsic property that they possess, such as chemical shift anisotropy, which increases with the field strength.

1.16 The high field instruments give the best results. For instance if you have the option to use 600 or 300 MHz instruments, then the 600 MHz instrument should be preferred, particularly if you have small sample quantities, as the *S/N* (sensitivity) ratio on the 600 MHz instrument would be better than that on the 300 MHz instrument by a factor of 600/300 squared, or four times better. The probe that you use is also important for optimum sensitivity. If you are employing an indirect detection probe, then this has the proton observe coil closer to the sample, on the inside, and it would give a better proton signal but a weaker carbon signal (if you use it for directly detecting carbon). A standard probe however has the carbon coil on the inside, and it gives a better carbon signal but a weaker proton signal than the indirect detection probe.

The *S/N* ratios are normally measured by a single scan on a concentrated sample (standard sample supplied by manufacturer). In actual experiments, one is normally measuring a dilute sample for several hours, and the relaxation times of the nuclei need to be taken into account. Also, nuclei take longer to relax at higher field strengths, so that the gain in the *S/N* may be less than expected. Carbons that do not have directly bonded protons (such as quaternary carbons, carbonyls, etc.) will take much longer times to relax than protonated carbons and hence their signals may be expected to be weaker. If one is interested in getting stronger signals for the quaternary carbons, then the D1 delay of 3 seconds or more should be employed to allow the quaternary carbons to relax after excitation before experiencing another pulse. Dramatic improvements in *S/N* ratios of carbon spectra can be obtained if one records a short or long-range 2D heteronuclear (proton, carbon) correlation experiment (Luciano, 1979).

1.17 The probe selection should be based on the actual requirements of a laboratory.

1. Small-diameter probes are generally suitable when the total availability of the sample is a limiting factor. For natural products or biochemical studies, a $^{13}C/^1H$ probe of 5 mm size is probably the best choice, since the major requirement here is to analyze small quantities of organic samples for the proton and carbon spectra. A 2.5 mm microprobe is also available for use with special sample tubes having a diameter of 2.5 mm, and it is highly recommended for small samples. Nanoprobes are available for very small samples quantities. An inverse probe is highly desirable for $^1H/^{13}C$ inverse-shift correlation experiments (e.g., HMQC/HSQC, HMBC).

2. For heteronuclear studies, where different types of nuclei are investigated routinely, the broad-band multinuclear probe is an excellent choice, since it can be tuned over a wide frequency range for various elements.

3. In organic-synthesis laboratories, where sample quantity is not a limiting factor, the larger probes provide a significant saving in time. A $^{13}C/^1H$ probe of 10–15 mm size is more appropriate.

4. A laboratory where nitrogen is the main nucleus to be analyzed, with occasional $^{13}C/^1H$ analysis, a broad-band probe specific for nitrogen should be acquired.

5. In a laboratory where structures of labeled proteins are studied, a triple resonance cryogenically cooled probe is ideal as it gives the capacity of measuring any of the labeled nuclei (^{2}H, ^{15}N, and ^{13}C), as well as decoupling any one of them with improved sensitivity.

1.18 An ideal probe should have the following properties:
   1. It should be made up of a material with a low magnetic susceptibility so that it does not distort the static magnetic field B_0 and adversely affect the line shape and resolution of the NMR signals.
   2. It should have a high field (B_1) homogeneity so that it can receive and transmit radiofrequency signals uniformly from all parts of the sample solution.
   3. It should fulfill the needs of a maximum number of users.

1.19 Cryogenically cooled probe technology has delivered the single largest increase in NMR sensitivity in the last four decades, after the advent of superconducting magnets. Injection of gaseous helium in probe electronics decreases the resistance and increases the conductance. This reduces the noise contribution and yields an enhancement of the S/N by up to five-fold, as compared to the comparable room temperature probe. This 3-5 fold jump in sensitivity by use of the cryogenically cooled probe enables one to analyze smaller quantities of sample that are impractical with conventional probes. Introduction of cryogenically cooled probe technology in the arsenal of NMR spectroscopist has had a direct impact on research in different fields. Most profoundly, it is now possible to analyze and record reasonably good quality NMR spectra of small quantities of proteins and other biomolecules, environmental contaminants, rare natural products, etc. Due to enhanced sensitivity, only a limited number of scans are required, increasing the sample throughput by up to 16-fold. Similarly, with cryogenically cooled probes, it is possible to record high quality NMR spectra of insensitive or low natural abundance nuclei, such as ^{15}N and ^{13}C.

1.20 The probe contains the electronics designed to detect the tiny NMR signal. The central component of the probe is a wire that receives the *Rf* pulse from the transmitter and dissipates it into the sample. It also receives the signal from the sample and transfers it to the receiver circuit. It is therefore necessary to tune the probe wire impedance to match it with those of the transmitter and the receiver. The optimum sensitivity of the NMR experiment will be realized only when this adjustment is made accurately. The probe tuning also minimizes the variable off-resonance effects, and it is therefore essential for the proper reproducibility of the pulse width.

1.21 Poor shimming would lead to poor line shape and resolution, as illustrated here.

(a) With proper shimming

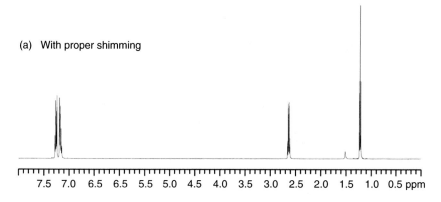

(b) With poor shimming

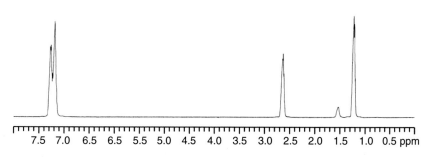

1.22 Shimming involves optimizing the homogeneity of the magnetic field by adjusting the resolution. Tuning involves minimizing the impedance at the frequency to be studied by adjusting the variable capacitors in the probe. If a probe is poorly tuned, it will result in the reflection of a lot of the power of the pulses, so that if we want to apply a 180° pulse, we may actually get only (say) a 130° pulse. This will not affect the resolution but it will lead to a poor *S/N* ratio (sensitivity). Experiments, such as COSY and DEPT (that depend on the use of accurate 90° pulses), may then not work properly and produce noise and artifact signals. If the *shimming* of an instrument is poor, it will result in a poor resolution, which will show up as a broadening of the NMR signals.

1.23 You should not try and tune the probe unless you have been trained as it is a specialized operation. The Bruker instruments have an automatic system of tuning. If you have an NMR spectrometer on which manual tuning is needed then you will need to bear in mind that tuning is affected by frequency, solvent, and sample peak heights.

1.24 1. This is probably because the lock is saturated. Go to the shim window and reduce the lock power by 4 in the lock power/gain/phase display. This should result in a drop of the lock power. If this does not happen, then keep reducing the lock power till the lock level has dropped.

2. You may need to load the standard shims again as the previous user may have disturbed the shims.

3. If you have not already done so, load the standard values for the shims. You do not know what sort of state the previous user left the shims in!

4. Make sure that the sample tube is not dirty or scratched, and that there is no material floating in the sample. If the concentration is too strong, then you may need to dilute the sample. The sample tube needs to be properly centered in the coil.

5. Shimming may also be difficult if you have paramagnetic ions in the sample which leads to line broadening. Broad lines may be caused by quadrupolar broadening with transition metals, or if chemical exchange is occurring.

1.25 Despite tremendous developments in the design of superconducting magnets, high field magnets have greater possibilities of instability and inhomogeneity. The shimming system is designed to adjust only very minor errors or inhomogeneities. In the case of high field NMR spectrometer, i.e., at 900 and 1000 MHz, the magnetic inhomogeneities are often much larger than what the shimming routine can adjust. Therefore, it is more challenging to shim high field magnets.

1.26 The deuterium lock prevents changes in the static field (B_0) and radio frequency (B_1) by maintaining a constant ratio between the two. It therefore ensures long-term stability of the magnetic field. If the ^2H lock is not applied, a drastic deterioration in the shape of the NMR lines is expected due to magnetic and radiofrequency inhomogeneities.

(a) With deuterium lock

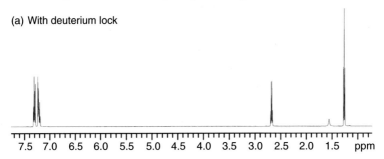

(b) Without deuterium lock

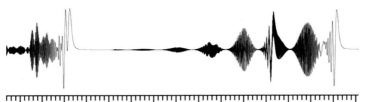

1.27 1. Make sure that you are using a deuterated solvent.

2. Make sure that the tube is spinning properly, and that you have not broken it when inserting it into the probe.

3. Switch off the lock in the lock display, increase the lock power and gain to the maximum value, and search for the sine wave by adjusting z^0. If a sine wave is found, then adjust z^0 to zero value. Now reduce the lock power, in order to avoid saturation, and click the lock button to "ON." If it loses lock when you lock on, then turn the lock off and adjust the lock phase (procedure should be described in the manual).

1.28 You will need to use about 0.7 mL of solvent if you are employing a 5 mm NMR tube. This can be reduced to about 0.5 mL if the sample quantity is low but then you may need to shim carefully. For ^{13}C spectra it will be worthwhile reducing the sample volume to get the best spectrum. If you have very small amounts of material and need to run an HMBC or HMQC spectrum, then the use of smaller sample tubes, or a higher sensitivity probe (cryogenically cooled probes give several fold increase in sensitivity) may be needed.

1.29 No. The signals from the lock transmitter would obliterate the signals we wish to observe.

1.30 The energy difference between LUMO and HOMO in UV spectroscopy ranges from 29.87 to 155.35 kcal mol^{-1}, while in IR spectroscopy the energy difference is of the order of 1.9–9.5 kcal mol^{-1}. In NMR spectroscopy, however, the energy difference between α and β states is 3×10^{-5} kcal mol^{-1} on a 300 MHz instrument. Therefore, the excitation and relaxation phenomena continue to occur simultaneously in NMR spectroscopy, leaving only a minor population difference (Boltzmann excess) in the lower energy state which is detected.

REFERENCES

Bascunan, J., Hahn, S., Park, D.K., Iwasa, Y., 2011. A 1.3-GHz LTS/HTS NMR magnet: a progress report. IEEE Trans. Appl. Supercond. 21 (3 Pt. 2), 2092–2095.

Bendet-Taicher, E., Mueller, N., Jerschow, A., 2014. Dependence of NMR noise line shapes on tuning, matching, and transmission line properties. Concepts Magn. Reson. B Magn. Reson. Eng. 44B (1), 1–11.

Chmurny, G.N., Hault, D.I., 1990. The ancient and honourable art of shimming. Concepts Magn. Reson. 2 (3), 131–149.

Daugaard, P., Jakobsen, H.J., Garber, A.R., Ellis, P.D., 1981. A simple method for NMR probe tuning and some consequences of improper probe tuning. J. Magn. Reson. 44 (1), 224–227.

Haase, J., Kozlov, M., Mueller, K.-H., Siegel, H., Buechner, B., Eschrig, H., Webb, A.G., 2005. NMR in pulsed high magnetic fields at 1.3 GHz. J. Magn. Magn. Mater. 290–291 (Pt. 1), 438–441.

Halse, M.E., Coy, A., Dykstra, R., Eccles, C.D., Hunter, M.W., Callaghan, P.T., 2009. Multidimensional earth's field NMR. In: Codd, S.L., Seymour, J.D. (Eds.), Magnetic Resonance Microscopy, WILEY-VCH Verlag GmbH&Co. KGaA, Weinheim, Germany, pp. 15–29.

Harris, R.K., 1989. Nuclear Magnetic Resonance Spectroscopy. Longman Scientific & Technical, Essex, England, p. 5.

Katz, I., Shtirberg, L., Shakour, G., Blank, A., 2012. Earth field NMR with chemical shift spectral resolution: theory and proof of concept. J. Magn. Reson. 219, 13–24.

Kovacs, H., Moskau, D., Spraul, M., 2005. Cryogenically cooled probes: a leap in NMR technology. Prog. Nucl. Magn. Reson. Spectrosc. 46, 131–155.

Liao, S.-H., Chen, M.-J., Yang, H.-C., Lee, S.-Y., Chen, H.-H., Horng, H.-E., Yang, S.-Y., 2010. A study of J-coupling spectroscopy using the earth's field nuclear magnetic resonance inside a laboratory. Rev. Sci. Instrum. 81, 10410-1–104104-6.

Luciano, M., 1979. Sensitivity enhanced detection of weak nuclei using heteronuclear multiple quantum coherence. J. Am. Chem. Soc. 101 (16), 4481–4484.

Melton, B.F., Pollak, V.L., 1971. Instrumentation for the earth's field NMR technique. Rev. Sci. Instrum. 42 (6), 769–773.

Pearson, G.A., 1991. Shimming an NMR magnet. 17-Dec-1991, http://nmr.chem.uiowa.edu/manuals/Shimming-GAP-NMR-magnet.pdf.

Poeschko, M.T., Schlagnitweit, J., Huber, G., Nausner, M., Hornicakova, M., Desvaux, H., Mueller, N., 2014. On the tuning of high resolution NMR probes. Chem. Phys. Chem. 15 (16), 3639–3645.

Ross, J.G., Holland, D.J., Blake, A., Sederman, A.J., Gladden, L.F., 2012. Extending the use of earth's field NMR using Bayesian methodology. Application of particle sizing. J. Magn. Reson. 222, 44–52.

Yanagisawa, Y., Piao, R., Iguchi, S., Nakagome, H., Takao, T., Kominato, K., Hamada, M., Matsumoto, S., Suematsu, H., Jin, X., Takahashi, M., Yamazaki, T., Maeda, H., 2014. Operation of a 400 MHz NMR magnet using a (RE:Rare Earth)Ba$_2$Cu$_3$O$_{7-x}$ high temperature superconducting coil: Towards an ultra-compact super-high field NMR spectrometer operated beyond 1 GHz. J. Magn. Reson. 249, 38–48.

Chapter 2

Creating NMR Signals

Chapter Outline

2.1 GENESIS OF NMR SIGNALS

As stated earlier, when placed in a magnetic field, the hydrogen nuclei adopt one of two different orientations, aligned either with or against the applied magnetic field B_0 (Section 1.1.1). There is slight excess of nuclei in the lower energy (α) orientation, in which the nuclei are aligned with the external field (B_0). There is a bulk or macroscopic magnetization M_z^0 corresponding to this difference that points in the direction of the applied field (z-axis, by convention) and that represents the sum of the magnetizations of the individual spins. As long as the bulk magnetization remains aligned along the z-axis, the precessional motion of the nuclei, though present, will not become apparent. This is the situation prevailing at the thermal equilibrium.

The NMR experiment is aimed at unveiling and measuring this precessional motion of the nuclei. One way to do this is to displace the magnetization from its position along the z-axis by application of a radiofrequency pulse. Once the magnetization M_0 is displaced from its position along the z-axis, it experiences a force due to the applied magnetic field B_0, causing it to "precess" about the z-axis at a frequency of γB_0 radians per second (or $\gamma B_0/2\pi$ Hz) (Fig. 2.1), a

Solving Problems with NMR Spectroscopy. http://dx.doi.org/10.1016/B978-0-12-411589-7.00002-4

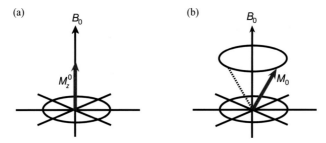

FIGURE 2.1 (a) Bulk magnetization vector, M_z^0 ~, at thermal equilibrium. (b) Magnetization vector M_0 after the application of a radiofrequency pulse.

motion called "*Larmor precession*." The aim of the NMR experiment is to measure this motion. When the magnetization M_0 is bent away from the z-axis, it will have a certain component in the xy-plane; it is this component that is measured as the NMR signal.

For a transition to occur from the lower energy state to the upper energy state, the precessing nuclei must absorb radiofrequency energy. This can happen only if the radiofrequency of the energy being emitted from the oscillator matches exactly the precessional frequency of the nuclei. Here is a simple analogy: imagine that you are holding a sandwich (an "*energy packet*," if you like) in your right hand while you are constantly rotating your left hand at a particular angular velocity. To pass this sandwich from your right hand to your left hand without stopping the rotation of your left hand, your right hand must also rotate at exactly the same angular velocity (*in-resonance*). The transfer of the sandwich (energy) between the two hands can take place only if they rotate *synchronously*. Similarly, only when the oscillator frequency *exactly matches* the nuclear precessional frequency can the nuclei absorb energy and flip-to the higher energy state. They can then relax back to their lower energy state *via* spin–lattice or spin–spin relaxations. The resulting change in the impedance of the detector coils during the relaxation process is recorded in the form of an NMR signal.

The z-component of the magnetization (M_z) corresponds to the population difference between the lower and upper energy states. Any change in this population difference − for instance, by nuclear Overhauser enhancement or polarization transfer − will result in a corresponding change in the magnitude of M_z, the NMR signal being strong when the magnitude of M_z is large.

Problem 2.1

Spin half nuclei ($I = \frac{1}{2}$), such as ^{1}H, ^{13}C, ^{15}N, ^{19}F, and ^{13}P adopt one of two different orientations aligned either with or against the applied magnetic field B_0. They are called dipole nuclei. Is this also correct for other nuclei, such as ^{2}H ($I = 1$), ^{14}N ($I = 1$), ^{17}O ($I = 5/2$), ^{35}Cl ($I = 3/2$), etc.?

What is the bulk or macroscopic magnetization in terms of nuclear spins?

2.1.1   Effects of Pulses

A pulse is a burst of radiofrequency energy that may be applied by switching on the *Rf* transmitter. As long as the pulse is on, a constant force is exerted on the sample magnetization, causing it to precess about the *Rf* vector.

The slight Boltzmann excess of nuclei aligned with the external magnetic field corresponds to a net magnetization pointing toward the +*z*-*axis*. We can bend this magnetization in various directions by applying a pulse along one of the axes (e.g., along the +*x*-, −*x*-, +*y*-, or −*y*-axis). If a pulse is applied along the *x*-axis, a linear field is generated along the *y*-axis that is equivalent to two vectors rotating in opposite directions in the *xy*-plane. However, interaction with the precessing nuclear magnetization occurs only with the vector that rotates in the same direction and with exactly the same frequency.

We can control the extent by which the +*z*-magnetization is bent by choosing the duration for which the pulse is applied. Thus, the term "90° pulse" actually refers to the *time period* for which the pulse has to be applied to bend the magnetization by 90°. If it takes, say, t μs to bend a pulse by 90°, it would require half that time to bend the magnetization by 45°, i.e., $t/2$ μs. A 180° pulse, however, will require double that time, i.e., $2t$ μs and cause the *z*-magnetization to become inverted so that it comes to lie along the −*z*-axis (Fig. 2.2). There will then be a Boltzmann excess of nuclei in the upper energy state, and the spin system is said to possess a "negative spin temperature." A subscript such as *x* or *y* is usually placed after the pulse angle to designate the direction in which the pulse is applied. A "$90°_x$" pulse therefore refers to a pulse applied along the +*x*-axis (in the direction +*x* to −*x*) that bends the nuclear magnetization by 90°, while a $40°_{-y}$ pulse is a pulse applied along the −*y*-axis (i.e., in the direction −*y* to +*y*) for a duration just enough to bend the nuclear magnetization by 40° from its previous position.

The duration for which the pulse is applied is inversely proportional to the bandwidth; i.e., if we wish to stimulate nuclei in a large frequency range, then we must apply a pulse of a short duration. Nuclear excitation will, of course, only occur if the magnitude of the B_1 field is large enough to produce the required tip angle. Typically, if the transmitter power is adjusted to 100 W on a high field instrument, then a 90° pulse width would have a duration of a few microseconds. This would also have a bandwidth of tens of kilohertz over which the nuclei could be uniformly excited. A "*soft*" *pulse* is one that has low power or a long duration (milliseconds rather than microseconds), and such pulses can be used to excite nuclei selectively in specific regions of the spectrum.

For ^{1}H-NMR experiments the pulse width is normally set at about 7–14 μs on the majority of instruments. The field width or bandwidth (bw) of excitation (in Hz) may be obtained from the formula:

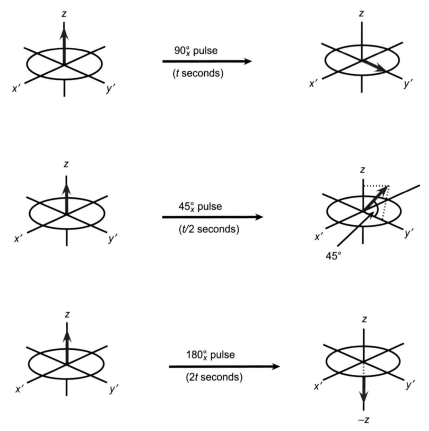

FIGURE 2.2   Effects of radiofrequency pulses of different durations on the position of the magnetization vector.

$$Rf = \frac{1}{(4 \times 90° \text{pulse})}$$

Thus for a 9 μs pulse, the *Rf* field corresponding to the bandwidth of excitation will be $1/(4 \times 0.000009) = 27{,}777$ Hz. Such a bandwidth of excitation is sufficient to excite the typical range of proton resonances in organic samples. On a 500 MHz instrument, these will be normally resonate in the bandwidth range of 5000–7000 Hz.

"*Hard*" *pulses* are short-duration pulses (duration in microseconds), with their power adjusted in the 1–100 W range. Figures 2.3 and 2.4 illustrate schematically the excitation profiles of hard and soft pulses, respectively. In a hard pulse the bandwidth of excitation is greater than the sweep width (sw). Thus, a 10 μs long rectangular pulse would excite in a bandwidth of $1/(4 \times 10$ μs)

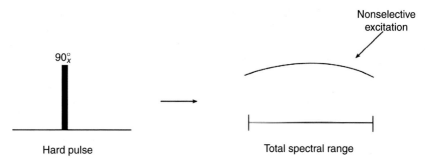

FIGURE 2.3 Time domain representation of a hard rectangular pulse and its frequency domain excitation function. The excitation profile of a hard pulse displays almost the same amplitude over the entire spectral range.

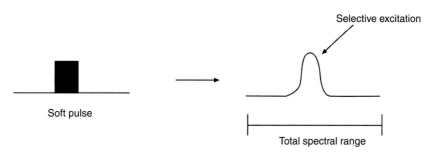

FIGURE 2.4 Time domain representation and frequency excitation function of a soft pulse. The soft pulse selectively excites a narrow region of a spectral range and leads to a strong offset-dependent amplitude of the excitation function.

= 250,000 Hz (i.e., 250 kHz). In the case of protons, the sweep width is say about 10–15 kHz so that this would be a hard pulse (since the bandwidth of excitation is much greater than the sweep width). In the case of a soft pulse, one may need to excite one particular proton selectively, say with a bandwidth of about 0.5 Hz. One would, therefore, need to apply the pulse for 500 ms since [(1/(4 × 0.5) = 0.5 s = 500 ms)].

Gaussian pulses are frequently applied as soft pulses in modern 1D, 2D, and 3D NMR experiments. The power in such pulses is adjusted in milliwatts. Readers wishing to know more about the use of shaped pulses for frequency-selective excitation in modern NMR experiments are referred to excellent reviews on the subject (Warren and Mayr, 2012; Kessler et al., 1991).

Problem 2.3

What is a pulse in NMR spectroscopy?

Problem 2.4

Bandwidth is inversely proportional to the pulse duration. Following is a computer simulation of a short pulse (left) and its calculated spectrum (right). Can you predict the shape of the pulse and its spectrum if a long pulse were employed?

The direction in which the nuclear magnetization is bent by a particular pulse is controlled by the direction in which the pulse is applied. *A pulse applied along a certain axis causes the magnetization to rotate about that axis in a plane defined by the other two axes.* For instance, a pulse along the x-axis will rotate the magnetization *around the x-axis in the yz-plane.* Similarly, a pulse along the y-axis will rotate the magnetization in the xz-plane. The final position of the magnetization will depend on the time t_1 for which the *Rf* pulse is applied (usually a few microseconds). The angle ("flip angle") by which the magnetization vector is bent may be calculated as $B = \gamma B_1 t_p$, where B_1 is the applied field and t_p is the duration of the pulse. Since the spectrometers are normally set up to detect the component lying along the y-axis, only this component of the total magnetization will be recorded as signals. The pulse, therefore, serves to convert longitudinal magnetization, or z-magnetization, to transverse, or detectable magnetization along the y-axis. The magnitude of the magnetization component along the y-axis is given by $B_0 \sin\theta$ where θ is the angle by which the magnetization is bent away from the z-axis (Fig. 2.5).

The direction in which the magnetization is bent by a particular pulse may be predicted by a simple "right-hand thumb" rule: if the thumb of your right hand represents the direction along which a pulse is applied, then the partly bent fingers of that hand will show you the direction in which the magnetization will be bent. Let us consider three different situations and see if we can predict the direction of bending of the magnetization vector in each case.

First, let us imagine that the net magnetization lies along the +z-axis and that we apply a 90°_x pulse. If you point your right-hand thumb along the direction +x to −x, then the bent fingers of that hand indicate that the magnetization will be bent away from the z-axis and toward the +y axis, adopting the position shown in Fig. 2.6. If we continue to apply this pulse for a second, identical time duration, i.e., if we apply a second 90°_x pulse immediately after the first 90°_x pulse (the two pulses actually constituting a 180°_x pulse) then the magnetization vector that has come to lie along the +y'-axis will be bent further by another 90° so that it comes to lie along the −z-axis. Continuous application of this pulse will effectively rotate this vector continuously around the x-axis in the y'z-plane.

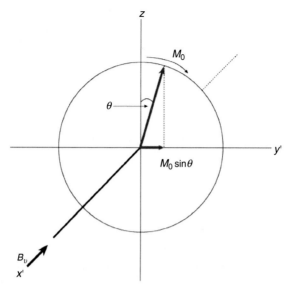

FIGURE 2.5   A cross-sectional view of the $y'z$-plane (longitudinal plane) after application of pulse B_v along the x'-axis (perpendicular to this plane), The pulse B_v applied along the x'-axis causes the equilibrium magnetization M_0 to bend by an angle θ. The magnitude of the component M along the y'-axis is $M_0 \sin\theta$.

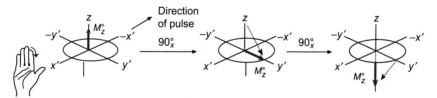

FIGURE 2.6   Effect of applying a 90° pulse on the equilibrium magnetization (M_0). Continuous application of a pulse along the x'-axis will cause the magnetization vector (M_0) to rotate in the $y'z$-plane. If the thumb of the right hand points in the direction of the applied pulse, then the partly bent fingers of the right hand point in the direction in which the magnetization vector will be bent.

Now assume that the vector lies along the $-y$-axis and a 90°_{-x} pulse is applied (i.e., a pulse in the $-x$ to $+x$ direction along the $-x$-axis). This pulse will now cause the magnetization vector to bend from the $-y$-axis to the $-z$-axis (Fig. 2.7). Applying another 90°_{-x} pulse without any intervening delay will rotate the magnetization vector by a further 90° so that it comes to lie along the $+y'$-axis. Continuous application of this pulse will cause the magnetization vector to rotate around the x-axis in the $y'z$-plane.

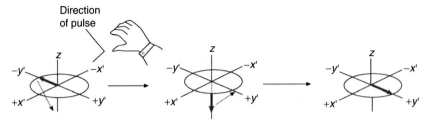

FIGURE 2.7   Applying a 90°_x pulse will bend the magnetization vector lying along the $-y'$-axis to the $-z$-axis, while continued application of the pulse along the x'-axis will cause the magnetization to rotate in $y'z$-plane.

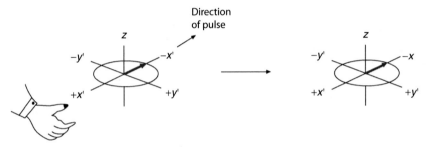

FIGURE 2.8   Applying a pulse along the x'-axis will not change the position of the magnetization vector lying along that axis.

Finally, let us assume that the magnetization vector lies along the $-x$-axis and that the magnetization pulse is also applied along the x-axis (i.e., in the $+x$ to $-x$ direction, Fig. 2.8). There is now no magnetization component that lies along the y or z axes, and so it will not have any effect. In other words, a pulse applied along a particular axis will have no effect on any magnetization component vector lying along that axis.

2.1.2   Rotating Frame of Reference

As stated earlier, when placed in a magnetic field, the bulk magnetization of the nuclei will precess about the applied magnetic field with a frequency γB_0. If this precessing magnetization is exposed to a second magnetic field B_1 produced by an oscillator perpendicular to B_0, then this causes it to tilt away from the z-axis and to exhibit an additional precessional motion *around* the applied *Rf* field at an angular frequency γB_1. This simultaneous precession about two axes (axes of B_0 and B_1 application) makes it quite difficult to visualize the movement of the magnetization. To simplify the picture, we can eliminate the effects of the precession about B_0 by assuming that the plane defined by the x- and y-axes is itself rotating about the z-axis at a frequency γB_0.

Problem 2.5

Based on the right-hand thumb rule described in the text, and assuming that only an equilibrium magnetization directed along the z-axis exists, draw the positions of the magnetization vectors after the application of:

1. a 90°_x pulse
2. a 180°_x pulse
3. a 90°_{-y} pulse
4. a 90°_y pulse and then a 90°_x pulse

Problem 2.6

Describe what is meant by "hard" and "soft" pulses.

To understand this simplification, imagine a man sitting on a merry-go-round with a large placard on which several sentences are written. As long as the merry-go-round is rotating rapidly, a stationary observer may not be able to read the words on the placard as it whirls round. However, if the observer were to jump onto the merry-go-round, then the apparent circular motion of the man with the placard would disappear to the observer since the two would now have the same angular velocity, and the words on the placard would be readable.

Another analogy is that of communication satellites in geostationary orbits. Imagine a satellite in the sky directly above us that is moving with an angular velocity that matches exactly the rotation of the Earth. From our "rotating frame of reference," the satellite will appear stationary to us, although it will be rotating with a much higher angular velocity than our own rotation on the Earth, since it will be in a larger orbit. If we could somehow change the angular velocity of the satellite, say, by firing its rockets so that it starts to travel faster, then it would appear to move slowly away from us; if it were made to travel more slowly, then it would appear to move away from us in the opposite direction. Only the *relative* motion between us and the satellite would be visible in such a "rotating frame of reference."

Considering the motions of nuclear magnetization in a "rotating frame" greatly simplifies their description. By rotating the plane defined by the xy-axes at the frequency B_0, we have effectively switched off the precession due to B_0, so that in the rotating frame only the precession due to the field B_1 of the oscillator will be evident. If the field B_1 is applied along the x-axis, it causes the bulk magnetization that was originally lying along the z-axis to precess about the x-axis (i.e., about B_1) in the yz-plane. A $\pi/2(90^\circ)$ pulse along the x-axis would bend the magnetization from the z-axis to the y-axis. Let us assume that the plane defined by the x- and y-axes is rotating at a frequency exactly equal to the Larmor frequency of the methyl protons of TMS. When a 90° pulse is applied along the x-axis, the magnetization vector (M) of the TMS protons will flip away from the z-axis by 90° and come to rest along the y-axis. Since the y-axis and the magnetization vector M are now rotating with identical frequencies, the vector M will appear static along the y-axis in this "rotating frame."

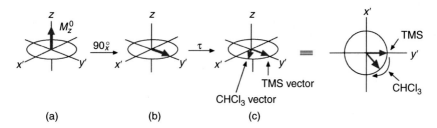

FIGURE 2.9 (a) Position of the magnetization vector at thermal equilibrium. (b) After the 90°_{x} pulse, the magnetization vector M^{0}_{z} will come to lie along the y'-axis on the $x'y'$-plane. (c) During the subsequent delay, the sample magnetization vector will move away from the TMS vector due to its (generally) higher precessional frequency on account of its chemical shift. Since the $x'y'$-plane is itself precessing at the precessional frequency of the TMS vector, the TMS vector appears to be stationary along the y'-axis in this rotating frame.

Now, let us consider a second case, in which we record the NMR spectrum of TMS in $CHCl_3$ and the xy-plane is again assumed to be rotating at the precession frequency of the TMS methyl protons. Application of a 90° pulse would superimpose both vectors (the one for the TMS methyl protons as well as the one for the $CHCl_3$ methine proton) on the y-axis. Because the TMS vector is rotating with the same frequency as the y-axis, the movement of the TMS vector becomes concealed in this rotating frame of reference, and it appears static along the y-axis. The frequency of precession of the C$\underline{H}$Cl$_3$ methine proton is, however, somewhat greater than that of the TMS vector, so in the time interval immediately after the pulse, it appears to move away clockwise with a *differential* angular velocity (Fig. 2.9). Similarly, if there are protons in a compound that resonate upfield from TMS (such as those in an organometallic substance), then after the 90° pulse their respective magnetization vectors would move away from TMS in the opposite (counter-clockwise) direction in the delay immediately after the 90° pulse. By convention, the x- and y-axis in the rotating frame of reference are differentiated by adding primes: x' and y'.

After the 90°_{x} pulse is applied, all the magnetization vectors for the different types of protons in a molecule will initially come to lie together along the y'-axis. But during the subsequent time interval, the vectors will separate and move away from the y'-axis according to their respective precessional frequencies. This movement now appears much slower than that apparent in the laboratory frame since only the *difference* between the reference frequency of the rotating frame and the precessional frequencies of the nuclei is observable in the rotating frame. If we consider the relative movement of two different vectors in the $x'y'$-plane, the angle between the faster vector and the slower vector will grow with time till it reaches 180°, then it starts decreasing. The two vectors thus repetitively come in phase and then go out of phase. The faster-moving vector corresponds to a downfield proton, while the slower-moving vector corresponds to an upfield proton. *Chemical shifts can therefore be considered in terms of differences between the angular velocities of the magnetization vectors of the nuclei, as compared to the angular velocity of the TMS vector.*

2.1.3   Free Induction Decay

As long as the bulk magnetization lies along the z-axis, there is no signal. Once a pulse is applied along the x'-axis, the magnetization M is bent away from the z-axis and a component is generated along the y'-axis. Since a pulse (a "hard" one) produces a wide band of frequencies with about the same amplitude, all the nuclei lying within the entire chemical shift range are simultaneously tipped by the same angle. If this is a 90°_x pulse which is applied, then the total magnetization M_0 will "flip" to the y'-axis; but if a pulse of a smaller pulse angle θ is applied, then the component aligned along the y'-axis will be correspondingly smaller ($M_0 \sin\theta$).

Problem 2.7

Following is a pictorial vector representation of a doublet. It shows the evolution of the components of a signal. Draw the positions of the vectors in the fourth and fifth frames, along with their directions of rotation.

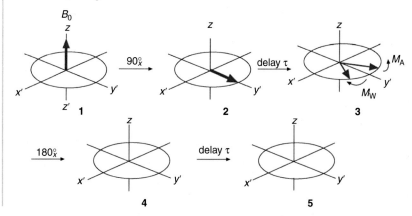

Problem 2.8

The effect of pulses on magnetization vectors is much easier to understand in the rotating frame than in the fixed frame. How do we arrive at the chemical-shift frequencies in the rotating frame?

Since the detector is conventionally regarded as being set up to detect signals along the y'-axis, the signal intensity will be at a maximum immediately

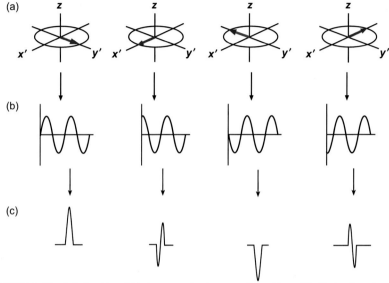

FIGURE 2.10   Relationship of (a) the magnetization vector's position with (b) the time-domain NMR signal and (c) the frequency-domain NMR signal (when detector is at y').

after the 90°_x pulse. During the subsequent delay, the magnetization vector will precess away from the $+y'$-axis and toward the $-x$-axis, and the signal will decay to a zero value when it reaches the x'-axis. As it moves further toward the $-y'$-axis, a negative signal appears that will reach its maximum negative amplitude when the vector points directly toward the $-y'$-axis. As the vector then moves toward the $-x'$-axis, the negative signal amplitude will again proceed to become zero and then grow to a maximum positive amplitude as it completes one full circle and reaches the $+y'$-axis. A sinusoidal variation of signal amplitude will thus occur. The relationship of the magnetization vector's position with the signal amplitude and signal phase is shown in Fig. 2.10.

Since the precessing magnetization vector will have a non-zero component in the $x'y'$ plane, this causes a corresponding oscillating voltage in the detection coils. Furthermore, the intensity of the signals does not remain constant but diminishes due to T_1 (spin-lattice) and T_2 (spin-spin) relaxation effects. The detector therefore records both the exponential decay of the signal with time and the interference effects in the form of a decaying beat pattern, the *pulse interferogram* or the *free induction decay* (indicating that it is *free* of the influence of the *Rf* field, that it is *induced* in the coils, and that it *decays* to a zero value). The time used to obtain the FID per scan is called the acquisition time (t_2), discussed in the following section.

Problem 2.9

In a decaying signal (FID) why does the amplitude decay asymptotically towards zero while the precessional frequency remains unchanged?

Problem 2.10

In practice, we sample the FID for a duration of about 2–3 seconds till most of it has been recorded. This means that a small portion of the "*tail*" of the FID will not be recorded. What effect would you expect this to have on the quality of the spectrum?

2.1.4 Fourier Transformation

The data on modern NMR spectrometers are obtained in the *time* domain (FIDs); i.e., they are collected and stored in the computer memory as a function of time. However, in order for such data to be used, they must be converted into the *frequency* domain, since the NMR spectrum required for interpretation must show characteristic line resonances at specific frequencies. The conversion of the data from the time domain to the frequency domain is achieved by *Fourier transformation*, a mathematical process that relates the time-domain data $f(t)$ to the frequency-domain data $f(w)$, described by the Cooley–Tukey algorithm (Eq. 2.1):

$$f(\omega) = \sum_{-t}^{+t} f(t)\exp\{i\omega t\}dt \qquad (2.1)$$

Apparently, the time-domain and frequency-domain signals are interlinked with one another, and the shape of the time-domain decaying exponential will determine the shape of the peaks obtained in the frequency domain after Fourier transformation. A decaying exponential will produce a Lorentzian line at zero frequency after Fourier transformation, while an exponentially decaying cosinusoid will yield a Lorentzian line that is offset from zero by an amount equal to the frequency of oscillation of the cosinusoid (Fig. 2.11).

Fourier transformation of the FIDs (which are in the time domain) produces frequency-domain components. If the pulse is long, then the Fourier components will appear over a narrow frequency range (Fig. 2.12), but if the pulse is narrow, the Fourier components will be spread over a wide range (Fig. 2.13). The time-domain signals and the corresponding frequency-domain partners constitute *Fourier pairs*.

Protons in different environments in a molecule will normally exhibit different chemical shifts (i.e., they will resonate at different frequencies) and different multiplicities (i.e., the individual components of each multiplet will resonate at their characteristic frequencies). Each of the individual components will contribute its decaying beat pattern to the FID, which will therefore represent a *summation* of the various decaying beat patterns. Fourier transformation

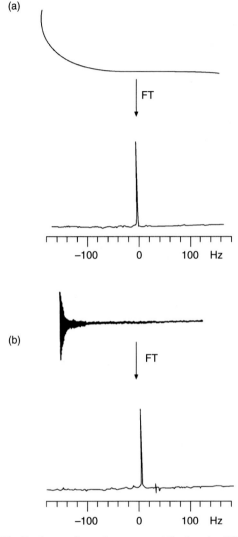

FIGURE 2.11 (a) The Fourier transform of an exponentially decaying FID yields a Lorentzian line at zero. (b) The FT of an exponentially decaying cosinusoid FID gives a Lorentzian line offset from zero frequency. The offset from zero is equal to the frequency of oscillation of the consinusoid. *(Modified from Homans, S. W. A dictionary of concepts in NMR, copyright © 1990. 127–29, by permission of Oxford University Press, Walton Street, Oxford OX2 6DP, U.K.)*

produces the NMR spectrum, in which the *positions* at which the individual signals appear will depend on the precessional frequencies of the respective nuclei, while the *signal widths* will depend on the life span of the decaying transverse magnetization. If the transverse magnetization decays slowly (i.e., if there is a long FID due to a long effective transverse relaxation time, T_2)

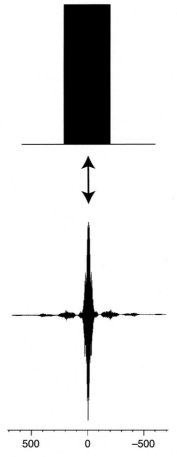

FIGURE 2.12 Fourier components of a long *Rf* pulse *(soft pulse)* are spread over a relatively narrow frequency range. *(Modified from Homans, S. W. A dictionary of concepts in NMR, copyright © 1990, 127–129, by permission of Oxford University Press. Walton Street, Oxford OX2 6DP. U.K.)*

then a sharp signal will be observed, but if there is a short FID, then a broad signal results (Fig. 2.14). If a molecule contains only one type of proton, then only a single peak will be observed after Fourier transformation. Such a simple FID will have only one decaying sinusoidal pattern. However, in a molecule containing several different types of nuclei, the FID obtained will be more complex, since it will be a summation of the various individual decaying beat patterns (Fig. 2.15).

Problem 2.11

Why is Fourier transformation essential in pulse NMR spectroscopy?

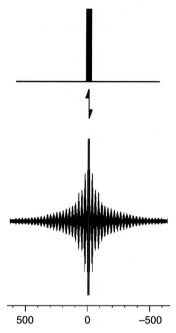

FIGURE 2.13   A short ("*hard*") *Rf* pulse has Fourier components spread over a relatively wide frequency range. *(Modified from Homans, S. W. A dictionary of concepts in NMR, copyright © 1990, 127–129 by permission of Oxford University Press. Walton Street, Oxford OX2 6DP. U.K.)*

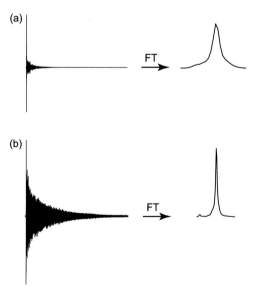

FIGURE 2.14   Free induction decay and corresponding frequency-domain signals after Fourier transformations. (a) Short-duration FIDs result in broader peaks in the frequency domain. (b) Long-duration FIDs yield sharp signals in the frequency domain.

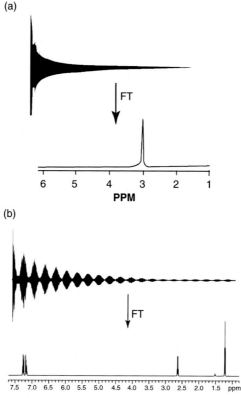

(a)

FT

6   5   4   3   2   1
PPM

(b)

FT

7.5 7.0 6.5 6.0 5.5 5.0 4.5 4.0 3.5 3.0 2.5 2.0 1.5 ppm

FIGURE 2.15   FIDs and Fourier-transformed frequency spectra. (a) Single-proton system display-ing only one decaying sinusoidal pattern. (b) Several different types of nuclei in a complex molecule yield a more complex FID pattern.

2.1.5   Data Acquisition and Storage

The free induction decay obtained in the NMR experiment is in the form of high-frequency electrical oscillations. These must be converted into numerical form before being stored in the computer memory. The electrical input is con-verted into binary output, representing the magnitude of the electrical voltage, by an analog-to-digital converter (ADC). To simplify calculations, the refer-ence frequency is subtracted from the observed frequency before the data are processed mathematically. For instance, on a 300-MHz (i.e., 300,000,000 Hz) instrument, the protons would normally resonate in the region between 0 and 12 ppm (0–3600 Hz, for 300 MHz instrument), so the frequencies to be pro-cessed would be between 300,000,000 and 300,003,600 Hz. Since it is only the difference between these two frequencies that concerns us, and since the use of such large frequencies would unnecessarily occupy computer memory, it is convenient first to subtract the reference frequency (300,000,000 Hz in this case) from the observed frequency, and to store the remainder (0–3600 Hz) in the computer memory.

After each pulse, the digitizer (ADC) converts the FID into a digital form and stores it in the computer memory. Ideally, we should keep on *sampling*, i.e., acquiring data till each FID has decayed to zero. This would require about $5T_1$ seconds, where T_1 is the spin-lattice relaxation time of the slowest relaxing protons. Since this may often take minutes, it is more convenient to sample each FID for only a few seconds. This means that the data obtained are from the larger-volume *"head"* of the FID, while a small portion of its *"tail"* is cut off. Since most of the required information is present in the *"head"* of the FID, we can thus avoid the time-wasting process of waiting for the *"tail"* to disappear before recollecting data from a new FID.

The next issue to consider is the rate at which the FIDs must be sampled. To draw a curve unambiguously, we need to have many data points per cycle. This becomes clear if we consider two different situations: in the first case, the signal is oscillating much faster than the rate at which the data are being collected; in the second case, the data are being collected faster than the *rate* of signal oscillation. Clearly, it is in the second case that we will obtain several data points per cycle, allowing a proper sinusoidal wave shape to be drawn. In the first case, the data collected will not correspond to any particular wave shape, and many different curves may be drawn through the data points.

To represent a frequency with accuracy, we must have *at least* two data points per cycle, so that a unique wave can be drawn that would pass through *both* data points. Thus, in Fig. 2.16a, only one data point has been registered per cycle, and the resulting ambiguity is apparent, since many curves can be drawn through these points. In Fig. 2.16, however, two data points have been registered per cycle, so only one curve can be drawn through them. This causes a problem: on a 500-MHz instrument, the signal will have to be sampled at twice its oscillator frequency, i.e., at 1000 MHz. In other words, we will need to collect one data point every $1/(10^6 \times 10^3)$ s, i.e., every 1 ns, for a second or longer.

Collecting such a large amount of data would require highly sophisticated and expensive instrumentation and is unnecessary, since we are really interested in the signals appearing in a rather narrow frequency range.

For instance, on a 500-MHz instrument, a 15 ppm range would represent 7500 Hz, and because we are interested only in the *difference* between the frequencies at which the signals of one sample appear and the frequencies of a reference standard, such as TMS, the sampling can be carried out at low frequencies (which can be digitized easily) even though detection is being done at much higher frequencies (radiofrequency range). In the earlier example, the minimum number of words of data storage per second required to cover the entire spectral width of 7500 Hz (0–15 ppm) would be $2 \times 7500 = 15,000$, and the *sampling rate* would need to be 1/15,000 s (0.066 ms). The relationship between the various experimental parameters when sampling NMR data is shown in Fig. 2.17.

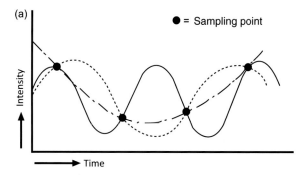

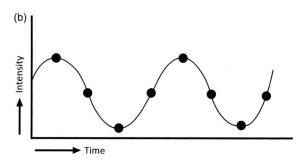

FIGURE 2.16   (a) If fewer than two data points are sampled per cycle, then more than one curve can be drawn through the points. This will result in an ambiguous wave shape. (b) Sampling of more than two data points per cycle removes the uncertainty and yields a particular wave shape.

The frequency at which the signals must be sampled is determined by the maximum separation between the signals in a spectrum, i.e., by the highest frequency to be characterized. If this is S Hz, then according to the Nyquist theorem, the signals will need to be sampled every $\frac{1}{2}S$ s. Moreover, to distinguish spectral features separated by $\Delta\delta$ Hz, the sampling must be continued for $1/\Delta\delta$ s. The total number of points that are sampled, stored in computer memory, and subjected to Fourier transformation is given by $2S/\Delta\delta$. In practice, we actually set the dwell time between the data points on the computer (which is equal to $\frac{1}{2}$ SW, where SW is the sweep width) and hence indirectly set the required sweep width. The design of the timers installed in NMR spectrometers does not allow them to be set at all possible time durations, so the computer program will select the dwell time that is slightly shorter than the required *dwell time*, and the spectral window will therefore be slightly larger than the required window.

Problem 2.12

Why is it necessary to subtract the reference frequency from the observed frequency before data storage and processing?

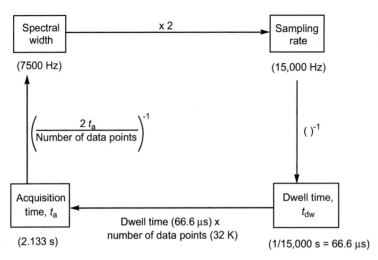

FIGURE 2.17 Interrelation of experimental parameters in FT NMR.

2.1.6 Digital Resolution

Even when the NMR spectrometer has been adjusted for optimum resolution, the spectrum may still have a poor resolution if enough computer memory was not allocated for the purpose, i.e., it will have poor *digital resolution*. A large allocation of computer memory (i.e., a greater number of words of data storage) requires a corresponding increase in acquisition time. The maximum acquisition time, in seconds, required after each pulse is given by Eq. 2.2:

$$\frac{Number\ of\ words\ of\ data\ storage}{2 \times spectral\ width\ in\ Hz} \qquad (2.2)$$

In the example given in the preceding section, the number of words of data storage was 15,000 and the spectral width was 7,500 Hz, so an acquisition time of $15,000/(2 \times 7,500) = 1.0$ s was required after each pulse.

The resolution is controlled not only by such factors as field homogeneity and the intrinsic nature of the compound (such as the presence of exchange or restricted rotation), but also by the *sampling rate*. With a spectral width of 7,500 Hz on a 500-MHz instrument, the sampling rate was $1/(2 \times 7,500)$ s, i.e., 0.066 ms. An acquisition time of 1.0 s requires 1.0 s/0.066 ms $\cong$ 15,000 data points of computer memory.

Suppose this gave a digital resolution of 1.0 Hz when the desired resolution was 0.2 Hz. A five-fold increase in digital resolution would now necessitate a corresponding five-fold increase in the number of data points (i.e., 75,000 instead of 15,000) to improve the line resolution from 1.0 to 0.2 Hz. Alternatively, we could maintain the same number of data points (15,000) but use a five-times smaller spectral width, i.e., scan in a 3 ppm region rather than the original spectral width of 15 ppm to improve the digital resolution from 1.0 to 0.2 Hz. As already indicated, there are other limiting factors that control the maximum achievable resolution, such as field inhomogeneity, and increase in the sampling rate would only improve the resolution to a certain point.

It is clear that the digital resolution (DR) depends on the amount of computer memory used in recording the spectrum. If the memory size is M, then there will be $M/2$ real and $M/2$ imaginary data points in the frequency spectrum, and the separation (in hertz) between these data points (DR) will be given by Eq. 2.3:

$$DR = \frac{2SW}{M} \tag{2.3}$$

where SW represents the sweep width. In a spectrum recorded with 16K ($2^{14} = 16,384$) data points, there will be 8K (8,192) real and 8K imaginary data points. For a spectral width of 6,000 Hz (12 ppm on a 500-MHz instrument), the digital resolution per point will be $6,000/8,192 = 0.73$ Hz (i.e., the accuracy of measurement will be ±0.73 Hz), so it will be impossible to differentiate between lines separated by less than 1.46 Hz.

If we wish to measure at a higher digital resolution, then the same spectrum could be recorded at, say, 32 K (16,384 real and 16,384 imaginary data points), giving a digital resolution of $6,000/16,384 = 0.365$ Hz. The same improvement could be achieved by halving the spectral width to 3,000 Hz and by measuring the spectrum at the same 16K digital resolution ($3,000/8,192 = 0.365$ Hz). Working at high digital resolutions increases the computing time, so spectra are normally not recorded beyond 32 or 64 K. If increase in digital resolution does not improve resolution, then field homogeneity, instrumental factors, or the intrinsic nature of the compound may be the limiting considerations.

We should decide in advance regarding the digital resolution at which we wish to acquire a spectrum and then set the acquisition time accordingly. The acquisition time AT (i.e., the product of the number of data points to be collected and the dwell time between the data points) is calculated as simply the reciprocal of the digital resolution (Eq. 2.4):

$$AT = \frac{1}{DR} \tag{2.4}$$

Problem 2.13

A poor digital resolution will result in loss of some of the fine structures of an NMR signal. To increase the digital resolution, we need either to maintain the same number of data points but reduce the spectral width or, alternatively, to maintain the spectral width but increase the number of data points. Which method would you prefer for achieving a better signal-to-noise ratio?

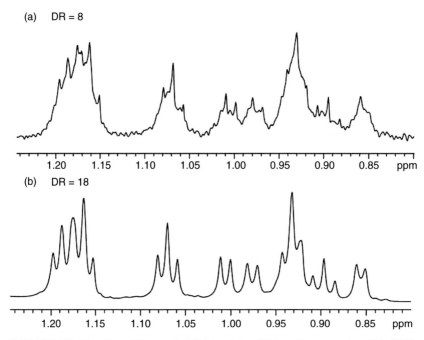

FIGURE 2.18 The effect of increased digital resolution (DR) on the appearance of the NMR spectrum. (a) The spectrum of cyclosporin recorded at a digital resolution of 0.1 Hz per point, allowing the spectral lines to be seen at their natural line width. (b) The spectrum of the same molecule recorded at a digital resolution of 0.4 Hz per point.

Figure 2.18 shows the effect of digital resolution on the appearance of signals. It is important to remember that *each* data point in the FID contains information about *every* peak in the spectrum, so that as more data points are accumulated and the FIDs acquired for a longer period, a finer digital resolution is achieved.

2.1.7 Peak Folding

If too small a spectral width is chosen, the signal lying outside the selected region will still appear in the spectrum, but at the wrong frequency. This is because the peaks lying outside the spectral width *fold over* and become superimposed on the spectrum (Fig. 2.19). The folded (or *aliased*) peaks can usually be recognized easily, since they show different phasing than the other peaks. To check if a peak is indeed a folded one, the spectral window may be shifted (say, by 500 Hz) to one side. All the "normal" peaks will then shift in the same direction by exactly this value, but the folded peaks will either shift in the wrong direction or will shift by the wrong value.

If too large a spectral width is chosen to avoid the folding of peaks, much of the computer memory will be wasted in storing noise data lying beyond the

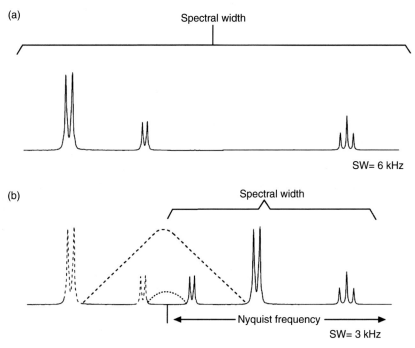

FIGURE 2.19 (a) Normal NMR spectrum resulting from the correct selection of spectral width. (b) When the spectral width is too small, the peaks lying outside the spectral width can fold over. Thus a' and b' represent artifact peaks caused by the fold-over of the a and b signals.

frequency range of the spectrum. The spectral width should therefore be chosen to be just wide enough to cover the region in which the signals are likely to appear. Even if the correct spectral width is chosen, the noise lying within the spectral region would still be digitized and processed. The magnitude of this problem can be reduced by filtering off the noise before digitization. Modern NMR spectrometers are fitted with filters that remove most of the noise, but they also somewhat reduce signal intensities within the spectral range while cutting them off outside the spectral range.

2.1.8 Dynamic Range Problem

The analog-to-digital converter (ADC) receives the FID signals in the form of electrical voltages and converts them into binary numbers proportional to the voltages stored in computer memory. To a great extent, the *digitizer resolution* (which is different from the *digital resolution* discussed earlier) determines the *efficiency* of this process. The digitizer resolution is dependent on the *word length* used in the ADC, i.e., the number of bits (binary 1 or 0) used to sample the electrical voltages. An 8-bit digitizer can process numbers from 00000000 binary (which is 0 in decimal) to 11111111 (which is 255 in decimal, or $2^8 - 1$). Thus, an 8-bit digitizer would accept the largest signal with an output of 255 and the smallest signal with an output of 1. The *dynamic range* of the 8-bit digitizer would then be described as 255:1.

The NMR signal should fit into the *dynamic range* properly; i.e., the gain of the instrument must be adjusted correctly. If the gain is too high, then the largest signal will not fall within the dynamic range of the digitizer and an ADC overflow will result, causing baseline distortions. If, however, some of the signals are too small, then they will be below the detection limits of the digitizer and will not be observed. To cope with this problem, modern NMR spectrometers are fitted with 12-bit digitizers with a dynamic range of 4095:1 or 16-bit digitizers with a dynamic range of 65,535:1.

If the receiver gain is set very low, then only a fraction of the available digitization levels of the ADC is utilized so that the intensity at each point of the FID has only a few possible values and the FID presents a "choppy" appearance. The NMR spectrum will then contain a lot of digital noise and a low signal-to-noise (*S/N*) ratio will result. As we increase the receiver gain, the FID becomes digitized with more digitization levels. This results in a greater increase in the level of the signal as compared to the noise, resulting in a better signal-to-noise ratio. By increasing the receiver gain setting, the signal-to-noise ratio would be improved till a point is reached when the signal exceeds the limit of the ADC. This will cause the FID to be clipped at the beginning, and the NMR spectrum will appear to be very distorted (Davis et al., 1985; Roth et al., 1980).

This occurs because the thermal noise in the FID at low settings of the receiver gain is smaller than the size of the digitization step of the ADC, and it will not be amplified by increasing the receiver gain till it is larger than the size of the digitization step of the ADC. Once this threshold is reached, it will be amplified in the same way as the signal.

Problem 2.14

What does the "ADC overflow" error message mean and how does one address the problem?

Problem 2.15

If the spectral width is inadequate to cover every peak in the spectrum, then some peaks in the downfield or upfield region may fold over and appear superimposed on the spectrum. How can you identify these folded signals?

The dynamic range problems become acute if we are trying to record a spectrum in which there is a large size difference between the largest and the smallest signals, such as in the spectra of an organic molecule in the presence of a signal for the H_2O protons due to large moisture content, present as an impurity. Attempts to fit the large hydroxylic signal within the dynamic range may make the signals for the protein molecule fall below the digitizer

(a) FID falls within the dynamic range

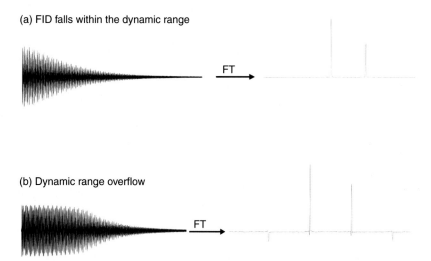

FT

(b) Dynamic range overflow

FT

FIGURE 2.20 (a) When the entire FID falls within the dynamic range of the digitizer, a regular-shaped signal results. (b) Dynamic range overflow (non-recording of the top and bottom of the head of the FID) reduces S/N ratio and distorts signal shapes.

detection limit. Careful adjustment of the gain and special water-suppression pulse sequences may be used to solve this problem.

One way to solve the dynamic range problem might be simply to adjust the gain in order to bring the smaller signals within the dynamic range and allow the intense FID signals to overflow, i.e., cut off a part of the "head" of the FID that contains the most intense signals. However, as stated earlier, *each* data point in the FID contains information about all parts of the spectrum. Cutting off the top and bottom sections of the "head" of the FID will therefore reduce the signal-to-noise ratio, distort the peaks (with negative peaks appearing on the sides of the signals), and give the appearance of new artifact peaks (Fig. 2.20).

With larger-capacity digitizers, the digitization rate is slower and the spectral width is also reduced. Thus, a 12-bit digitizer operates at a maximum speed of 300 kHz, requiring 3 ms for each sampling to characterize a spectral range of 150 kHz, while a 16-bit digitizer would reduce the spectral range to 50 kHz. In spite of the slower digitization rate, it is still preferable to use larger-capacity digitizers to detect the smaller signals, which would otherwise be lost in the quantization noise caused by digitization at smaller word lengths (Fig. 2.21).

Problem 2.16

The dynamic range of the digitizer is important. The NMR signal should fit into the digitizer appropriately; i.e., the gain must be adjusted properly. What will happen if there is a large solvent signal that makes the gain too high?

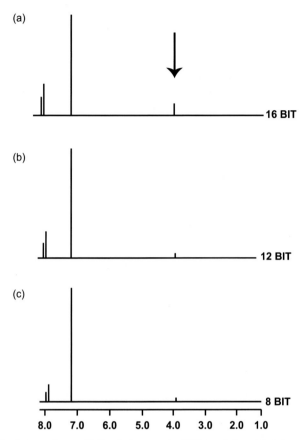

FIGURE 2.21 Increase in sensitivity *via* increase in ADC resolution. (a) A tiny signal of an impurity at δ 3.90 appears clearly with a 16-bit digitizer. (b) The signal at δ 3.90 is smaller when a 12-bit digitizer is used. (c) The signal at δ 3.90 is hard to see when the spectrum is recorded with an 8-bit digitizer.

2.1.9 Oversampling

It is important therefore to take care that the receiver gain is neither set too high nor too low. A problem can arise, particularly when one is recording the NMR spectrum of a mixture. If one wishes to observe the small signals of the minor components in the presence of much stronger signals from major components, then if one sets the receiver gain too high, it would cause an overload with accompanying distortions but if the gain is too low, then the signals of the minor components would be too weak.

This difficulty can be addressed by *oversampling* the FID. This is achieved by digitizing the spectral data at a much faster rate than that required to satisfy the Nyquist condition. This results in a substantial reduction of the digitization noise. If the oversampling factor is N, then the digitization noise will be reduced by a factor of $N^{1/2}$. Oversampling factors for ^{1}H-NMR are often kept at 16 or 32.

2.1.10   Quadrature Detection – Digital Signal Detection

So far, we have been concerned with a single detector measuring only the y-component of the magnetization. In such a single-detection system, we can measure only the *frequency difference* between the reference frequency and the frequency of the signal, but not its *sign*; i.e., it is impossible to tell whether the signal frequency is greater or less than the reference frequency. If the *Rf* pulse is placed at one end of the spectrum, then all the signals will appear on one side of the pulse frequency while only noise will lie on the other side. Because the positive and negative frequencies cannot be distinguished, this noise will fold over and come to lie on the top of the spectrum, thereby decreasing the *signal-to-noise ratio* by a factor of $\sqrt{2}$ or 1.4.

Clearly, this is not acceptable. In modern NMR spectrometers, the irradiating frequency is therefore placed in the middle of the spectrum rather than at one end. To distinguish between the signals lying above and below the reference frequency, the so-called *quadrature detection* method is employed (Gengying and Haibin, 1999; Villa et al., 1996; Samuelson et al., 1977). This involves detecting not only the magnetization component along the y-axis, but also those along the x-axis by employing two detectors to detect the signals from the same coil, with their reference phases differing by 90°. This allows us to discriminate between the sign of the detected frequencies. The x-magnetization M_x results in a signal of magnitude $\cos \Omega t$ in the x-channel, while the y-magnetization M_y creates a signal of magnitude $\sin \Omega t$ in the y-channel.

One of these FIDs is termed a "real" FID while the other is termed an "imaginary" FID. Fourier transformation of these FIDs produces a "real" spectrum and an "imaginary" spectrum. This allows us to phase one of the NMR spectra (e.g., the "real" one) to afford an absorption mode spectrum. The real and imaginary spectra can then be added together to afford an increase in the signal-to-noise ratio by $\sqrt{2}$ (Hoult and Richards, 2011). A block diagram of how the data is collected in the quadrature mode is given below (Fig. 2.22):

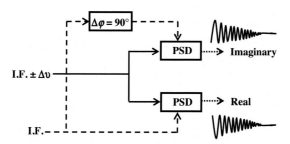

FIGURE 2.22   Block diagram of data collection in quadrature mode.

The broken line (signal) corresponds to the intermediate frequency (IF) of the spectrometer. The solid line (signal) represents the intermediate frequency of the spectrometer which undergoes modulation by the NMR spectral data (IF ± Δv). The real and imaginary data are detected by two phase sensitive detectors (PSDs) that differ in phase by 90 degrees. These devices produce a difference signal between the two inputs. The inputs to the first phase sensitive detector are the IF and the IF plus or minus the spectral data (Δv). The output is therefore the information content of the NMR spectrum (± Δv). The second PSD has the same inputs as the first PSD except that the phase of the IF is shifted by 90°. The output of the second PSD is therefore similar to that of the first PSD except that it has a 90° phase difference. The two outputs correspond to the "real" and "imaginary" FIDs which are subjected to Fourier transformation.

Two different methods have been employed for quadrature detection. In the first method, the two signals along the x- and y-axes are collected *simultaneously*, while in the other method they are collected *sequentially*. This has implications in two-dimensional spectra in which the spectral widths may be defined differently in the two dimensions. The sequence of 90° phase shifts used in the quadrature phase-cycling routine (also known as *cyclops*) results in the magnetization being successively generated along the y-, x-, $-y$-, and $-x$-axes, corresponding to the 0°, 90°, 180°, and 270° phase shifts, respectively. The data from the four separate phase shifts are co-added in the computer.

The underlying principle of this Redfield technique is illustrated in Fig. 2.23. Since the spectral width is now reduced, the noise lying outside the spectral region cannot fold back onto the spectrum, resulting in an improved signal-to-noise ratio (Fig. 2.24). For simplicity, we can assume that one of the two phase-sensitive detectors is set up correctly to detect the cosine (odd) component of the magnetization while the other detects the sine (even) component (actually each detector detects both components). The two signals, corresponding to the real and imaginary parts of a complex spectrum, are digitized separately and subjected to Fourier transformation so that one line in the frequency domain is reinforced while the other is cancelled (Fig. 2.25). This allows us to distinguish between the signs of the signals, i.e., whether they are at higher or lower frequencies than the reference frequency.

Since quadrature detection involves the cancellation of an unwanted component by adding two signals that have been processed through different parts of the detection system, they will cancel out completely only if their phases differ by precisely 90° and if their amplitudes are exactly equal.

In practice, this may not be achieved perfectly, so weak *quad images* may be produced. They can be readily recognized as they show different phases than the rest of the spectrum.

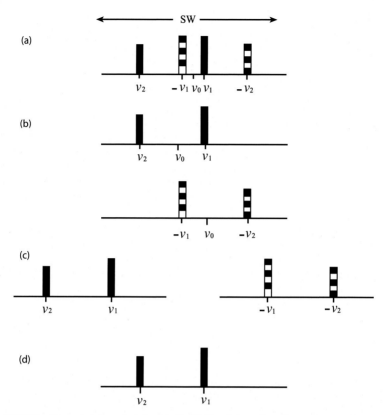

FIGURE 2.23 The underlying principle of the Redfield technique. Complex Fourier transformation and single-channel detection gives spectrum (a), which contains both positive and negative frequencies. These are shown separately in (b), corresponding to the positive and negative single-quantum coherences. The overlap disappears when the receiver rotates at a frequency that corresponds to half the sweep width (SW) in the rotating frame, as shown in (c). After a real Fourier transformation (involving folding about $n_1 = 0$), the spectrum (d) obtained contains only the positive frequencies.

2.1.11 Signal-to-Noise Ratio

As stated earlier, the spectral width of the frequencies is determined by the rate at which the data are collected; so in order to have a spectral width of W Hz, the data must be sampled at $2W$ Hz. In other words, to collect data over wider spectral widths, we must collect data more *slowly*. However, because the *accuracy* with which the frequencies are measured depends on the length of time spent in collecting data points *in each* FID, to obtain a high resolution spectrum we need to collect data points for a greater *length of time*. Fourier transformation converts the FID, which is in the time domain, to the spectrum, which is in the frequency domain (2^N data points in the time domain giving 2^{N-1} points in the frequency domain).

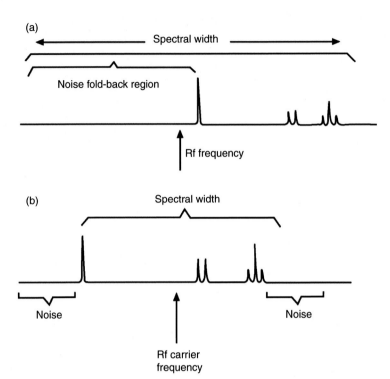

FIGURE 2.24   (a) Reduced *S/N* ratio resulting from noise folding. If the *Rf* carrier frequency is placed outside the spectral width, then the noise lying beyond the carrier frequency can fold over. (b) Better *S/N* ratio is achieved by quadrature detection. The *Rf* carrier frequency in quadrature detection is placed in the center of the spectrum. Due to the reduced spectral width, noise cannot fold back onto the spectrum.

Problem 2.17

A serious problem associated with quadrature detection is that we rely on the cancellation of unwanted components from two signals that have been detected through different parts of the hardware. This cancellation works properly only if the signals from the two channels are exactly equal and their phases differ from each other by exactly 90°. Since this is practically impossible with absolute efficiency, some so-called "image peaks" occasionally appear in the center of the spectrum. How can you differentiate between genuine signals and image peaks that arise as artifacts of quadrature detection?

In the older CW instrument it was necessary to scan slowly from one end of the spectrum to the other, and record excitations of nuclei as they came to resonance sequentially. However, since at any one time only nuclei in a specific region of the spectrum are being subjected to excitation while for the remainder of the scan time only the baseline noise or other spectral regions are being scanned, acquisition of spectra on continuous-wave instruments involves the time-wasting process of scanning one region at a time.

(a)

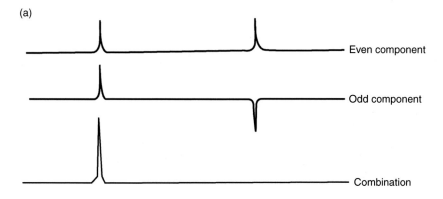

(b)

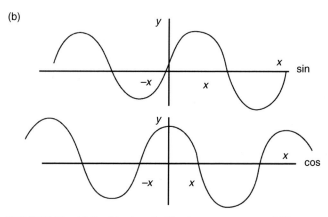

FIGURE 2.25   (a) Combination of odd and even components yields a spectrum with cancellation of signals through the addition of opposite-phased signals. (b) Sine waves are odd, and cosine waves are even. *(Reprinted from Derome, E. Modem NMR techniques for chemistry research, copyright © 1987, 78–79, with permission from Pergamon Press Ltd., Headington Hill Hall, Oxford OX3 OBW, U.K.)*

Fourier transform NMR spectroscopy overcame this problem because all the nuclei in the spectral range of interest are excited simultaneously by a short, sharp burst of radiofrequency energy (pulse). On the modern PFT instruments, the FIDs can therefore be obtained much more rapidly, with only a certain relaxation delay being inserted between successive scans. Since the sample signals appear at exactly the same frequency while the noise signals vary in their positions, the noise tends to grow at a slower rate. If n scans are accumulated, the signal will grow n times while the noise will grow by a factor of $\sqrt{n}$, resulting in an improvement of the signal-to-noise ratio by $\sqrt{n}$. Thus, if a certain signal-to-noise ratio is achieved with 64 scans ($\sqrt{64} = 8$), then to double this ratio we would need to accumulate 256 scans ($\sqrt{256} = 16$), which would take four times longer than did accumulating the original 64 scans. Simply doubling the scan time, therefore, does not double the signal-to-noise ratio, a common misconception. With small sample quantities, a point is soon reached beyond which it is no longer feasible to devote more instrument time to accumulate additional scans for improving the signal-to-noise ratio. It is also possible to obtain an improved signal-to-noise ratio by manipulating the spectrum through digital filtering and other apodization techniques that will be discussed later (Moskau, 2002).

The maximum signal intensity is obtained if the magnetization vectors point directly to the y'-axis, i.e., if a 90° pulse is employed. However, the nuclei would require a certain time, depending on the relaxation rate R, to relax back to their thermal equilibrium state before a new pulse is applied. If the pulses are applied too rapidly so that the nuclei cannot relax sufficiently, then a state of saturation would soon be reached and the signal will disappear. Moreover, if all the nuclei do not have the same relaxation rates (as is usually the case), and if the spectrum is recorded before they have relaxed to their respective equilibrium states, the signal intensities will not correspond to their respective integrations. If, however, we wait long enough between the pulses so that the z-magnetization of all the nuclei is fully restored (normally, $5T_1$ of the slowest-relaxing nuclei in the molecule), then much time would be spent between successive scans that an insufficient number of scan may be accumulated in a given time. A compromise between these two extremes is therefore desirable. It is more time-efficient to tip the nuclei by a smaller angle so that they take less time to relax back to their respective equilibrium states, allowing the spectra to be recorded with smaller time delay intervals between successive scans. The optimized pulse angle α (Ernst angle) is given by the equation $\cos\alpha = \exp(-t_r/T_1)$ were t_r is the repetition time between the individual pulses and T_1 is the open-lattice relaxation time.

The relaxation rates of the individual nuclei can be either measured or estimated by comparison with other related molecules. If a molecule has a very slow-relaxing proton, then it may be convenient *not* to adjust the delay time with reference to that proton and to tolerate the resulting inaccuracy in its intensity but adjust it according to the *average* relaxation rates of the other protons. In 2D spectra, where 90° pulses are often used, the delay between pulses is typically adjusted to $3T_1$ or $4T_1$ (where T_1 is the spin-lattice relaxation time) to ensure that no residual transverse magnetization from the previous pulse (that could yield artifact signals) remains. In 1D proton NMR spectra, the tip angle θ is usually kept at $30°-40°$.

The signal-to-noise ratio (S/N), commonly referred to as sensitivity, is a key factor in the development and applications of NMR spectroscopy. Due to its central importance in NMR spectroscopy, it is separately discussed in Chapter 3.

Problem 2.18

What is a signal-to-noise (S/N) ratio, and how can it be improved by acquiring a large number of FIDs?

2.1.12 Apodization

Having recorded the FID, it is possible to treat it mathematically in many ways to make the information more useful by a process known as *"apodization"* or window function (Lindon and Ferrige, 1980; Ernst, 1966). By choosing the right "window function" and multiplying the digitized FID by it, we can improve either the signal-to-noise ratio or the resolution. Some commonly used apodization functions are presented in Fig. 2.26.

Problem 2.19

The signal-to-noise ratio can be increased by treating the data so as to bias the spectrum in favor of the signals and against the noise. This can be done by multiplying the FIDs by the proper apodization functions. What would happen if the spectrum is recorded without apodization?

Problem 2.20

Apodization is likely to change the relative intensities of signals with different line widths. Can it also affect the chemical shifts of the signals?

2.2 SENSITIVITY ENHANCEMENT

To increase the signal-to-noise ratio, we need to multiply the FIDs by a window function that will reduce the noise and lead to a relative increase in signal strength. Since most of the signals lie in the "head" of the FID while its "tail" contains relatively more noise, we multiply the FID by a mathematical function that will emphasize the "head" of the FID and suppress its "tail" (Spencer and Traficante, 1998; Pacheco and Traficante, 1996).

This can be done simply by multiplying each data point in the FID by an exponential decay term that starts at unity but decays to a negligible value at its end. Such an exponential multiplication (EM) is a simple and effective way to increase the signal-to-noise ratio at the expense of added line broadening (LB). The LB term in this function can be altered by the operator, the larger the value of LB, the more rapid will the apodized FID decay, leading after Fourier transformation to some broadening of lines (i.e., reduction in resolution) but often with significant improvement in the signal-to-noise ratio. If the lines are already rather broad, then LB should be chosen to correspond to the existing line widths; if the lines are narrow, then LB may be set to correspond to the digital resolution. If a negative LB value is chosen, then an opposite effect will be observed after Fourier transformation, and resolution will be improved, i.e., the lines will be sharper but there will also be more noise. Such a "resolution-enhanced" spectrum will have a lower signal-to-noise ratio. Figure 2.27 shows the effects of multiplying an FID when various LB values are included in the exponential multiplication.

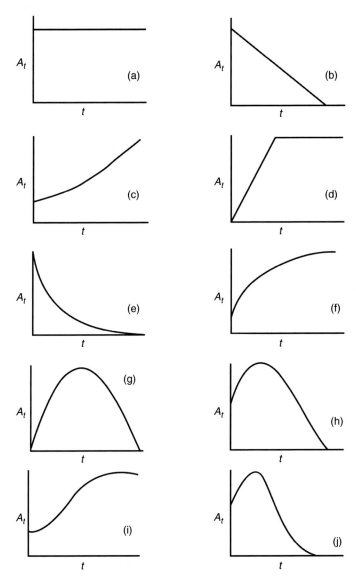

FIGURE 2.26 Various selected apodization window functions: (a) an unweighted FID; (b) linear apodization; (c) increasing exponential multiplication; (d) trapezoidal multiplication; (e) decreasing exponential multiplication; (f) convolution difference; (g) sine-bell; (h) shifted sine-bell; (i) LIRE; (j) Gaussian multiplication. *(Reprinted from Lindon and Ferrige, Progress in NMR Spectroscopy, copyright © 1980, 27–66, with permission from Pergamon Press Ltd., Headington Hill Hall, Oxford OX3 OBW, U.K.)*

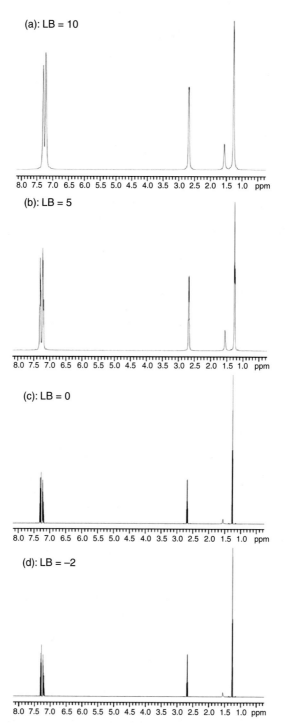

FIGURE 2.27 The effect of line broadening (LB) multiplication on the appearance of ¹H-NMR spectra. ¹H-NMR spectra recorded after multiplying the FIDs with positive LB values: (a: LB = 10), (b: LB = 5), ¹H-NMR spectra and the same ¹H-NMR spectrum recorded without line broadening (c: LB = 0). Sharper signals are obtained when the FID is multiplied by negative LB values (d: LB = − 2).

2.3 RESOLUTION ENHANCEMENT

The process of exponential multiplication just described produces a rapid decay of the FID and the production of broad lines; suppressing the decay of the FID gives narrow lines and better resolution, with increased noise level. An alternative approach to resolution enhancement is to reduce the intensity of the earlier part of the FID. Ideally, we should use a function that reduces the early part of the FID, to give sharper lines, as well as reduces the tail of the FID, to give a better signal-to-noise ratio.

A simple way to do this is to multiply by a symmetrical shaping function, such as the *sine-bell* function (Marco and Wüthrich, 1976), which is zero in the beginning, rises to a maximum, and then falls to zero again, resembling a broad inverted cone (Fig. 2.26g). One problem with this function is that we cannot control the point at which it is centered, and its use can lead to severe distortions in line shape. A modification of the function, *the phase-shifted sine bell* (Wagner et al., 1978) (Fig. 2.26h), allows us to adjust the position of the maximum. This leads to a lower reduction in the signal-to-noise ratio and improved line shapes in comparison to the sine-bell function. The *sine-bell squared* and the corresponding *phase-shifted sine-bell squared* functions have also been employed (see Section 3.5.1 also).

Gaussian multiplication (Marco and Wüthrich 1976; Ernst, 1966) has been used widely for resolution enhancement without significant loss of sensitivity in 1D NMR spectra. There are two parameters altered by the operator: Lorentzian Broadening (LB) controls the line widths (negative values lead to ·sharper lines) (Fig. 2.27), and Gaussian broadening (GB) determines the position of the maximum of the function. LB initially may be set at -2 or at 3DR (where DR is digital resolution) and GB should be located at the point where the FID begins to merge into the noise. Increasingly negative values of LB lead, up to a point, to sharper lines, but with corresponding decreases in the signal-to-noise ratio; smaller values of GB tend to prevent the lowering of the signal-to-noise ratio.

The existence of a high *signal-to-noise* ratio in the collected data therefore allows the use of resolution enhancement functions with greater freedom, since it offers the operator greater flexibility in sacrificing some of the signal-to-noise ratio to obtain the desired resolution. If the signal-to-noise ratio in the original data is low to start with, then multiplication by a resolution enhancement function may lead to weakening of the signals to the extent that they become indistinguishable from the noise.

Problem 2.21

Define sensitivity and resolution in NMR spectroscopy.

Problem 2.22

What changes in the line shape of an NMR spectrum occur after resolution enhancement?

Problem 2.23

Why are pulse Fourier transform (PFT) NMR experiments preferred over continuous wave (CW) NMR techniques?

Problem 2.24

How does saturation affect the sensitivity of an NMR experiment?

Problem 2.25

Summarize the common methods for enhancing sensitivity in NMR spectroscopy.

2.4 PULSE WIDTH CALIBRATION

The power (strength) of a pulse depends directly on its frequency. Thus pulses that are weak (or "soft") are those that have a low frequency, that is less cycles per second. Strong (or more powerful) pulses are those that have a high frequency i.e. a larger number of cycle per second. A pulse with a lower "pulse width" (t_p), (a weak or "soft" pulse), has a pulse width in the millisecond range while a hard ("strong") pulse will have a pulse width (t_p), in the microsecond range.

It is important that the pulse widths be calibrated accurately, since the NMR experiments cannot be performed properly if the wrong flip angles are chosen. The flip angle may not be critical in simple one-dimensional NMR experiments, but it is still important to know the duration of the 180° pulse. In two-dimensional NMR experiments, however, an accurate knowledge of pulse widths is essential. In some heteronuclear 2D NMR experiments in which a proton or X-band decoupler is used as a pulse transmitter, it may be necessary to determine the pulse widths of the decoupler amplifiers. In 2D multiple quantum coherence NMR experiments, even a slight deviation from the required pulse angle can adversely affect the outcome of the experiment, making the accurate determination of the relationship between spin flip angle and pulse duration even more critical. Also, the pulse calibration must be repeated regularly. Since pulse widths depend on the power output of the transmitter and decoupler, and since power levels can vary with time, pulse widths determined on one day may not be valid on the following day. In certain 2D NMR experiments, it is also advisable to calibrate the pulse width immediately before the start of the experiment.

It is important to avoid saturation of the signal during pulse width calibration. The Bloch equations predict that a delay of $5*T_1$ will be required for complete restoration to the equilibrium state. It is, therefore, advisable to determine the T_1 values; an approximate determination may be made quickly by using the inversion-recovery sequence (see the next paragraph). The protons of the sample on which the pulse widths are being determined should have relaxation times of less than a second to avoid unnecessary delays in pulse width calibration. If the sample has protons with longer relaxation times, then it may be advisable to add a small quantity of a relaxation reagent, such as $Cr(acac)_3$ or $Gd(FOD)_3$ to induce the nuclei to relax more quickly.

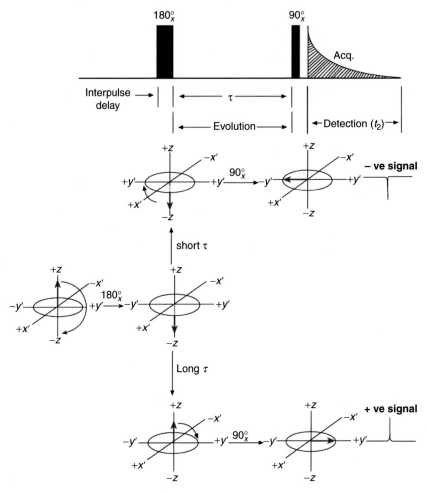

FIGURE 2.28 The effect of the duration of the evolution delay τ on the spin-lattice relaxation in an inversion-recovery experiment is shown. A short duration produces only minimal relaxation along the z-axis. Thus, when a 90°_x pulse is applied to the evolved magnetization, the z-magnetization vector (which points to the $-z$-axis) will be rotated to the $-y'$-axis, ultimately giving rise to a *negative* signal in the spectrum (assuming the detector lies along y'). Longer duration of the evolution delay results in a restoration of the $+z$-magnetization so that the 90°_x pulse will cause the magnetization to rotate to the $+y'$-axis, producing a *positive* signal.

The inversion-recovery pulse sequence used to determine T_1 values is shown in Fig. 2.28. The first 180° pulse causes the magnetization of the protons to be inverted so that it comes to rest along the $-z$-axis. During the subsequent evolution period τ, the magnetization would relax from the $-z$-axis back toward its original equilibrium position along the $+z$-axis. The return to equilibrium is not instantaneous, but usually takes place with a first-order

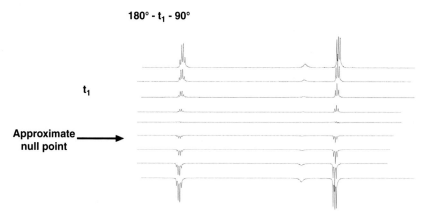

180° - t₁ - 90°

t₁

Approximate
null point

FIGURE 2.29   Representation of an inversion-recovery T_1 experiment.

rate constant R_1. In the study of nuclear relaxations involving time-dependent measurements, however, it is more convenient to consider the relaxation-time constant $T_1(= 1/R_1)$ rather than the rate constant R_1. To convert the longitudinal magnetization into a signal, it must first be converted into transverse magnetization (i.e., brought in the $x'y'$-plane) by the application of a subsequent 90°_x pulse. If this 90°_x pulse is applied very soon after the original 180°_x pulse (i.e., if the evolution period or delay τ is very short), then the $90°x$ pulse will "catch" the magnetization while it is still on the $-z$-axis and cause it to rotate to the $-y'$-axis, thereby giving a negative signal. As the evolution period τ is increased progressively, the 90°_x pulses will "encounter" the z-magnetization as it recedes along the $-z$-axis to zero value. With still longer τ intervals, it would grow along the $+z$-axis toward its original equilibrium value, the one that existed before application of the 180°_x pulse. This would produce a series of corresponding spectra in which the signals initially have decreasing negative amplitudes and then, after going through a null point, increasingly positive amplitudes. The null points may sometimes appear as tiny out-of-phase signals and be readily recognized (Fig. 2.29).

The whole sequence of successive pulses is repeated n times, with the computer executing the pulses and adjusting automatically the values of the variable delays between the 180°_x and 90°_x pulses as well as the fixed relaxation delays between successive pulses. The intensities of the resulting signals are then plotted as a function of the pulse width. A series of "stacked plots" are obtained (Fig. 2.30), and the point at which the signals of any particular proton pass from negative amplitude to positive amplitude is determined. This zero transition time τ_0 will vary for different protons in a molecule, depending on their respective spin-lattice relaxation times, and it is related to T_1 by the equation $T_1 = \tau_0/\ln 2 = \tau_0/0.693$. Determining τ, in seconds, at which the amplitude of a particular nucleus is zero and dividing it by 0.693 thus gives its spin-lattice relaxation time T_1.

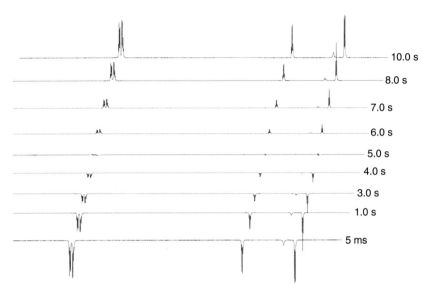

FIGURE 2.30   Stacked plots of ^{1}H-NMR spectra for ethylbenzene. This experiment can be used to measure the spin-lattice relaxation time, T_1.

There are other ways to determine the spin-lattice relaxation times, but this is the most convenient.

As already indicated, determination of the T_1 times helps in setting the inter-pulse delays to be used during pulse calibrations. Having derived the T_1 values, we next acquire an FID with a very short pulse (say, 250 μs). Fourier transformation and phasing produces a spectrum. The process is repeated to give a series of FIDs with the appropriate intervening relaxation delays, and the pulse widths are incremented successively in small steps (say, 5 μs). The spectra thus collected are arranged as a stacked plot, and each signal is seen to undergo an oscillation that reaches its maximum positive amplitude at 90° and its maximum negative amplitude at 270°. The 180° pulse width (zero amplitude or very weak signal) can thus be determined.

It is convenient to use a sample of p-dioxane in CDCl₃, having enough concentration to give an acceptable signal-to-noise ratio in one acquisition. The 90° pulse width is not read directly, since it is often difficult to see when the peaks reach their maximum amplitude but easier to see the null point. The 90° pulse is, therefore, calculated by halving the duration determined for the 180° pulse. We should check that inadequate intervening delays between successive pulses are not producing an erroneous value due to saturation. We can do this by repeating the experiment, with increased intervals between successive FIDs and checking that the signal intensity does not increase. If it does, then the plot should be repeated, with longer delays between successive FIDs. Once the approximate null point is known, to get a more accurate reading the experiment is repeated, with smaller incrementation of pulse durations, say, 1 or 2 μs.

A closely related procedure involves setting the pulse width to the approximate value of the 180° pulse and then recording many spectra. Before doing this, the approximate phase constants must first be determined. The transmitter frequency is placed close to the signals being used for calibration. We should obtain a reasonable signal-to-noise ratio with a single transient. But if the signal obtained with a single transient is too weak, then we can accumulate several transients, with a relaxation delay of at least $5T_1$ s between successive transients. The transmitter offset is placed close to the signal in the sample of interest, and the FID is acquired with a small flip angle (10°−20°). The sensitivity is enhanced by exponential line broadening, the FID subjected to Fourier transformation in the absolute intensity mode, and the phase corrected for the pure absorption mode across the spectrum. The phase correction constants thus defined are carefully noted.

The pulse width is next adjusted to the expected value of the 180° pulse, and a new FID recorded. This is again subjected to exponential multiplication, Fourier transformation, and phase correction using the phase constants defined in the earlier experiment. The phasing of the signal (i.e., positive or negative) depends on whether the pulse is longer or shorter than 180°. Many spectra are thus recorded, each with a slightly different value of the pulse width, till the 180° condition is reached, when the residual signal will have a symmetric dispersive (positive/negative) phasing with minimum amplitude. A broadened positive hump often also remains, due to poor shimming, but only the positions of the central narrow dispersive region should be used for judging the 180° pulse width.

There are many other methods known for accurate calibration of pulse widths, but such discussion is beyond the scope of this text (see Nielsen et al., 1986; Wesener and Günther, 1985; Bax, 1983; Lawn and Jones, 1982; Thomas et al., 1981).

In practice, it is usually unnecessary to determine exact pulse widths for each sample; we can use approximate values determined for each probehead, except in certain 2D experiments in which the accuracy of pulse widths employed is critical for a successful outcome. Proper tuning of the probehead is advisable, since pulse widths will normally not vary beyond ±10% with well-tuned probeheads.

Problem 2.26

I am having difficulty in correctly phasing my spectrum. What do I do?

2.4.1 Composite Pulses

Since there is a slight delay between the time when a pulse is switched on and when it reaches full power, an error may be introduced when measuring 90° or smaller pulses directly. If the 90° pulse width is required with an accuracy of better than ±0.5 μs, then it may be determined more accurately by using

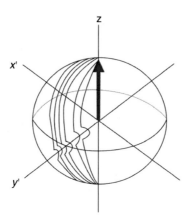

FIGURE 2.31   Applying the first incorrectly adjusted " 90_x° " pulse (actually, 85_x° pulse) bends the z-magnetization vector 5° above the y'-axis. The 180_y° pulse at this stage will bring the magnetization vector 5° below the y'-axis (to the mirror image position). Applying another similarly maladjusted " 90_x°" pulse causes a further bending of the magnetization vector precisely to the $-z$-axis. The composite pulse sequence (i.e., $90_x^{\circ} - 180_y^{\circ} - 90_x^{\circ}$) is thus employed to remove imperfections in the 90_x° pulse.

self-compensating *pulse clusters* that produce accurate flip angles even when there are small ($<10\%$) errors in the setting of pulse widths (Levitt, 1986; Levitt et al., 1984).

For example, to bend the equilibrium magnetization from the z-axis to the $-z$-axis, we need to apply a 180_x° pulse. Suppose the " 180_x°" pulse is wrongly adjusted so that in practice it rotates the z-magnetizations by 170° instead of the desired 180°. One simple way to compensate for this error is to apply a cluster of pulses comprising a $90_x^{\circ} - 180_y^{\circ} - 90_x^{\circ}$ *composite pulse*. The first " 90_x°" pulse in this cluster actually produces a similar error factor, so it will in reality be an 85_x° pulse, causing the magnetization to come to lie 5° above the y'-axis in the $y'z$-plane. The 180_y°-pulse causes the magnetization vector to jump across the y'-axis and adopt a mirror image position* on the other side of the y'-axis, i.e., 5° below the y'-axis in the $y'z$-plane. Note that a 180° pulse about any axis would cause the magnetization vector lying on one side of that axis to adopt a position that is the exact mirror image of its earlier position, across the axis about which the pulse was applied. The angle traversed may *appear* to be smaller than 180° if viewed in a particular plane. For instance, in the example given, the vector has traveled only about 10° in the $y'z$-plane but has actually moved by a *full semicircle* (Fig. 2.31). The magnetization now lies as far below the y'-axis as it was above the y'-axis before the application of the " 180_y° " pulse; i.e., an

---

* Actually, it will not be quite the mirror image position: the 180_y° pulse now applied will really be only a 170_y° pulse. But since the vector is now only moving a very short distance on the circular surface in the $y'-z$-plane (about 10°), the error thus produced will be significantly less.

equal and opposite error has been produced by the "$180°_y$" pulse. Applying the final "$90°_x$" (actually $85°_x$) pulse now moves the vector an additional 85°, so that it comes to lie fairly accurately on the $-z$-axis.

Problem 2.27

What problems would you expect to encounter in the case of incorrect alignment of the pulse width during an NMR experiment?

Composite pulses are usually described in abbreviated notations. For instance, a $\pi_x/2$ pulse (i.e., $90°_x$ pulse) is represented as X, and a π_x pulse (i.e., $180°_x$ pulse) as 2X. Phase shifts of 180° are represented by a bar over the X or Y. Hence, a $180°_{-y}$ pulse (i.e., a 180° pulse applied along the $-y$-axis) will be described as $2\bar{Y}$ (since it is 180° phase-shifted from the 180° pulse applied along the $+y$-axis). A composite pulse sequence known as GROPE-16 represents the following cluster of pulses: $3\bar{X}4XY3\bar{Y}4YX$ (i.e., $270°_{-x} - 360°_x - 270°_{-y} - 360°_y - 90°_x$). It compensates for up to 20% error in pulse width and an offset error of $0.5\,B_1$ (Shaka and Freeman, 1983).

Composite pulses have also been used in overcoming problems due to sample overheating during broad-band decoupling experiments. A widely used pulse sequence is Waltz-16 (Shaka et al., 1983), which may be represented as $A\ \overline{AA}\ A$, where $A = 3\,4\,\bar{2}\,3\,\bar{1}\,2\,\bar{4}\,2\,3$. The subsequently modified GARP I has larger bandwidth (Shaka et al., 1985). MLEV-16, comprising the composite 180° pulse $ABBA\ BBAA\ BAAB\ AABB$ (where $A = 90°_{-y} - 180°_x - 90°_{-y}$, and $B = 90°_y - 180°_{-x} - 90°_x$), was originally developed for more efficient heteronuclear decoupling experiments (Levitt et al., 1982); later it found wide application in homonuclear Hartmann Hahn echo spectroscopy ($HOHAHA$) (Bax and Davis, 1985; Davis and Bax, 1985). A modified version of MLEV-16 is MLEV-17, which contains another $180°_x$ uncompensated pulse at the end of the MLEV-16 sequence (Bax and Davis, 1985; Davis and Bax, 1985).

Problem 2.28

What are the advantages of using composite pulses instead of a unified pulse?

2.4.2 Pulsed Field Gradients

The use of pulsed field gradients (PFG) has radically transformed modern NMR spectroscopy (Matsukawa, 2006; Espinoza et al., 2004). Combined with selective excitation using shaped pulses it has become possible to obtain clean spectra of compounds with excellent sensitivity. If one has sufficient sample of an organic compound (say 15 mg of a compound of molecular weight below 1000), then most 2D spectra can be recorded within about 20 minutes using pulse field gradients, instead of the hours that it would take if pulsed field gradients were not applied. Water suppression is also far more effective than was possible with the older presaturation methods. Application of pulsed field gradients also allows one to effectively suppress any signals in the spectrum in which one is not interested, thereby allowing the other interesting signals to be recorded with improved sensitivity. It is now also possible to automatically adjust and optimize all 28 shims of a spectrometer within a few minutes (Sarkar et al., 2008).

Application of pulsed field gradients allows the magnetic field to be made very homogeneous across the entire sample volume. Such a homogeneous magnetic field is essential to record a good spectrum with sharp peaks (Jerschow and Muller, 1998; Zhu and Smith, 1995). This is achieved by first deliberately destroying the homogeneity across the sample by the application of a gradient in a linear and predictable manner and then by restoring it. Thus if a z-gradient is applied it would reduce the magnetic field in the lower part of the sample and increase it in the upper part of the sample in a linear manner (Barker and Freeman, 1985). The magnitude of the magnetic field experienced by the organic molecules in the sample will be a function of the position of the molecules in the NMR tube in the z-axis as described by the following equation:

$$B_{g(z)} = B_0 + G_z * Z \qquad (2.5)$$

where B_g is the magnetic field strength when the gradient is being applied, G_z is the magnitude of the field gradient, and z represents the position of the organic molecule on the z-axis.

The zero of the z-axis is assumed to be in the center of the sample. Molecules above the center, therefore, experience a slightly greater magnetic field while those below the center experience a somewhat decreased magnetic field, the difference between them being tiny, only about 50 Gauss in a magnetic field B_0 of magnitude 117,440 Gauss in a 500 MHz NMR spectrometer. This gradient can be switched off and on very rapidly, for durations of 1 millisecond. Since the resonance frequency of a nucleus is always proportional to the applied magnetic field, molecules above the center of the z-axis will experience a slightly greater magnetic field, and they will therefore precess faster than those that are below the center of the z-axis which experience a decreased magnetic field. Hence, molecules will experience differing magnetic fields depending on their positions in the NMR tube. It is assumed that diffusion is so slow that their relative positions do not change. The magnetization now exists in a twist or helical form which rotates at the end of the gradient pulse as a function of the z-axis throughout the sample. Because different molecules will be experiencing different magnetic fields and have correspondingly different precessional frequencies. If we tried to observe an NMR signal at this point of time after the first pulse, no signal will be visible as the nuclei would be pointing in all possible directions. We would thus have essentially destroyed the NMR signal. Why would we want to do that? The reason is that there may be artifacts or some other inhomogeneities that we would want to eliminate by this process. In order to regenerate the clean signal, free of artifact peaks, we now apply another gradient pulse of the same magnitude but of opposite sign in order to create the

"*Gradient Echo.*" This results in a magnetic field which is now slightly less above the center of the z-axis, and greater below the center of the z-axis. The vectors that were rotating slower after the application of the first pulse now rotate correspondingly faster after the second pulse, while those that were rotating faster after the first pulse now rotate at correspondingly slower precessional frequencies. The "untwisting" of the helix and refocusing of the magnetization vectors now occurs, leading to a clean signal free of artifact peaks. *We can express this phenomenon by stating that the first gradient pulse leads to the encoding of the position of each molecule into its magnetization, thereby scrambling the net magnetization of the sample, while the second gradient pulse leads to the decoding of this information through an unscrambling process.* So the first pulse destroys all the signals while the second pulse reverses the process, and regenerates the genuine sample signals but not the artifact signals.

Practical Considerations: The gradient coils that produce the pulses surround the sample tube in the NMR probe. The gradient magnetic field is created by an electric current of 10 amperes provided by the amplifiers in the console. This current is delivered to the gradient coils in the probe by a cable.

Gradient pulses can be switched on and off abruptly (rectangular pulses) or they can be "shaped" to turn off gradually. "Sine" shaped gradient pulses are normally used. The parameters that are adjusted when producing pulsed field gradients are (a) the strength of the gradient (% of maximum gradient current) in each of the three axes x, y, and z, (b) the gradient shape (normally sine) in each of these three directions, (c) the duration of the gradient pulse (usually 1 ms), and (d) the duration of the recovery delay (usually 0.2 ms). Experiments in which gradients are commonly used include gradient COSY (or DQF-COSY), gradient TOCSY, and gradient HSQC or HMQC.

Gradient COSY (DQF-COSY): In the double quantum filtered version of COSY, gradients are employed to select the double quantum pathway. This blocks signals such as methyl singlets, water signals, or signals of other protons that are not coupled to any other protons from appearing in the spectrum. It is usually possible to increase the receiver gain significantly during the recording since the double quantum filter eliminates the strongest signals lying along the diagonal. The sensitivity of the experiment can be further enhanced by recording an echo-anti-echo version of the experiment (Lin et al., 2000).

Gradient TOCSY: This is similar to the normal TOCSY spectrum (see Section 7.5), except that cleaner and more sensitive spectra are obtained. The use of gradients allows the phase information to be encoded in the F1 (indirect) dimension. In a sensitivity enhanced version of the experiment, an echo-anti-echo mode is employed in which the sign of the final gradient is changed (selection of $p = +1$ coherence instead of $p = -1$). This leads to an enhancement of sensitivity by about 40%.

Gradient HSQC or HMQC: The gradients in this experiment are chosen so that the coherence pathway selected involves transfer of coherence from ^1H to ^{13}C and then back to ^1H. This allows the signals of protons bound to ^{12}C to be eliminated and the gain can hence be increased considerably. Phase cycling is additionally used to suppress protons bound to ^{12}C. HSQC is more sensitive than HMQC and is preferred for this reason. In a further variant of HSQC, the CH_2 cross peaks appear inverted, relative to the CH and CH_3 cross-peaks which appear positive (Section 8.2.1.1). This then allows the 2D HSQC spectrum to be obtained with information from a DEPT spectrum built into it (Rinaldi and Keifer, 1994).

Problem 2.29

What are the key objectives of the pulsed-field gradients (PFG) technique?

Problem 2.30

Constant efforts are being made in NMR spectroscopy to achieve a homogenous magnetic field throughout the sample volume in order to get sharper peaks for various nuclei. In contrast, in the PFG technique, the magnetic field homogeneity is deliberately destroyed. What is the logic behind this?

2.4.3   Shaped Pulses

The earlier NMR spectrometers could only apply rectangular pulses which were delivered at high power but for a very short duration. These were non-selective in nature as they led to the excitation of all the nuclei of a particular type, such as ^1H, irrespective of their chemical shifts. For instance if a high powered 90° pulse is applied for a short duration, the magnetizations of protons at all chemical shifts will undergo a rotation by 90°. If we wish to achieve the same rotation at lower power, then we could do this by increasing the time duration for which a weaker pulse is applied. By lowering the power of the *Rf* signal to the probe, it is possible to achieve greater selectivity: the lower the power the greater the selectivity. Protons located at the center of the spectrum (corresponding to the carrier frequency) may be bent by 90° but protons away from the center may be bent to a lesser extent, say 30° or 40°. This would be problematic if we wanted to excite just one proton and look at its effects (by *J*-coupling or NOE) on other protons, without affecting other protons in the spectrum. This is what is achieved in a 2D NMR spectrum, but if we could selectively excite individual protons, it would provide us with the same result except that it would essentially correspond to a slice of the 2D NMR spectrum in a very short time.

Low powered rectangular pulses of long duration achieve a certain level of selectivity, but they also excite other protons that are not at the center of the spectrum, leading to unwanted signals. This is where *shaped pulses* can be very useful. Special hardware is required to generate pulses with shapes other than rectangular, and it is now becoming available on the latest NMR spectrometer systems. For simple applications, the Gaussian (sine bell) shape is the one most widely used. Such pulses are applied for longer durations (e.g., 90 ms) and with power that is low but enough to cause a 90° rotation of the selected proton without significantly affecting other nearby protons. Hundreds of other pulse shapes are also now available.

To apply a shaped pulse, a number of parameters must first be selected. These include the duration of the pulse (pulse width), the shape of the pulse (such as Gaussian), and the maximum power at the center of the pulse (at the top of pulse shape). If one wishes to excite a peak that is not at the center of the spectral window, then the offset frequency will also need to be selected. One needs to carry out a calibration first in order to determine the power levels that correspond to 180° or 90° pulses. For experiments described below, the instrument would have built-in packages with the parameters already pre-determined, and one normally needs only to set the offset frequency (in Hz, relative to the center of the spectral window in Hz) so that it corresponds to the chemical shift of the nucleus that one wants to excite. Shaped pulses have found many applications:

Selective TOCSY: A selective TOCSY spectrum is a one-dimensional spectrum that corresponds to a slice through the conventional 2D TOCSY spectrum. A single peak in the spectrum is selectively excited and the spread of magnetization to other coupled protons in a time-dependent manner is then recorded. Those protons near the excited proton appear first while those farther away but part of the same coupled system appear at greater durations of the mixing time. This allows one to discern close lying protons from those located farther away within a spin system. Such a selective 1D TOCSY can be recorded far more quickly than the conventional 2D TOCSY spectrum.

Selective Transient NOE: This is a selective version of the NOE difference spectrum. The experiment involves the application of a 180° pulse on a single peak, followed by a certain delay (mixing time) during which the perturbation is carried to nearby protons in the molecule by the NOE effect. The FID is then recorded, Fourier transformation of which produces the spectrum in which the peak that was subjected to the excitation appears inverted while the protons that were perturbed by the NOE appear as weak positive signals. There is no subtraction needed as this is not a difference spectrum, and artifacts due to subtraction are thus avoided.

Selective HMBC: In this heteronuclear experiment, a specific peak in the ^{13}C NMR spectrum is selected and protons connected to it through two or three bonds are identified. This therefore corresponds to a slice taken from the conventional HMBC experiment. Since it is a 1D spectrum that is obtained through the selective excitation, the time taken is much shorter. If a quaternary carbon is selectively excited, then the conformational or stereochemical information may be obtained from the heteronuclear *J* values.

Selective HMQC: This again affords a 1D NMR spectrum, corresponding to a slice taken through a conventional 2D HMQC spectrum. A carbon is selectively excited and the spectrum provides peaks of the proton(s) directly attached to that carbon by a single bond.

2.5 PHASE CYCLING

Most NMR experiments use combinations of two or more of the four pulse phases shown in Fig. 2.32. The phase of the pulse is represented by the subscript after the pulse angle. Thus, a $90°_{-y}$ pulse would be a pulse in which the pulse angle is $90°$ and $-y$ is its phase (i.e., it is applied in the direction $-y$ to $+y$), so that it will cause the magnetization to rotate about the y-axis in the xz-plane.

Phase cycling is widely employed in NMR spectroscopy to suppress artifact signals due to field inhomogeneities or imperfect pulse settings. We have already discussed how phase cycling can be used to suppress image peaks in quadrature detection. As mentioned earlier, there are two separate sections of the computer memory, designated A and B, which digitize the signals received. Let us assume that the signals of $0°$ phase go into section A while signals with $90°$ phase go into section B. To eliminate imbalances between the two receiver channels, the signals are cycled so that both channels contribute equally to the data. Because each receiver channel is receiving signals of only one phase type ($0°$ or $90°$), the switching of the receiver channels is

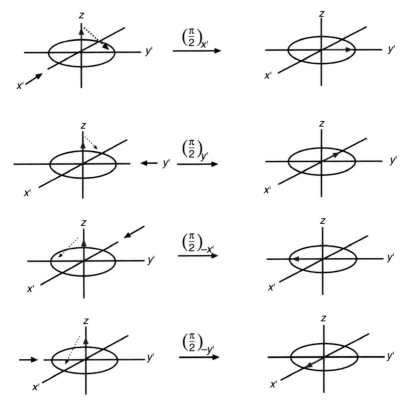

FIGURE 2.32 A $90°$ pulse brings the equilibrium magnetization to the $x'y'$-plane. Its orientation in the $x'y'$-plane depends on the direction of the pulse. Applying the pulse along one axis causes the magnetization to rotate in a plane defined by the other two axes.

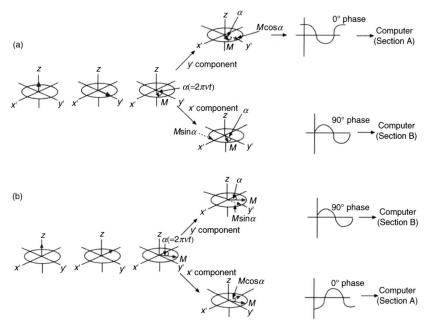

FIGURE 2.33 The first two steps of the CYCLOPS phase cycling scheme. Any imbalance in receiver channels is removed by switching them so that they contribute equally to the regions A and B of the computer memory.

carried out simultaneously with the changing of the phases (*phase cycling*) so that absorptive and dispersive signals are recorded on successive scans (Ivchenko et al., 2003; Pennino, 1987).

A simple, two-step phase cycling scheme may, therefore, be employed: the signals of 0° phase and 90° phase pass through signal channels (1) and (2) to sections A and B, respectively, of the computer memory during the first cycle; in the second cycle the receiver channels are switched so that both channels contribute equally to the signals (Fig. 2.33).

To suppress other interference effects, the phase of the transmitter pulse is also shifted by 180° and the signals subtracted from sections A and B, leading to the CYCLOPS (*CYCL*ically *O*rdered *Pha*Se cycle) phase cycling scheme shown in Table 2.1 In this the two different receiver channels differing in phase by 90° are designated as 1 and 2 and the four different receiver pulses ($90^\circ_x, 90^\circ_y, 90^\circ_{-x}, \text{and}\, 90^\circ_{-y}$) are called x, y, $-x$, and $-y$, respectively.

Problem 2.31

What is phase cycling, and why is it used in NMR spectroscopy?

Problem 2.32

What is the effect of phase cycling on the appearance of NMR signals?

TABLE 2.1 CYCLOPS Phase Cycling*

Scan	Pulse	Sign A	Sign B	Receiver Phase Code	Receiver Mode	Pulse Phase	A	B
1	$90°_x$	$\cos \omega t$	$\sin \omega t$	R0	x	x	+1	+2
2	$90°_y$	$\sin \omega t$	$-\cos \omega t$	R1	y	y	−2	+1
3	$90°_{-x}$	$-\cos \omega t$	$-\sin \omega t$	R2	−x	−x	−1	−2
4	$90°_{-y}$	$-\sin \omega t$	$\cos \omega t$	R3	−y	−y	+2	−1

*The two memory blocks in the computer are designated A and B; the two receiver channels differing in phase by 90° are shown as 1 and 2.

2.5.1 Phase Cycling and Coherence Transfer Pathways

Coherence may be considered a generalized description of transverse magnetization, and it corresponds to a transmission between two energy levels. The magnitude of coherence in each spin system is governed by the coherence level. The
difference in magnetic quantum number m_z of the two energy levels connected
by the same coherence represents the *coherence order, p*. The path describing the
progress of a coherence order in a pulse sequence is called a *coherence transfer
pathway*. It is possible to change the coherence level by applying a pulse, while
in the time interval delay between the pulses the coherence level does not change.

The transitions between energy levels in an *AX* spin system are shown in
Fig. 2.34. There are four single-quantum transitions (these are the "normal"

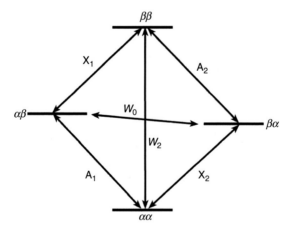

FIGURE 2.34 Transitions between various energy levels of an *AX* spin system. A_1 and A_2 represent the single-quantum relaxations of nucleus *A*, while X_1 and X_2 represent the single-quantum
relaxations of nucleus *X*. W_2 and W_0 are double- and zero-quantum transitions, respectively.

transitions A_1, A_2, X_1, and X_2 in which changes in quantum number of 1 occur), one double-quantum transition W_2 between the $\alpha\alpha$ and $\beta\beta$ states involving a change in quantum number of 2, and a zero-quantum transition W_0 between the $\alpha\beta$ and $\beta\alpha$ states in which no change in quantum number occurs. The double-quantum and zero-quantum transitions are not allowed as excitation processes under the quantum mechanical selection rules, but their involvement may be considered in relaxation processes.

Transverse magnetization represents a particular type of coherence involving a change in quantum number p of ±1. Each coherence σ_{rs} is equal to the difference in magnetic quantum numbers of the nuclei r and s, i.e., the coherence order is $M_r - M_s$, and pulses cause transitions to occur between the different coherence orders. At thermal equilibrium, the change in quantum number p is zero, and it is at this point that the coherence transfer pathway is initiated by the application of a pulse. For signals to be detected by the receiver, the coherence transfer must end with single quantum coherence (i.e., with $p = -1$).

When a 90° pulse is applied to the sample at equilibrium, the *longitudinal* (z) magnetization vanishes (i.e., the population difference between the α and β states decreases to zero) and transverse magnetization is created in the $x'y'$-plane. A phase coherence is now said to exist between the α and β states of the nucleus, since they precess coherently with the same phase (a property conveyed to them by the pulse). The coherence that now exists, termed *single-quantum coherence*, causes a precessing net magnetization of the nucleus, which can be detected in the form of a signal. It is possible to transfer this coherence by applying additional pulses to other states.

Suppose the first pulse resulted in the creation of a phase coherence across the A_1 transition between the $\alpha\alpha$ and $\alpha\beta$ states (Fig. 2.34). It is possible to transfer this phase information from the $\alpha\beta$ state to the $\beta\beta$ state by applying a selective π pulse across the X_1 transition. The two successive pulses would, therefore, transfer the phase of the $\alpha\alpha$ state to the $\beta\beta$ state, with the two states now becoming phase coherent with one another.

Since the $\alpha\alpha$ and $\beta\beta$ states are separated by a quantum number difference of 2, *double-quantum coherence* is said to have been created. Similarly (in more complex spin systems), coherence can be created between states differing by 2, 3, 4, or higher quantum numbers. The quantum mechanical selection rules stipulate that such *multiple-quantum coherences* do not give rise to detectable magnetization, the detectable signals resulting only from single-quantum transitions. Double-, triple-, and other multiple-quantum transitions must, therefore, be converted into single-quantum coherence before detection. Many different coherence transfer pathways can be used, and by selecting appropriate phase cycling procedures, a particular coherence transfer pathway can be detected exclusively. Some typical coherence transfer pathways in common 2D NMR experiments are shown in Fig. 2.35.

When designing the phase cycling procedure it is necessary to tailor them to the coherence pathways by which the signals reach the receiver. The application of the first pulse causes the z-magnetization to be transferred to some coherence

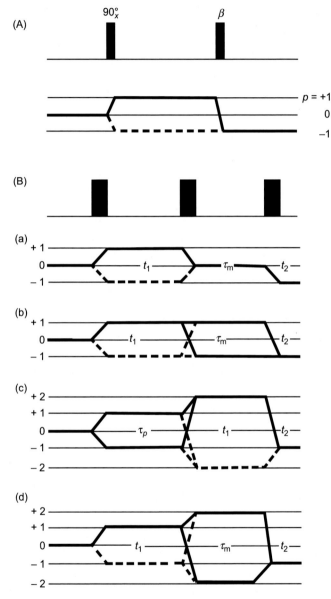

FIGURE 2.35   Coherence transfer pathways in 2D NMR experiments. (A) Pathways in homonuclear 2D correlation spectroscopy. The first 90° pulse excites single quantum coherence of order $p = \pm 1$. The second mixing pulse of angle β converts the coherence into detectable magnetization ($p = -1$). (B:a) Coherence transfer pathways in NOESY/2D exchange spectroscopy; (B:b) relayed COSY; (B:c) double-quantum spectroscopy; (B:d) 2D COSY with double-quantum filter ($\tau_m = 0$). The pathways shown in (B:a,b, and d) involve a fixed mixing interval (τ_m). *(Reprinted from Bodenhausen, G. et al., J. Magn. Reson. 58, 370, copyright © 1984, Rights and Permission Department, Academic Press Inc., 6277 Sea Harbor Drive, Orlando, Florida 32887).*

level. The coherence level may then be systematically altered by subsequent applications of one or more pulses, with the sequence of coherence levels by which the signals reach the receiver representing the *coherence pathway*. The phase cycling procedures adopted determine which signals are to be coadded and which signals canceled, according to the coherence pathway chosen. Coherence levels may be negative or positive. For instance in a nucleus of spin 1/2, applying a pulse can create a single-quantum coherence of level +1 and another coherence of level −1 corresponding to the clockwise and counterclockwise rotating magnetization vectors obtained from the x and y Bloch magnetizations. If the phase of a pulse is shifted, then the behavior of a coherence subjected to such a pulse will be governed by the change in the coherence level. For instance, if a pulse is shifted by angle θ and the change in the coherence level is Δ_m, then the effect on the coherence is obtained by multiplying $e^{i\Delta m}$ by the coherence. It is possible thereby to derive a simple rule that would predict which signals following a certain coherence pathway will survive the particular phase cycling procedure and which signals would be eliminated (Bain, 1980).

Before the application of a pulse, only equilibrium magnetization exists, directed toward the z-axis corresponding to the zero coherence level for all coherence pathways. When the pulse is applied, two coherence levels, +1 and −1, are created during the evolution period that evolves into $Me^{i\omega t}$ and $Me^{-i\omega t}$, respectively, where ω is the absolute Larmor frequency of the nucleus and t is the evolution time. Before detection, the signal is mixed with the carrier frequency W_0 so that it comes within the audio frequency range to be picked up by the receiver. In quadrature detection, the carrier frequency may be considered to be circularly polarized, and mathematically the mixing of the carrier frequency is equivalent to multiplication of each of the two magnetization components by $e^{-i\omega_0 t}$. As a consequence, the magnetization component corresponding to the coherence level of +1 is detected, since it is the only signal $Me^{-i(\omega-\omega_0)t}$ in the audiofrequency range. Apparently, all coherence pathways will therefore start at zero coherence levels and end at +1 coherence levels; since the quadrature receiver is sensitive only to the +1 polarization, only the single-quantum coherence is detected.

The signals reaching the receiver from each coherence pathway will, in general, depend on the *pulse phase factor* (and hence on the phase) as well as on the receiver phase factor (i.e., the data routing). To analyze the phase cycling sequence, it is therefore necessary to calculate the change in the coherence level and the phase factor for each pulse in the cycle, including the receiver phase. By multiplying all the phase factors together and coadding the results over the cycle, it is possible to check whether the result is zero or not. If the results add up to zero, then that pathway does not contribute to the signal; if the answer is not zero, then that pathway may lead to a signal.

Problem 2.33

Summarize the different events during a pulsed NMR experiment in the form of a flow diagram.

SOLUTIONS TO PROBLEMS

2.1 Only the dipole or spin ½ (I) nuclei have two energy states (α and β). Other nuclei may have three or more energy states, half of them aligned with B_0 and half opposed to it. These nuclei are called quadrupole nuclei. Examples include ^2H (deuterium, $I = 1$) with three states (+1, 0, −1), ^{14}N ($I = 1$) with three states (+1, 0, −1), ^{17}O ($I = 5/2$) with six states (−5/2, −3/2, −1/2, +1/2, +3/2, +5/2) and ^{35}Cl ($I = 3/2$) with four states (−3/2, −1/2, +1/2, +3/2). These nuclei are often difficult to detect, with the exception of a few because of the very low energy differences between various energy states resulting in a low population difference.

2.2 Bulk or macroscopic magnetization (M_0) is the sum of the z-components of all the nuclear magnetic moments of individual spins in a sample. M_0 exists because of the difference in population between the energy states, $N_\alpha > N_\beta$.

2.3 The pulse is a burst of radiofrequency radiation of a specific power, duration, and width which is used to excite all nuclei of one type simultaneously.

2.4 A long duration pulse results in a narrow NMR signal.

2.5

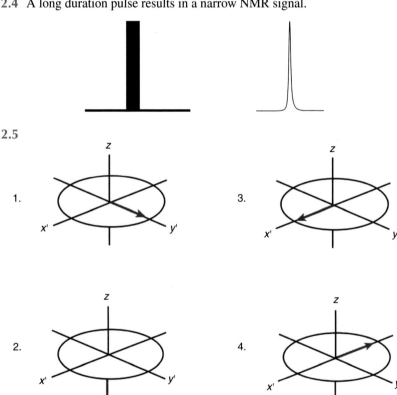

2.6   *"Hard"* and *"soft"* pulses depend on the pulse width: the shorter the width of the pulse, the "harder," or the more powerful, it will be. Similarly, a long width pulse is "soft" and less powerful. For example, a pulse that requires 25-kHz B_0 for a duration of 20 μs will be "hard," while the same pulse of 20-ms width will be considered "soft."

2.7

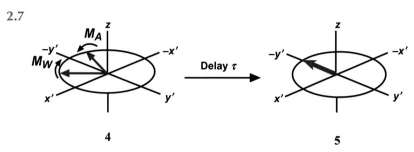

4   5

2.8   We can use the angular frequency of TMS as the reference frequency of the rotating frame. Deducting this from the Larmor frequency of the signal will leave only the *differential* frequency (or, in other words, the chemical shift) associated with the magnetization vector of the signal.

2.9   The amplitude represents a circular motion of the magnetization vector along the *xy*-plane, which slowly relaxes back towards the *z*-axis; the Larmor frequency of the nucleus is inherent to it and would therefore remain unchanged throughout the FID.

2.10 The "tail" of the FID contains very little information; most of the relevant information is in the initial large-volume portion of the FID envelope. The loss of the "tail" of the FID, should not, therefore, significantly affect the quality of the data. Another manipulation to compensate for the lost information is exponential multiplication, in which the FID is multiplied by a negative exponential factor.

2.11 In pulse NMR we measure in the time domain; i.e., the variation of signal amplitude with time (FID) is recorded. The time-domain data are then subjected to Fourier transformation to convert it into the frequency domain.

2.12 If we do not subtract the reference frequency, we will have to process a very large amount of data. For example, on a 500-MHz NMR spectrometer, the frequencies to be processed would be 500,000,000–500,006,000 Hz for protons (since protons normally resonate within 0–12 ppm, i.e., 0–6000 Hz).

2.13 It is better to maintain the same number of data points and reduce the spectral width as far as possible. In the alternative case, the improved digital resolution will be at the cost of sensitivity, since it will produce a corresponding increase in acquisition time, AT. Either a greater time period would then be required or a lesser number of scans would be accumulated in the same time period, with a corresponding deterioration in the *signal-to-noise* ratio.

2.14   The sample generates an NMR signal which is first amplified by the receiver before being digitized by the analog-to-digital converter (ADC). If the signal intensity that is received is too strong, then a message *"Receiver Overflow"* or *"ADC Overflow"* is displayed. If no action is taken, then the FID will be clipped and a distorted spectrum will result. To avoid this one needs to adjust the receiver gain to a lower value or to do it automatically by employing autogain (one types in "gain = n"). If this does not solve the problem, then the observe pulse width (PW) should be reduced to half the value being used. If this still does not resolve the problem, then one can dilute the sample. If the ADC overflow is being caused by a solvent signal, then the solvent suppression technique can be used.

2.15   The folded peaks are easy to identify since they show different phases than the "normal" signals. On shifting the spectral window to one side, all the "normal" signals will shift in the same direction, by the same value, as the spectral window. In contrast, the folded signals will move either in the opposite direction or by a different value in the same direction, so that their *relative* disposition to other signals in the spectrum will change.

2.16   If the gain is too high, then the largest signal of solvent in the spectrum will cause ADC overflow, resulting in severe distortions.

2.17   The image peaks resulting from quadrature detection can be easily distinguished from small genuine peaks since they show different phases and move with changes in the reference frequency.

2.18   The ratio of the height of an NMR signal to the noise is called the *S/N* (*signal-to-noise*) ratio. The simplest way to improve *S/N* is through signal accumulation. Recording a large number of spectra and combining them together will increase *S/N*. Since the noise contribution is random whereas the signals occur in exactly the same place each time, the signals will build up over a number of scans. This means that over n scans of the experiment, the signal will increase n times, while the noise amplitude will increase by $\sqrt{n}$ to give an overall increase in *S/N* of $\sqrt{n}$. As a result, increasing the number of scans by four-fold will double in the signal-to-noise-ratio.

2.19   Following are examples of spectra recorded with and without apodization.

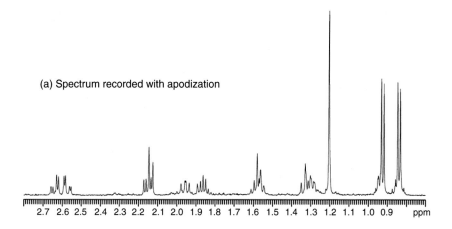

(a) Spectrum recorded with apodization

(b) Spectrum recorded without apodization

2.20 Apodization (exponential multiplication) is used to improve the *signal-to-noise* ratio, and it does not affect the chemical shifts of the NMR signals.

2.21 The simplest definition of *sensitivity* is the signal-to-noise ratio. One criterion for judging the sensitivity of an NMR spectrometer or an NMR experiment is to measure the height of a peak under standard conditions and to compare it with the noise level in the same spectrum. *Resolution* is the extent to which the line shape deviates from an ideal Lorentzian line. Resolution is generally determined by measuring the width of a signal at half-height, in hertz.

2.22

(a) Spectrum before resolution enhancement

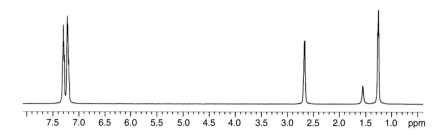

(b) Spectrum after resolution enhancement

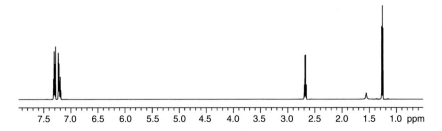

2.23 Greatly enhanced sensitivity within a short measuring time is the major advantage of pulse Fourier transform (PFT) experiments. In the continuous-wave (CW) experiment, the radiofrequency sweep excites nuclei of different Larmor frequencies, one by one. For example, 500 s may be required for excitation over a 1-kHz range, while in a PFT experiment a single pulse can simultaneously excite the nuclei over 1-kHz range in only 250 μs. The PFT experiment, therefore, requires much less time than the CW NMR experiment, due to the short time required for acquisition of FID signals. Short-lived unstable molecules can only be studied by PFT NMR.

2.24 If the radiofrequency power is too high, the normal relaxation processes will not be able to compete with the sudden excitation (or perturbation), and thermal equilibrium will not be achieved. The population difference (Boltzmann distribution excess) between the energy levels (α and β) will decrease to zero, and the intensity of the absorption signal will also therefore become zero.

2.25 Following are some common methods for sensitivity enhancement:
1. *Sample concentration*: When the sample concentration in a given volume of solvent increases, the number of NMR active nuclei also increases resulting in an improved signal-to-noise ratio.
2. *Temperature*: Slight lowering of the sample temperature increases the energy difference between energy levels and therefore increases the population difference based on the Boltzmann distribution equation.
3. *Magnetic field strength* (B_0): The population difference (Boltzmann excess) increases with increasing magnetic field.
4. *Rf power* (B_1): The signal strength increases with the increase in *Rf* power. Care must be taken, however, to avoid population saturation.
5. *Number of scans*: In the pulse Fourier Transform NMR experiment, the *signal-to-noise* ratio increases as the square root of the number of accumulated scans.
6. Polarization transfer and nOe effects also contribute to sensitivity enhancement.

7. *Relaxation rate*: Slow-relaxing nuclei cannot attain thermal equilibrium within the pulse interval, which results in a lower sensitivity. One way to cope with this problem is to introduce a sufficiently large delay time between the pulses. Addition of small quantities of relaxation reagents can also result in a better signal-to-noise ratio, but this may also produce a change in chemical shifts due to complexation.

8. Others, such as use of cryogenically cooled probes, use of specialized NMR tubes, and dynamic nuclear polarization (DNP) methods.

2.26 Certain types of spectra, such as DEPT or APT, will afford some peaks in positive phase and others in negative phase. This is how the spectrum is supposed to look, so do not try to get all positive peaks through rephasing such spectra. You also need to use a sufficiently large spectral width as otherwise resonances outside the spectral region will "fold-back" and you may see phase errors. This can easily be corrected by doubling or tripling the spectral width and recording the spectrum again.

2.27 The pulse angle controls the extent to which the magnetization vectors are bent. A misalignment of the pulse would lead to various artifact signals. Including 180° spin-echo pulses can, to some extent, compensate for setting pulses incorrectly. But in certain experiments (e.g., inverse NMR experiments), it is important for the success of the experiment that the proper pulse angles be determined and employed.

(a) ^{1}H-NMR spectrum with proper alignment of pulse width

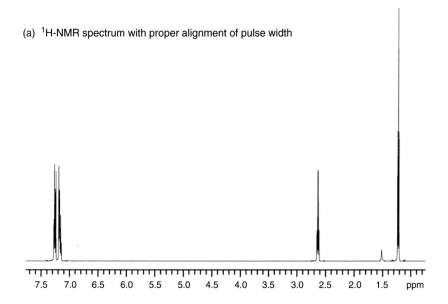

(b) ^{1}H-NMR spectrum with poorly set pulse width

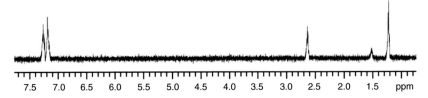

2.28 Composite pulses reduce the errors introduced due to the delay between the start of the pulse and when it reaches full power. They have also been used to overcome problems of sample overheating during broad-band decoupling particularly in experiments in which pulses have to be applied for long durations.

2.29 In the pulsed-field gradients (PFG) technique, the homogeneity of the magnetic field B_0 is deliberately disturbed in a predictable manner and then restored to achieve the following objectives:

1. To automatically optimize all shims simultaneously in a few minutes.
2. To substantially reduce the total time required for 2D-NMR experiments.
3. To suppress the water signal (even in 90% H_2O), for which this technique is far superior to conventional presaturation methods.
4. To eliminate artifacts in 2D spectra, and thus to improve the sensitivity by selecting only the signals which belong to the samples, which are suppressing all others.

2.30 In PFG, first the z-axes gradient alters the magnetic field in such a way that the molecules in the lower half of the NMR tube experience a lesser magnetic field, as compared to those in the upper half. Since the Larmor frequency of a nucleus is directly proportional to the magnetic field strength, this means that if we have magnetization vectors in the xy-plane (e.g., after a 90° pulse), the magnetization vectors will rotate in the xy-plane at different rates, depending on the position of the molecules in the NMR sample tube. Magnetization vectors in the upper half of the tube will precess a little faster, while magnetization vectors in the lower part will precess a little slower than the average. This results in a "twisted" magnetization in the sample. Attempts to acquire an NMR spectrum at this point, after the gradient was turned off, will not lead to any observable signal as the vectors will point in all possible directions in the xy-plane, and no net magnetization will exist. This first step in PFG sequence thus destroys all the NMR signals. These signals may belong to your sample or they may belong to other unwanted sources, such as artifacts, solvent signals, or they may be done to other features of the spectrum that you wish to eliminate. In this way, the PFG technique can be used to "clean up" or the remove "unwanted" NMR signals.

In the next stage, the twisted magnetization in the sample can be "*untwisted*" by applying a second (second) *z*-gradient pulse of exactly the same magnitude, *but of opposite sign*. This gradient now does the opposite, i.e., it decreases the magnetic field strength in the upper half of the NMR sample tube, and increases it in the lower half of the NMR tube. Thus, the magnetization vectors that rotated slower in the first gradient pulse are now rotating faster by the same magnitude and the ones that rotated faster will be rotating slower. At the end of the second gradient pulse, all the magnetization vectors are lined up throughout the sample. If we start collecting the FIDs at this stage, we will get a clean NMR spectrum, devoid of unwanted artifact signals. The two-stage gradient change is done fairly rapidly, leading to substantial saving in the NMR experimental time.

2.31 Phase cycling is widely employed in multipulse NMR experiments. It is also required in quadrature detection. Phase cycling is used to suppress artifact peaks, to correct pulse imperfections, and to select particular responses in 2D or multiple-quantum spectra.

2.32

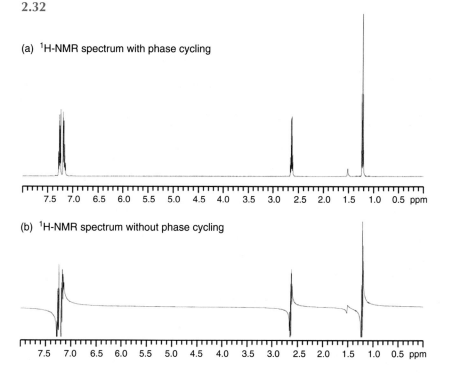

(a) ^{1}H-NMR spectrum with phase cycling

7.5 7.0 6.5 6.0 5.5 5.0 4.5 4.0 3.5 3.0 2.5 2.0 1.5 1.0 0.5 ppm

(b) ^{1}H-NMR spectrum without phase cycling

7.5 7.0 6.5 6.0 5.5 5.0 4.5 4.0 3.5 3.0 2.5 2.0 1.5 1.0 0.5 ppm

2.33

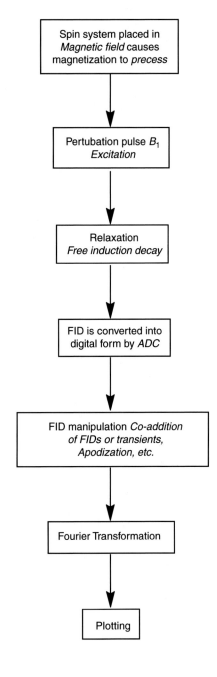

REFERENCES

Bain, A.D., 1980. Superspin in NMR: application to the ABX system. J. Magn. Reson. 37 (2), 209–216.

Barker, P., Freeman, R., 1985. Pulsed field gradients in NMR. An alternative to phase cycling. J. Magn. Reson. 64 (2), 334–338.

Bax, A., 1983. Determination of heteronuclear coupling constants *via* semiselective two-dimensional *J* spectroscopy. J. Magn. Reson. 52, 330–334.

Bax, A., Davis, D.G., 1985. MLEV-17-based two-dimensional homonuclear magnetization transfer spectroscopy. J. Magn. Reson. 65, 355–360.

Davis, S., Bauer, C., Barker, P., Freeman, R., 1985. The dynamic range problem in NMR. J. Magn. Reson. 64 (1), 155–159.

Davis, D.G., Bax, A., 1985. Simplification of one ^{1}H NMR spectra by selective excitation of experimental sub-spectra. J. Am. Chem. Soc. 107, 7197–7198.

Ernst, R.R., 1966. Sensitivity enhancement in magnetic resonance. Adv. Magn. Reson. 2, 1–135.

Espinoza, L., Garbarino, J.A., Chamy, M.C., Piovano, M., 2004. Pulsed field gradients in NMR high resolution experiments. J. Chil. Chem. Soc. 49 (2), 189–195.

Gengying, L., Haibin, X., 1999. Digital quadrature detection in nuclear magnetic resonance spectroscopy. Rev. Sci. Instrum. 70 (2), 1511–1513.

Hoult, D.I., Richards, R.E., 2011. The signal-to-noise ratio of the nuclear magnetic resonance experiment. J. Magn. Reson. 213 (2), 329–343.

Ivchenko, N., Hughes, C.E., Levitt, M.H., 2003. Multiplex phase cycling. J. Magn. Reson. 160 (1), 52–58.

Jerschow, A., Muller, N., 1998. Efficient simulation of coherence transfer pathway selection by phase cycling and pulsed field gradients in NMR. J. Magn. Reson. 134 (1), 17–29.

Kessler, H., Mronga, S., Gemmecker, G., 1991. Multidimensional NMR experiments using selective pulses. Magn. Reson. Chem. 29, 527–557.

Lawn, D.B., Jones, A.J., 1982. Pulse length measurement in N.M.R. spectroscopy: a convenient digital method using second-dimension transformation. Aust. J. Chem. 35 (8), 1717–1722.

Levitt, M.H., Rance, M., Frey, M., Suter, D., Counsell, C., Ernst, R.R. Composite pulses and multiple-quantum coherence. In: Mueller, K.A., Kind, R., Roos, J. (Eds.), Congr. AMPERE Magn. Reson. Relat. Phenom., Proc., 22nd, 1984, 420–421.

Levitt, M., Freeman, R., Frenkiel, T., 1982. Broadband heteronuclear decoupling. J. Magn. Reson. 47 (2), 328–330.

Levitt, M.H., 1986. Composite pulses. Prog. Nucl. Mag. Res. Sp. 18 (2), 61–122.

Lin, G., Liao, X., Lin, D., Zheng, S., Chen, Z., Wu, Q., 2000. Noise and sensitivity in pulsed field gradient experiments. J. Magn. Reson. 144 (1), 6–12.

Lindon, J.C., Ferrige, A.G., 1980. Digitization and data processing in Fourier transform NMR. Prog. NMR Spectrosc. 14, 27–66.

Marco, A.D., Wüthrich, K., 1976. Digital filtering with a sinusoidal window function: an alternative technique for resolution enhancement in FT-NMR. J. Magn. Reson. 24, 201–204.

Matsukawa, S. Webb, G. A. (Ed.), 2006. NMR measurements using field gradients and spatial information, Mod. Magn. Reson., 1, 129–134.

Moskau, D., 2002. Application of real time digital filters in NMR spectroscopy. Concept Magn. Reson. 15 (2), 164–176.

Nielsen, N.C., Bildsöe, H., Jakobsen, H.J., Sörensen, O.W., 1986. Pulse techniques for calibration of the decoupler radiofrequency field strength. J. Magn. Reson. 66 (3), 456–469.

Pacheco, C.R., Traficante, D.D., 1996. Two new window functions for long acquisition times. J. Magn. Reson. 120 (1), 116–120, Series A.

Pennino, D.J., 1987. The use of phase cycling in multipulse Fourier transform-nuclear magnetic resonance experiments. Spectroscopy 2 (3), 36–40.

Rinaldi, P.L., Keifer, P., 1994. The utility of pulsed-field-gradient HMBC for organic structure determination. J. Magn. Reson. 108 (2), 259–262, Series A.

Roth, K., Kimber, B.J., Feeney, J., 1980. Data shift accumulation and alternate delay accumulation techniques for overcoming the dynamic range problem. J. Magn. Reson. 41 (2), 302–309.

Samuelson, G.L., Obenauf, R.H., Albright, M.J., 1977. American Laboratory (Shelton, CT, USA). 9(9), 85–86, 88–90, 92, 94.

Sarkar, R., Moskau, D., Ferrage, F., Vasos, P.R., Bodenhausen, G., 2008. Single or triple gradients? J. Magn. Reson. 193 (1), 110–118.

Shaka, A.J., Freeman, R., 1983. Composite pulses with dual compensation. J. Magn. Reson. 55 (3), 487–493.

Shaka, A.J., Barker, P.B., Freeman, R., 1985. Computer-optimized decoupling scheme for wide-band applications and low-level operation. J. Magn. Reson. 64 (3), 547–552.

Shaka, A.J., Keeler, J., Freeman, R., 1983. Evaluation of a new braoad band decoupling sequence: WALTZ-16. J. Magn. Reson. 53 (2), 313–340.

Spencer, L.R., Traficante, D.D., 1998. New window function for very short acquisition times. Appl. Spectrosc. 52 (1), 139–142.

Thomas, D.M., Bendall, M.R., Pegg, D.T., Doddrell, D.M., Field, J., 1981. Two dimensional ^{13}C, ^{1}H polarization transfer J spectroscopy. J. Magn. Reson. 42 (2), 298–306.

Wagner, G., Wüthrich, K., Tschesche, H., 1978. A ^{1}H nuclear-magnetic-resonance study of the conformation and the molecular dynamics of the glycoprotein cow-colostrum trypsin inhibitor. Eur. J. Biochem. 86 (1), 67–76.

Warren, S., Mayr, S. M., 2012. In: Harris, R. K., Wasylishen, R. E. (Eds). Encyclopedia of NMR, 8, 4394-4402.

Wesener, J.R., Günther, H., 1985. Steady state pulse calibration. J. Magn. Reson. 62 (1), 158–162.

Villa, M., Tian, F., Confrancesco, P., Halamek, J., Kasal, M., 1996. High-resolution digital quadra-ture detection. Rev. Sci. Instrum. 67 (6), 2123–2129.

Zhu, J.-M., Smith, I.C.P., 1995. Selection of coherence transfer pathways by pulsed-field gradients in NMR spectroscopy. Concept Magn. Res. 7 (4), 281–291.

Chapter 3

Sensitivity Enhancement

Chapter Outline

3.1 SENSITIVITY PROBLEM

Nuclear magnetic resonance spectroscopy is a powerful, robust, and noninvasive analytical technique which provides information about the structure, dynamics, and chemical kinetics of complex molecular entities. It is a key tool in the fields of organic, medicinal, analytical, and environmental chemistry, as well as in structural biology and material sciences. NMR is also widely used in medical diagnosis (magnetic resonance imaging). However, as compared to other spectroscopic techniques, such as ultraviolet (UV), infrared (IR), or

Solving Problems with NMR Spectroscopy. http://dx.doi.org/10.1016/B978-0-12-411589-7.00003-6

electron spin resonance spectroscopy, NMR is significantly *insensitive* largely because of the small energy and population differences between the ground and excited spin states (Section 1.1.1). Therefore, an important aim of new experimental and instrumental developments has been to minimize this disadvantage by improving the sensitivity. Improvements in instrumentation design and development of new pulse sequences in the last three decades have led to significant improvements in sensitivity and performance of NMR spectrometers. This has allowed NMR spectroscopy to be applied in many diverse fields where the sample quantity and the insensitive nature of the nuclei to be detected are serious constraints.

The key objectives of this chapter are to highlight the importance of the sensitivity issue in NMR spectroscopy, as well as to place the major developments in the field of sensitivity enhancement together under one heading. It is important to note that almost all of the techniques covered in this chapter are discussed in greater detail in some other chapters, and some level of repetition is, therefore, unavoidable.

Initial observation of NMR signals by physicists was made on simple and abundant molecules, such as water, paraffin, and ethanol in the early 1940s. However, with time the need to analyze minor quantities of structurally complex samples (Molinski, 2010), initially by chemists and then by biologists, challenged the then-existing design and concept of NMR spectroscopy. The low sensitivity of NMR, due to reasons mentioned below, was a major impediment in the development of this new technique. However, numerous benefits of the technique greatly overweighed this disadvantage. Improving the sensitivity and resolution of NMR signals understandably became the major driving force in many earlier developments in the concepts and instrumentation.

The early users of NMR spectroscopy had experienced sensitivity problems in NMR spectra due to several reasons, including:

1. Low natural abundance of the nucleus to be analyzed, such as ^{13}C or ^{15}N (resulting in a small number of NMR active nuclei in a given sample)
2. Low magnetogyric ratio of the nucleus under study, such as ^{13}C (resulting in weak magnetic susceptibility of the target nuclei)
3. High noise levels (due to the imperfect *Rf* electronics of the first generation of NMR spectrometers)
4. Low sample concentration (resulting in a small number of NMR active nuclei in a sample)

Sensitivity is generally measured by dividing the size of the signal by the size of the background noise—the *signal-to-noise* (*S/N*) ratio. The ^{13}C-NMR spectrum is about 5,800 times lower in sensitivity as compared to a ^{1}H-NMR spectrum. This is due to the low natural abundance (0.11%) and the low magnetogyric ratio of the ^{13}C nucleus. The *S/N* ratio is the ratio of the highest peak in the spectrum to the magnitude of the noise (difference between the most positive and most negative peaks corresponding to the noise level). This meant that for analysis involving experiments with nuclei having relatively low natural abundance and low magnetogyric ratio, such as ^{13}C or ^{15}N (for instance Hetero-COSY and INADEQUATE experiments described in Sections 7.1.2 and 7.7), large quantities of sample are required, often not accessible to the users. Such challenges have been the biggest driver of technical advances in NMR spectroscopy.

Problem 3.1

Why *sensitivity* and *signal-to-noise (S/N)* ratio terms are used interchangeably in NMR spectroscopy?

Problem 3.2

The *S/N* ratio depends on the number of NMR active nuclei in a given volume of solvent in the NMR tube (i.e., on the sample concentration).
1. Does this mean that the more molecules we have in our sample, the better will be the *S/N* ratio?
2. What will be the effect of molecular weight (mass) of the sample on the sensitivity of the NMR signal?

The *sensitivity test* is routinely performed for all NMR spectrometers to assess the performance of the instruments. This involves the measurement of the *S/N* ratio. These tests are performed using standard compounds in sealed tubes, provided by the instrument manufacturer. For protons (^1H-NMR) 0.1% ethylbenzene in $CDCl_3$ is used, for ^{13}C-NMR experiments 10% ethylbenzene in $CDCl_3$ is normally used, while for ^{15}N-NMR 90% formamide dissolved in d_6-benzene is employed. The sensitivity test is the measurement of *S/N* ratio in which the peak height of the signal in the sample is compared with the average height of the noise signals in the spectrum (Fig. 3.1).

The signal intensity is directly proportional to the population difference between the ground (α) and excited (β) energy states, represented by the Boltzmann distribution excess of the lower state over the higher energy state at thermal equilibrium. *This slight excess between the two states for protons or ^{13}C nuclei (spin ½ nuclei) (or more states for quadruple nuclei) gives birth to the NMR signal.* This energy difference is, in turn, proportional to the strength of the applied magnetic field (B_0). The earlier attempts to improve the sensitivity of the NMR signal were therefore focused on developing more powerful magnets. The *S/N* ratio, as measured for the quartet of the methylene protons (in ethylbenzene), is shown in Fig. 3.1. Other things being equal to the gain in *S/N* ratio on going from a smaller field NMR spectrometer (B_0 (old)) to a higher field NMR spectrometer (B_0 (new)) are given by Eq. 3.1:

$$Gain\ in\ \frac{S}{N} = \left[\frac{B_0(new)}{B_0(old)} \right]^{1.5} \tag{3.1}$$

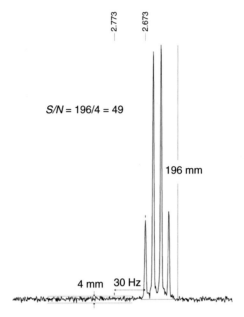

FIGURE 3.1   Sensitivity test in ^{1}H-NMR spectroscopy using 1% ethyl benzene solution in CDCl$_3$. The quartet of methylene protons is used to measure the S/N ratio. The S/N ratio is defined as the ratio of the average signal height, S, and the average noise level, N, which is measured as N = noise height/2.5.

where B_0 is the strength of the magnetic field of the magnet. This means that if we go from a 500 MHz NMR spectrometer to an 800 MHz NMR spectrometer, the gain in sensitivity would be $[800/500]^{1.5} = 2.02$, so that the sensitivity would be about doubled. Such gain in sensitivity has been the driving force behind the development of high field superconducting magnets (Section 1.2.1).

3.2   FIRST MILESTONE IN THE NMR SENSITIVITY ENHANCEMENT

Historically, the first major advance in improving the S/N ratio in NMR spectroscopy was the advent of *pulse Fourier transform (PFT) NMR spectrometers*. The departure from *CW* (continuous wave) NMR spectrometers represented a major milestone and a fundamental change in the field (Hyberts et al., 2014).

It was recognized that a further enhancement in sensitivity can be achieved by spectral accumulation. By recording spectrum many times and then adding them together, the S/N ratio improves because noise signals are random, varying both in intensity and in sign. So repeated addition of such noise signals leads to a sharp decrease in their magnitude. However, signals originating from absorptions of Rf energy by sample nuclei always occur at their respective chemical shifts with constant intensity and sign. Repeated addition of spectra therefore results in an increase in the *S/N* ratio which is dependent on the *square root of the number of scans*. The need for spectral accumulation over long time periods to achieve sensitivity enhancement was one of the reasons that led to the complete displacement of the CW method by pulse Fourier transform (PFT) NMR. This is explained in the following paragraph.

In the older *CW* NMR spectrometers, now only of historical interest, one swept the field of the electromagnet (B_0) or the transmitter frequency (*Rf*) through the resonance. Signals of chemically distinct nuclei (e.g., three different types of protons in ethyl alcohol) were recorded *one by one* when their Larmor frequency exactly matched the transmitter frequency and absorption of *Rf* took place. For a decent resolution, the sweep rate had to be rather slow, say of the order of 1 Hz/second. For a 1000 Hz wide spectrum, a single scan would therefore take 1000 seconds or over 16 minutes! By scanning repeatedly and adding the resulting spectra on a <u>C</u>omputer for <u>A</u>verage <u>T</u>ransients (CAT) one could obtain a reasonable spectrum but this would often require several hours of NMR time for the processing of a single sample, particularly if the sample quantities were small, making CW NMR spectroscopy an inefficient and *insensitive* technique for solving practical problems.

The advent of PFT NMR made it possible to record a large number of free induction decays (FIDs) (Sections 2.1.3–2.1.5) in the shortest possible time which led to a substantial increase in the *S/N* ratio. Through PFT NMR, it became possible to improve the *S/N* ratio by recording a large number of FID acquisitions and then processing them through signal averaging. Signal averaging basically involves the addition of a large number of scans together in order to enhance the signal strength and facilitate the distinction between signal and noise peaks. It is based on the assumption, as indicated above, that the signal peaks will always overlap and gain in strength with the addition of spectra, while the noise signals being random, will decrease in relative strength with the addition of spectra. Signal averaging required a great deal of computing capacity and <u>A</u>nalog-to-<u>D</u>igital <u>C</u>onversion (ADC). When signal averaging is performed, the magnitude of the NMR signal grows linearly, but the magnitude of the noise grows as well, though at a slower rate. One cannot simply double the *S/N* ratio by just doubling the number of scans, because the signal and the noise do not build up at the same rate. Instead, the signal increases as $NS^{1/2}$, where NS is the number of scans. Since the *S/N* ratio depends on the square root of the number of scans, in order to double the *S/N*, one will need to increase the number of scans by four times. Since the length of the experiment is directly proportional to the number of scans, this approach of enhancing the signal can quickly become extremely time-demanding.

The *S/N* ratio depends on many factors, as presented in Eq. 3.2.

$$\frac{S}{N} \propto n\gamma_e \sqrt{\gamma_d^3 B_0^3 t} \tag{3.2}$$

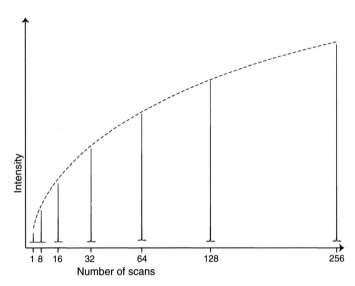

FIGURE 3.2 Signal intensity (~*S/N* ratio) *versus* number of scans. Spectra are plotted on an absolute intensity scale against the number of scans collected.

where "*n*" is the number of nuclear spins observed, γ_e is the magnetogyric ratio of the nuclear spin being excited, γ_d is the magnetogyric ratio of the spins being detected, B_0 is the magnetic field strength, and *t* is the acquisition time for the experiment.

With the advent of PFT NMR spectroscopy, ^{13}C, a nucleus with low natural abundance and a weak spin (low magnetogyric ratio) could be detected routinely in natural abundance. ^{15}N- and ^{17}O-NMR spectroscopy also emerged as well, though only on enriched/labeled samples due to their very low natural abundance (0.368% and 0.038%, respectively), opening the field for a whole range of biological applications. ^{31}P-NMR spectroscopy emerged for metabolomic studies and biofluid analysis, despite the relatively low magnetogyric ratio of phosphorus (108.291 for ^{31}P *versus* 267.513 for ^{1}H).

Interestingly, sensitivity enhancement by signal averaging through recording a large number of scans has inherent limitations (Fig. 3.2). For example if a particular *S/N* ratio is achieved in 5000 scans, just to double this ratio, the number of scans needs to be increased to 20,000 scans and to double it again, it would require 80,000 scans, and so on, thereby requiring long NMR instrument measurement times. For insensitive nuclei, such as ^{13}C, acquiring a large number of scans is necessary, and one needs also to ensure that there is a suitable relaxation period between the successive scans. A delay of at least $5T_1$ (five times the longitudinal relaxation time) is required between the scans.

Problem 3.3

Why is the advent of the PFT considered to be a *conceptual revolution* in NMR spectroscopy?

Problem 3.4

How many scans do we need to increase in order to double the *S/N* ratio every time?

Problem 3.5

Assume that the ^{13}C-NMR spectrum of 100 mg of a sample, dissolved in a solvent of choice, requires only 10 scans to afford a certain *S/N* ratio. In order to get the *same S/N* ratio with 50 and 25 mg of the same compound, how many scans will be required in each case?

Problem 3.6

Before running an NMR experiment, the number of scans (transients) needs to be decided. What factors are important in deciding the number of scans?

Problem 3.7

What is the most appropriate number of scans for various NMR experiments?

Problem 3.8

Why does not doubling the number of scans lead to a doubling of the *S/N* ratio?

Key challenges in NMR sensitivity and relevant answers and solutions are presented in Table 3.1, along with references.

Before a detailed discussion of various options for sensitivity enhancement, it is important to briefly explain the difference between sensitivity and resolution, two important properties which determine the quality of an NMR spectrum. The resolution measures how closely the discrete frequencies are spaced and it reflects the ability to distinguish between signals that are close in frequency. The sensitivity however reflects how strong the signal is as compared to the noise.

3.3   ADVANCES IN NMR HARDWARE FOR IMPROVING THE SENSITIVITY AND DETECTION

3.3.1   High Field Magnets

The population difference between the energy states is directly proportional to the magnetic flux density which, in turn, depends on the strength of the applied magnetic field B_0. This means that the NMR *S/N* ratio increases non-linearly with B_0. The use of high field NMR spectrometers has, therefore, been recognized as the most straightforward means to enhance NMR sensitivity. NMR spectroscopy began with low field permanent magnets and later electromagnets became common. Both permanent magnets and electromagnets, however, suffered from low field strengths with correspondingly low sensitivity, as well as

TABLE 3.1 Problems Commonly Encountered in NMR Measurements, and Their Possible Solutions

Number	Questions	Solutions	Reference
1.	I wish to record the NMR spectrum of an element in my sample which exists in low natural abundance. What do I do?	Use a high field NMR spectrometer which will improve sensitivity. Improve the S/N ratio by increasing the number of scans followed by signal averaging. Use an NMR spectrometer equipped with cryogenically cooled probe, or use more sensitive dedicated probes suitable for the nucleus under study. Use specially designed NMR tubes, such as Shigemi tubes. These tubes allow the use of smaller volume of NMR solvents (more concentrated samples) and thus "*concentrate*" the molecules in front of the detection coils. Use samples artificially enriched with greater abundance of the nuclei in the NMR tube under study.	Jarrell, 2009; Kovacs et al., 2005
2.	I have a sample for NMR analysis but it has organic impurities. What are my options?	Organic impurities are genuine compounds which will appear in NMR spectra with an intensity proportional to their concentration. Sensitivity enhancement techniques can not remove the signals of impurities in samples. Try using the diffusion order spectroscopy (DOSY) (Section 9.3.1) or TOCSY (Section 7.5) experiments. This might give separate ^{1}H-NMR spectra of your samples and impurities therein. Use of pulsed field gradients allows one to select coherences of interest or destroy unwanted coherences thereby dramatically reducing measurement times.	Hanabusa, 1971
3.	I have sufficient quantity of sample, dissolved in a suitable deuterated solvent; even then my NMR spectrum does not look so promising. What is wrong?	In such cases, the problem may well be in the NMR instrument. This may include poor shimming and wrong pulse calibration. This can quickly be checked by recording a spectrum of a standard, such as ethylbenzene or menthol and calculating the sensitivity and resolution. A comparison of this data with the instrument specifications allows one to identify if the problem exists in the spectrometer.	

poor magnetic field stability over long measurement times. The advent of stable and powerful superconducting magnets set the stage to develop NMR spectroscopy as a robust technique for a variety of purposes, including structural elucidation of complex chemical and biological samples, detection of low abundant nuclei in biological samples, quantitative analysis, medical diagnosis, etc. Today, magnets up to 1000 MHz (1 GHz, 23.5 T) are available in commercially available NMR spectrometers, while even more powerful magnets have been developed for specialized research purposes in more advanced laboratories of the world. The process of development of high field magnets continues with emerging new technologies related to superconducting magnets.

The use of high field NMR magnets provides the dual benefits of sensitivity enhancement and increased signal dispersion. Signal dispersion is a term used to define how well the resonances (signals) in an NMR spectrum are separated. For small molecules, signal dispersion may not be a major issue. However for large organic and biological macromolecules, both sensitivity and signal dispersion are important. Here, high field NMR spectrometers contribute in a major way. For example, on a 300 MHz spectrometer 1 ppm corresponds to 300 Hz, whereas on a 800 MHz spectrometer, 1 ppm corresponds to 800 Hz. This leads to a much greater dispersion of signals on the 800 MHz instrument which helps in the interpretation of the NMR data as it also leads to reduction of second-order distortions (Fig. 3.3).

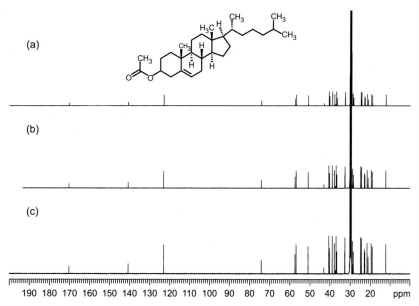

FIGURE 3.3 ^{13}C-NMR spectrum of cholesteryl acetate (1 mg): (a) at 300 MHz, (b) 400 MHz, and (c) 500 MHz, instruments RT in 18,432 scans. The dual benefits of sensitivity enhancement and signal dispersion are observed.

3.3.2 Advances in Probe Technology

The performance, sensitivity, and flexibility of probes are essential elements in the overall performance and utility of NMR spectrometers. Over the years, major developments in probe technology have contributed to high performance with good sensitivity and specificity in NMR experiments. For limited sample amounts, one should use the smallest diameter probe available. Thus, one can get ~50% improvement in S/N ratio if a 3 mm probe is used instead of a 5 mm probe for identical sample amounts. The use of susceptibility plugs from Shigemi results in the solution containing the sample to be located only opposite to the *Rf* coil length, resulting in a 2.5 to 4 fold increase in sensitivity (see later).

3.3.2.1 Cryogenically Cooled NMR Probes

Over the last two decades, the sensitivity of NMR signal detection has greatly increased due to the availability of more powerful magnets, improved *Rf* electronics, and the advent of cryogenic probe technology (Lee et al., 2014). A substantial increase in sensitivity is achieved if cryoprobes are used as the level of noise is substantially reduced and 3-4 fold enhancement in sensitivity is obtained as compared to the conventional probe. This allows small sample quantities to be analyzed relatively quickly (Fig. 3.4). The sensitivity of the NMR spectrometer is limited by the thermal noise in the process of signal detection. The probe is placed in the middle of the magnet bore and possesses two coils,

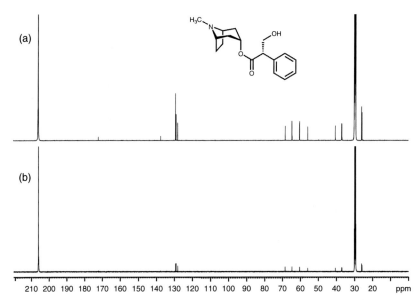

FIGURE 3.4 ^{13}C-NMR spectrum of hyoscyamine (1 mg) using: (a) cryogenically cooled (500 MHz) (helium gas at 20 K) probe, and (b) room temperature probe (500 MHz) at 298 K, in the same number (18,432) of scans.

one for sending pulses to the sample for excitation and a second coil to detect the signal generated by the relaxation process that is being received from the sample. In addition to this, a cryogenically cooled probe also houses a preamplifier to enhance the detected signal. The enhancement in sensitivity is achieved by lowering the temperature of both these coils since the thermal noise in the receiver coil is substantially decreased on cooling. The concept of decreasing the temperature of the probe for sensitivity enhancement was first presented by Hoult and Richards in 1976. Later, Styles and his coworkers made a probe in which detector coil and amplifier were cooled by liquid helium.

The technology involves cooling the probe coil and/or built-in signal pre-amplifier with a stream of helium (He) gas at ∼20 K, or with liquid nitrogen (N) (Styles et al., 1984). This increases the conductance and decreases the resistance within signal acquisition. The direct effect is sensitivity enhancement of the probe coil, and reduction in the levels of thermal noise generated by electronic circuitry of signal receiver, tuning and matching units, preamplifier, filters, transmit-receive-switch, etc.

Cryogenic probe technology was developed in the 1990s and the first installation took place in 1999. Since then, the developments have continued for high field NMR spectrometers and today this technology is available for all spectrometer ranges from 300 MHz to 1 GHz. These can be fitted with LC systems and flow injection modes with the option of 1.7 micro, 3 and 5 mm standard and shaped sample tubes (Behnia and Webb, 1998; Subramanian et al., 1998). Innovations in the development of a variety of probes are rapidly taking place to meet the demand of users. Cryogenic probes are available today for ^{1}H, ^{13}C detection and inverse probes with simultaneous decoupling of ^{13}C, ^{15}N, and ^{31}P. Broad range probes are available that include the widest range of nuclei both for direct and inverse detection. Cyroprobes are also available for tritium ^{3}H detection.

In the modern NMR spectrometers, the probe is cooled by using cold gaseous helium (20 K) instead of liquid helium (since that would need efficient thermal insulation). The cryogenic system includes a helium compressor and cryogenic cooling unit. This cryo unit comprises a closed-loop cooling system in which helium is compressed in one chamber and then cooled in another chamber by heat exchange following the ideal gas law (Eq. 3.3) since temperature is directly proportional to pressure and volume.

$$T = \frac{PV}{nR} \tag{3.3}$$

The cooled helium gas is then transferred to the probe through vacuum-insulated lines. The probe is also contained inside an insulated Dewar containing both the receiver and the preamplifier. The receiver coil is cooled to ∼20 K and preamplifier to ∼80 K. This close fitted assembly results in the thermal noise being kept to a minimum level (Styles et al., 1984). The warming up and cooling down of the system takes about a couple of hours.

Currently, cryogenically cooled probes are of two types (Kovacs et al., 2005). One is based on a closed cycle helium gas cryo cooled system. The other type is the open cycle liquid nitrogen cooled probes. Both types, though based on the same principle, have different sensitivity gains, costs, and ease of operation and maintenance.

3.3.2.1.1 Helium Cooled Cryoprobes

Closed cycle helium probes provide an increase in *S/N* ratio of about four-fold, in comparison to room temperature (RT) probes (Section 1.2.2). This has the following impact on the sensitivity of an NMR experiment:

1. NMR experiments can be run in about 1/16th the time duration as compared to an RT probe with the same level of *S/N* ratio.
2. NMR experiments can be performed on samples with 1/4 of the concentration in the same amount of time as on a conventional RT probe.

Problem 3.9

What is the fundamental difference in sensitivity enhancement between the use of higher field magnets or using cryogenically cooled NMR probes?

Problem 3.10

What is the benefit of sensitivity enhancement in terms of *NMR instrument time*?

The typical components of a cryogenically cooled probe are:

1. The main cooling system which contains a cold head which cools the helium gas to ~10 K.
2. A helium compressor which provides highly compressed helium gas at room temperature.
3. A cooling system for the compressor. The compressor can be either air-cooled or water-cooled.
4. A high purity helium gas cylinder.
5. A cryoprobe – the electronics is bathed in ~20 K helium gas.

Since cold probes have several interacting systems, special care and routine maintenance is regularly required to optimally maintain all of the components. The helium compressor operates at high temperatures. Even under the best conditions, compressors can last about 3–5 years before requiring expensive maintenance. In addition, operating the compressor at higher temperatures may cause the oil to leak and release contaminants into the cryo cooling unit. This can cause the whole system to fail or to develop chronic problems. The compressor should therefore be operated at the lowest temperature possible. Moreover, if the compressor cooling system fails even briefly (ca. 30 seconds), the compressor will shut off and the probe will warm up, leading to diminished sensitivity and possible damage to the system.

Contamination of the helium can be a major issue because any contaminants will freeze out in the cold head and disrupt the whole cooling system. Therefore, only very high purity helium gas should be used. The main contaminants come from the inner walls of the cylinder itself. Optimally, only grade 6 helium cylinders which have been individually tested and certified should be used.

Vibrations are often an issue because the movement of the cold head creates mechanical vibrations throughout the system. These vibrations can show up as noise in the NMR spectra. Great care must be taken in placing anti-vibration stands, making sure that the stands are flat on the floor and fixed tight. If the stands can move at all, the probe will sense the vibrations.

3.3.2.1.2   Liquid Nitrogen Cooled Probes

A more recent development in cryogenic probe technology is the development of liquid nitrogen cooled probes (Section 9.1.4). Liquid nitrogen cools down the temperature of Rf coils and preamplifiers. This technology delivers a considerable boost in sensitivity, i.e., two- to three-fold increase for insensitive nuclei such as ^{15}N to ^{31}P. The lower cost and ease of operation makes this a truly attractive option for many NMR laboratories. The cost is significantly less than that of cryoprobes. In addition to this, the use of liquid nitrogen in place of high purity helium gas lowers the cost of operation.

The open cycle liquid nitrogen probe comprises the following components:

1. Control unit
2. Liquid nitrogen vessel and
3. Cryoprobe electronics bathed liquid nitrogen

Cryogenic probes have greatly facilitated NMR studies of insensitive nuclei (Fig. 3.5) and re-introduced the prospect of using modern versions of the inherently insensitive ^{13}C-NMR experiments, such as INADEQUATE (^{13}C–^{13}C direct/one bond coupling) (incredible natural abundance double quantum transfer experiment). The INADEQUATE experiment (Section 7.7) greatly assists in determining the carbon frameworks of unknown structures, especially of small organic molecules.

Cryogenically cooled NMR probes have also facilitated biological NMR studies. The big sensitivity gains in ^{15}N spectra have enabled scientists to work on the structures and dynamics of proteins, nucleic acids, etc.

3.3.2.2   NMR Probeheads

The NMR probehead is a key component in an NMR spectrometer. The NMR probe contains the radiofrequency (Rf) coils, tuned at particular frequencies for the detection of specific nuclei in a given applied magnetic field B_0. The probe also contains the necessary hardware to control the sample temperature (when

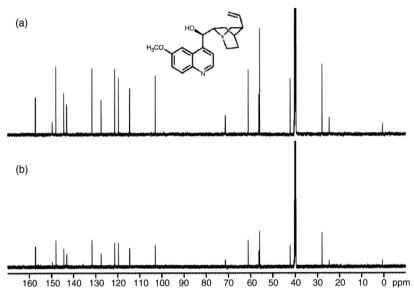

FIGURE 3.5 ^{13}C-NMR spectrum of quinine by using: (a) gaseous helium cooled probe, i.e., cryo-probe (20 K) and (b) open cycle cooled liquid nitrogen probe, using 64 scans on 600 MHz instrument.

connected to an external temperature controller). To optimize the sensitivity (S/N ratio), the probehead must be adapted for the specific application.

The probehead geometry is central to probe performance. Probeheads are usually constructed with two observe coils, one closest to the sample (the *inner* coil), and another further from the sample (the *outer* coil). This allows the probe to respond to multiple frequencies, and to allow the excitation/irradiation of multiple nuclei. The nuclei that use the *inner* coil are detected with the highest sensitivity. For example, in ^{13}C selective dual probehead, the inner coil is tuned to observe ^{13}C nuclei, while the outer coil is tuned to irradiate ^{1}H. Such probes are ideal for high sensitivity detection of ^{13}C through polarization transfer from the more sensitive coupled ^{1}H nuclei. This probehead affords the ^{13}C–{^{1}H} spectrum.

For solution NMR spectroscopy, several types of specific high resolution probeheads are used. Each has a specially tuned inner coil for direct detection of specific nucleus/nuclei and an outer coil for the irradiation of another nucleus/nuclei (Subramanian et al., 1999; Subramanian and Webb, 1998). Dozens of such probeheads are now available for solution NMR spectroscopy. Only the most commonly used probes are presented below:

1. *^{1}H selective normal probes* **SEL** (room temperature). In this probe, the inner coil is tuned to observe ^{1}H. It is highly sensitive for the detection of ^{1}H.
2. *^{19}F selective normal probes* **SEF** (room temperature). In this probe, the inner coil is tuned to observe ^{19}F.

3. *Selective probes for specific nuclei* **SEX** (room temperature). The inner coil in such probes is tuned to observe a specific nucleus selectively (e.g., ^{13}C, ^{31}P, ^{11}B, ^{29}Si, ^{27}Al). There is a decoupling channel for ^1H as outer coil. The probe is optimized for low background noise on the observe channel.

4. *Dual probes for $^1H/^{13}C$ selective* **DUL** (room temperature). These probes are tuned for observing ^1H and ^{13}C nuclei. The probe contains a ^1H decoupling channel, which is designed in such a manner that it can be used for decoupling, as well as for observation. This probe is specially suited for the routine analysis of organic compounds.

5. *Dual probes* **DUX** (room temperature). These probes are designed for observing a given nucleus (e.g., ^{31}P) in addition to ^1H. The probe features a ^1H decoupling channel, which can be used for decoupling, as well as for observation. This type of probe is ideal for the analysis of a given nucleus, such as ^{31}P in biofluids.

6. *Selective inverse or indirect probes* **SEI** (room temperature). The inner coil in these probes is tuned to observe ^1H, while the outer coil is tuned for selective decoupling on one nucleus (^{13}C, ^{15}N, etc.).

7. *Triple resonance inverse probes* **TXI** (room temperature). In this probe, the inner coil is tuned to observe ^1H, while the outer coil is double tuned for simultaneous decoupling of two nuclei (e.g., ^{13}C and ^{15}N).

8. *Triple resonance observe probes* **TXO** (room temperature). In this probe, the inner coil is double tuned to two nuclei (e.g., ^{13}C and ^{15}N) and the outer NMR coil is tuned for decoupling of ^1H.

9. *Broad-band probes* **BB** (room temperature). The probe has an *inner*-coil tuned to observe a range of nuclei, so-called equipped with a "broad-band channel" (i.e., ^{13}C, ^{15}N, ^{31}P, ^{109}Ag, etc.). The broad-band channel allows a continuous coverage of a very large frequency band. This probe is designed to analyze any NMR active nucleus, from ^{31}P all the way down to ^{109}Ag. Because of the non-specific nature of the inner coil (broad-band channel), the sensitivity is often much lower than for the above-cited dedicated probes (Fig. 3.6).

Most of the above probeheads are also available as cryogenically cooled probes for further enhancement in sensitivity, as discussed in Section 1.2.2.

Problem 3.11

Which probe(s) can give the highest proton (^1H) sensitivity in NMR spectroscopy?

Problem 3.12

I have a molecule with CF_3 groups present. Which probe is the most suited one for this purpose?

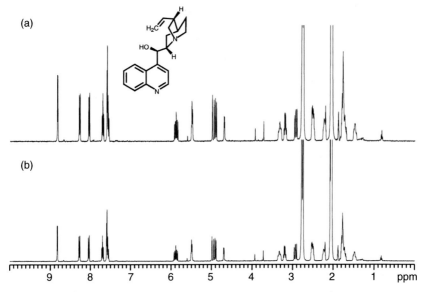

FIGURE 3.6 ^{1}H-NMR spectra of cinchonidine (1 mg) in 64 scans on: (a) ^{1}H-detected selective probe, and (b) broad- band probe.

Problem 3.13

Quite often, the inner coils of probes are tuned for high sensitive detection of a specific nucleus. For example, SEX probe may be used for the direct detection of ^{13}C. Can the same probe be used for the detection of ^{1}H? If yes, what would be the sensitivity in such a measurement?

There can be 2, 3, 4, or even more independent *Rf* channels installed in the instrument, making it possible to simultaneously process various nuclei (such as ^{1}H, ^{13}C, ^{15}N, ^{19}F, etc.).

3.4 NMR TUBES

The sensitivity of the NMR spectrum is dependent on a number of factors (Lee et al., 2014). These include the number of NMR active nuclei in the region of detection coil, the filling factor used, the magnetic field B_0 strength, the coil geometry, the magnetogyric ratio γ, natural abundance of the nuclei under observation, and the absolute temperature (Olson et al., 1995). The number of magnetic nuclei is proportional to the molar concentration and the volume of the solution in the active region. It is important to measure the NMR spectrum with as high concentration of the sample as conveniently possible, and to ensure that the sample is close to the detection coil by using the appropriate sample tubes and having probeheads of the optimum geometry.

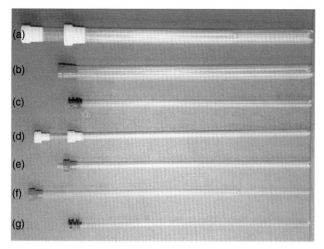

FIGURE 3.7 Various kinds of NMR tubes. (a) 10 mm coaxial NMR tube, (b) 10 mm NMR tube, (c) 5 mm NMR tube, (d) 5 mm co-axial tube, (e) 5 mm capillary NMR tube, (f) 2.5 mm NMR tube, and (g) 3 mm NMR tube.

The use of the proper NMR tubes can play an important role in enhancing the sensitivity (Section 1.2.7) (Jarrell 2009). Selecting the right kind of NMR tubes (Fig. 3.7), which ensure maximum number of NMR active nuclei (spins) per volume of the detection coil, may be a more convenient initial option to obtain some enhancement in sensitivity for users as compared to procuring expensive cryogenically cooled probe technologies.

The NMR tube diameter is a key factor in determining the number of NMR active nuclei in the detection coil. For example, if we keep the concentration of the sample solution constant (say, 10 mM), then an NMR measurement in a 5 mm diameter sample tube will afford a certain S/N ratio. If we place the same sample solution in a 3 mm tube in the same NMR spectrometer and record with the same conditions, the signal will be much weaker. The reason is that fewer nuclei will now be exposed to the Rf excitation frequency. If, however, the same solution is placed in a 10 mm NMR sample tube, then stronger signals will be obtained as there will be more NMR active nuclei exposed to the Rf excitation frequency. If, however, the sample is not sufficiently soluble or sample quantities are very low, then smaller NMR tubes are preferred.

A standard NMR tube is a thin-walled tube, made of borosilicate glass. There are several key elements in the architecture of NMR tubes, which affect their performance. These include the length, diameter, concentricity (variation in the radial centers, thickness), and camber (straightness or linearity). These tubes are available in 7 and 8 inch lengths. The ones with 5 mm outer diameter tubes are most commonly used, but tubes of 3 mm and 10 mm outer diameters are also available. Borosilicate glass NMR tube cannot be used for ^{11}B-NMR spectroscopy. For this purpose, quartz NMR tubes, containing low boron content, are used.

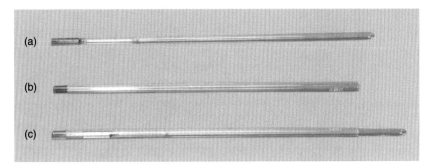

FIGURE 3.8 (a) Inner plunger of Shigemi NMR microtube, (b) outer Shigemi tube, (c) 5 mm Shigemi NMR microtube.

The Shigemi NMR microtube is an excellent addition in NMR hardware for high precession quality NMR measurements. This is a microscale NMR tube typically used for specialized NMR experiments. Shigemi microtubes are most appropriate for protein NMR experiments on high field NMR spectrometers, where only a smaller sample quantity is available. The thinner glass wall in the sample area, as compared to symmetrical microtube (0.20 mm *versus* 0.40 mm), increases the sample quantity held in the *Rf* region by a factor of 1.2 (about 20%). This provides 16% increase in the *S/N* ratio. A corresponding smaller solvent volume used in the Shigemi tube ensures a higher sample concentration. The reduced sample depth in Shigemi tubes is compensated by having a solid glass plug in the NMR tube beneath the level of the sample. Once air bubbles have been removed, the plunger is placed on the tube, and secured by parafilm. Ideally, the tube building materials are "matched" with the deuterated solvent used to have better spectrum resolution. Shigemi tubes are available with their construction optimised for various solvents, including D_2O, $CDCl_3$, CD_3OD, DMSO-d_6, etc. The tubes contain a bottom glass plug in the outer tube, and a glass plug in the plunger both of which are made of a material which is optimized for the specific deuterated solvent that is to be used in the tube (Fig. 3.8).

Shigemi tubes allow the restriction of the active volume and hence reduce the amount of solvent used without causing line shape problems. There are also susceptibility plugs available from different suppliers, that achieve a similar reduction in solution volumes. These can be employed for the ordinary NMR tubes (Fig. 3.9).

Following are some NMR tube options available for various types of NMR measurements:

1. 10 mm outer diameter borosilicate tubes for routine analysis
2. 5 mm outer diameter borosilicate tubes
3. 3 mm outer diameter borosilicate tubes
4. 1.7 mm outer diameter borosilicate tubes
5. 1.0 mm outer diameter borosilicate tubes
6. Microtubes
7. Nanotubes

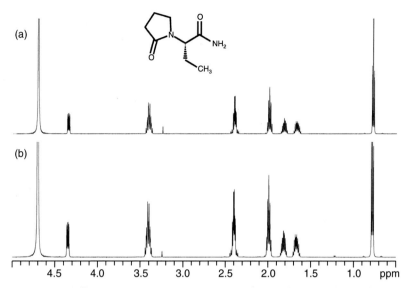

FIGURE 3.9   (a) ^{13}C-NMR spectrum of levetiracetam (0.5 mg) on 600 MHz NMR spectrometer (room temperature dual probe) in 16 scans, using 5 mm NMR tubes, and (b) same using Shigemi microtubes.

8. Capillary tubes
9. Shigemi NMR microtubes optimized for various deuterated solvents

Problem 3.14

What is in the architecture of the Shigemi microtubes which makes them more sensitive than the corresponding ordinary microtube?

Problem 3.15

Why are Shigemi NMR microtubes considered to be particularly suited for protein/biomolecular NMR measurements?

Problem 3.16

Which hardware options would afford the best sensitivity for recording ^{13}C-NMR spectra?

3.5   INNOVATIVE PROCESSING AND PULSE TECHNIQUES FOR IMPROVING THE SENSITIVITY AND DETECTION

The sensitivity of an NMR spectrometer is, apart from various hardware options (Sections 1.2.1, 1.2.2, and 3.3), also dependent on the innovative pulse sequences and data processing techniques that have been developed to improve the sensitivity of NMR experiments (Donovan and Frydman, 2015; Fujiwara and Ramamoorthy, 2006).

Since the strength of the NMR signal increases as the population difference between the energy levels increases, a number of direct and indirect techniques have been developed that increase this population difference, resulting in increased sensitivity. *Direct* methods for sensitivity enhancement are based on increasing the population difference using high field magnets, polarization transfer, nuclear Overhauser effects, Dynamic Nuclear Polarization (DNP), as well as the use of appropriate probeheads which can concentrate NMR active nuclei in the detection area (Cai et al., 2006; Merkle et al., 1992; Granger, 1990).

Indirect methods for sensitivity enhancement are based on increasing the contribution of the sample signal and decreasing the level of noise, either by using cryogenically cooled probes or by data manipulation techniques.

3.5.1 Sensitivity Enhancement by Data Manipulation

Let us first discuss sensitivity enhancement by an easy indirect method, that involving mathematical data manipulation. The nuclear excitations caused by application of pulses afford a large number of FIDs (Sections 2.1.3–2.1.5). The acquired FIDs are then converted from the time domain to the frequency domain signals by Fourier transformation. There are various ways through which the acquired FID data can be manipulated before the Fourier transformation. These methods involve manipulating the data using certain "window functions" (Section 2.1.12). The main objective of multiplying the acquired data by a suitable window function is to improve the *S/N* ratio (sensitivity) and resolution.

In a typical time domain signal (FID), the contribution of the signal is strongest at the beginning (in the "head" of the FID), but due to the decay, it decreases over time; however, the noise remains constant throughout. This means that the early part of a FID is *more important* in terms of contribution to the signal as it is here where the signal is the strongest, whereas the latter parts are *less important* as the signal has decayed, leaving the noise as the dominant contributor in the "tail" of the FID. We can favour the early parts of a FID by multiplying it with an exponentially decaying function which starts with the value one and then steadily decrease to zero (Fig. 3.10).

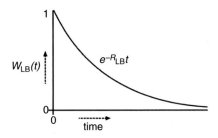

FIGURE 3.10 A graph of the decaying exponential function commonly employed to enhance the contribution of the signal in the "head" of the FID and decrease the contribution of noise in the "tail" of the FID.

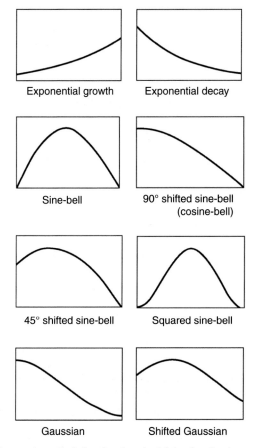

FIGURE 3.11   Commonly used window functions (weighting functions).

The window functions are mathematical weighting functions which are used to de-emphasize the latter parts of the FIDs. This forces the tail toward zero. The careful use of window functions, such as exponential, Lorentz-Gauss, Traficante, sine-bell, etc., can lead to a spectrum with enhanced sensitivity (Fig. 3.11). However, the increased decay rate of the NMR signal may cause the lines to broaden by decreasing the resolution. A detailed discussion on the use of window functions is presented in various texts (Zerbe and Jurt, 2014; Claridge, 2009). Figure 3.12 illustrates the effect on the S/N ratio due to the use of different weighting functions.

3.5.2   Sensitivity Enhancement by NOE through Decoupling

The presence of spin–spin coupling between a ^{13}C nucleus and ^{1}H nuclei bonded to the ^{13}C splits the carbon-13 peaks and results in weaker signals of the

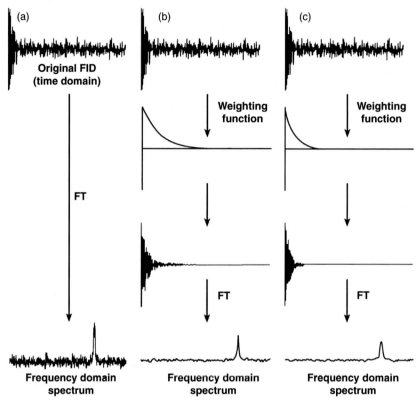

FIGURE 3.12   The effect of multiplying an FID with various decaying exponential functions (weighing functions) on the *S/N* ratio of the spectrum.

multiplet. This problem can be addressed by decoupling, which leads to NOE based enhancement of the ^{13}C-NMR signal.

To understand the effect of decoupling on sensitivity, the ^{13}C-NMR spectrum of fenchone is presented (Fig. 3.13a). The signals for protonated carbons appear as split signals due to their (one-bond) couplings with the attached ^{1}H nuclei. The splitting effect due to the long-range ^{13}C and ^{1}H coupling has been removed for simplicity, through off-resonance decoupling (Section 4.1.4). Decoupling is achieved with the help of a saturation pulse, tuned to the ^{1}H frequency. A decoupling saturation pulse is a relatively low power B_1 field applied for long enough for all proton magnetization effects to disappear. A saturation pulse, applied on the ^{1}H nuclei, rotates their magnetization vectors clockwise about the x' axis constantly. By saturating the methylene protons, the splitting disappears. When the effects of the ^{1}H-^{13}C coupling are removed, a fully decoupled ^{13}C-NMR spectrum is obtained, where each signal appears as a clean singlet, representing a specific carbon at its corresponding chemical shift, which depends on its electronic environment.

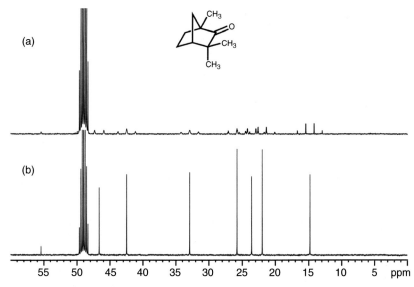

FIGURE 3.13   (a) ^{13}C-NMR spectrum of fenchone with direct (one bond) ^{13}C–^{1}H coupling split-ting of the signals of all protonated carbons, (b) broad-band decoupled ^{13}C-NMR spectrum with enhanced sensitivity due to NOE effect.

The effect on signal intensity due to the decoupling is dramatic, as shown in Fig. 3.13b. One would expect a stronger decoupled signal due to the collapse of the split (doublet, triplet, or quartet) signal. However, the enhancement in intensity is more than due to the simple collapse of split signal. What causes this additional enhancement?

The additional enhancement is caused by the *nuclear Overhauser effect* (NOE). The concept and the theory of the NOE are described in Chapter 6 of this book. Broad-band decoupled ^{13}C-NMR spectra are now routinely recorded for the assignments of carbon chemical shifts, and they result in enhanced signal intensity achieved through homonuclear and heteronuclear decoupling. Selective heteronuclear decoupling can also be carried out on selected ^{1}H nuclei, leading to signal enhancement of the specific attached carbons.

3.5.3   Sensitivity Enhancement by Polarization Transfer

3.5.3.1   Polarization Transfer

After nuclear excitation, there is a small excess of spins in the lower energy state than the upper energy state according to Boltzmann distribution. As stated earlier, the signal intensity depends on the difference in population of the two states (in case of spin ½ nuclei). One of the most exciting develop-ments in the field of NMR spectroscopy is the advent of polarization transfer to enhance sensitivity.

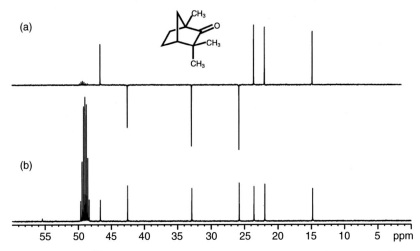

FIGURE 3.14   (a) ^{13}C-NMR spectrum of fenchone with polarization transfer (DEPT) (only protonated carbons are seen), (b) broad-band decoupled ^{13}C-NMR spectrum of the same compound, at 125 MHz.

Polarization transfer techniques are generally used to enhance the sensitivity of ^{13}C- and ^{15}N-NMR spectra, by transferring the polarization from sensitive or highly polarized nuclei to the less sensitive spin systems. Polarization transfer in liquids is often achieved through J-coupling between insensitive and sensitive nuclei, using the INEPT experiment (Section 4.2.1), while in solids dipolar coupling is used for cross polarization transfer. The most widely used polarization transfer NMR experiments include _D_istortionless _E_nhancement _P_olarization _T_ransfer (DEPT) and _I_nsensitive _N_uclei _E_nhancement by _P_olarization _T_ransfer (INEPT) for ^{13}C- and ^{15}N-NMR spectroscopy. Figure 3.14 shows the sensitivity gain in a ^{13}C-NMR spectrum by polarization transfer (DEPT) (only protonated carbons are visible) in comparison to broadband decoupled ^{13}C-NMR spectrum.

The concept and applications of polarization transfer for sensitivity enhancement are presented in Chapter 4, Section 4.2 of this book.

3.5.3.2   Inverse Polarization Transfer

The discovery and use of inverse polarization transfer was another breakthrough in the field. In the ^{1}H-detected NMR experiments, such as _H_eteronuclear _M_ultiple _Q_uantum _C_oherence (HMQC), _H_eteronuclear _S_ingle _Q_uantum _C_oherence (HSQC), and _H_eteronuclear _M_ultiple _B_ond _C_onnectivity (HMBC) advantage is taken of the greater sensitivity of protons (or ^{19}F). The insensitive nuclei are detected through the more sensitive nuclei after so called "inverse polarization transfer." The technique is extensively discussed in Section 7.6. Sensitivity gains in HMQC spectrum *versus* conventional HETCOR (Hetero-COSY) experiments can be seen in Fig. 3.15.

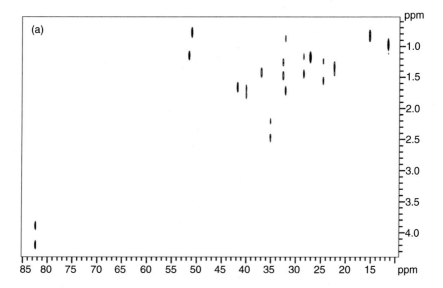

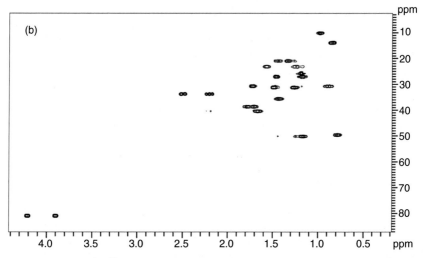

FIGURE 3.15   (a) ^{1}H–^{13}C hetero COSY NMR spectrum of compound oxandrolone (25 mg), (b) HSQC spectrum (involving inverse polarization transfer) of the same compound (500 MHz). Difference in sensitivity is clearly seen.

3.6 OTHER METHODS FOR SENSITIVITY ENHANCEMENT

3.6.1 Dynamic Nuclear Polarization (DNP)-NMR

NMR spectroscopy is a key tool for structure determination in chemical and biochemical sciences, as well as in structural biology. However, the inherently low sensitivity of certain nuclei (^{13}C, ^{15}N, etc.), due to their low magnetogyric ratios and natural abundance of certain nuclei, is a major constraint. It results in a small difference in the number of spins in lower and upper energy states. Therefore, the signal intensities in NMR experiments are weak so that extensive signal accumulation/averaging and polarization transfer are needed in order to get acceptable *S/N* ratio. In polarization transfer techniques, polarization transfer is carried out from the more sensitive nuclei (^{1}H or ^{19}F) to the low sensitive (insensitive) nuclei (^{13}C, ^{15}N, etc.) in order to improve the sensitivity of the insensitive nuclei. However, the nuclear spin reservoir for even the so-called sensitive nuclei, such as ^{1}H, is limited. As a result, the sensitivity enhancement through nuclear polarization transfer still proves to be inadequate in many cases. Interestingly the Boltzmann distribution excess even for the most sensitive nucleus ^{1}H (with one of the biggest magnetogyric ratios) is low (only an excess of 21 nuclei in the lower energy state in a total population of two million nuclei at 500 MHz).

However, at the same magnetic field B_0, the electron spin reservoir (e.g., in a free, stable paramagnetic polarizing agent) is significantly larger than the nuclear spin reservoir, due to the much higher magnetic moment of the electron spins. This difference in population between the two reservoirs is reflected by the corresponding ratio of the magnetogyric ratios of electron and proton. Dynamic nuclear polarization (DNP)-NMR spectroscopy for solution and solid state currently makes use of the polarization transfer from electron spins to nuclear spins. It is governed by the ratio γ_e/γ_n with γ_e being much higher than γ_n. This makes it possible to transfer the large Boltzmann polarization of the electron spin reservoir to the nuclear spin reservoir to provide a major boost in NMR signal intensities by several orders of magnitude. This dramatically increases the signal intensity (sensitivity) in an NMR experiment (Nanni et al., 2011).

There are essentially two types of DNP methods. The first method (continuous-wave DNP) relies on continuous irradiation of electrons with microwaves at or near the electronic resonance. It becomes inefficient at high magnetic fields. Ironically it is at these high magnetic fields that the resolution of NMR spectroscopy is the highest.

The second method involves pulsed DNP, which allows a greater flexibility in the manipulation of both the electronic and the nuclear spins. Among the DNP experiments, the *N*uclear spin *O*rientation *V*ia *E*lectron spin *L*ocking (NOVEL) sequence has special significance, since the enhancement it can generate is not expected to decline with the strength of the applied magnetic field strength B_0. Some polarizing agents used in DNP experiments are presented in Fig. 3.16. However for technical reasons, the DNP experiments are still not widely used in NMR spectroscopy.

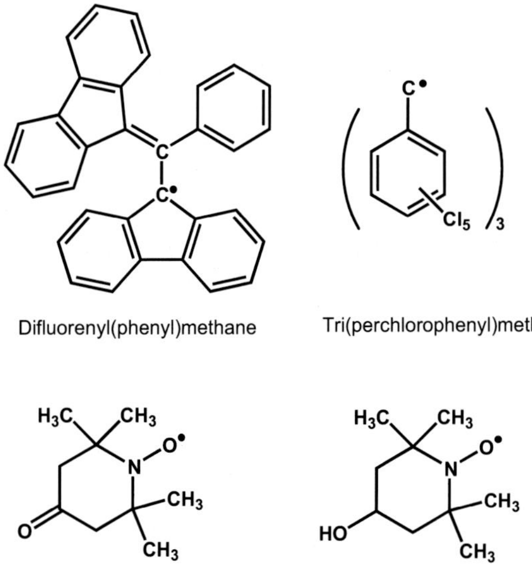

Difluorenyl(phenyl)methane   Tri(perchlorophenyl)methyl

TEMPONE   TEMPOLE
(4-Oxytetramethylpiperidine oxide)   (4-Hydroxytetramethylpiperidine oxide)

FIGURE 3.16   Some polarizing agents for DNP-NMR experiments.

Problem 3.17

What is the most likely mechanism of the transfer of polarization from the electron reservoir to the nuclear reservoir in DNP-NMR spectroscopy?

3.7 KEY APPLICATIONS OF SENSITIVITY ENHANCED NMR SPECTROSCOPY

The use of cryogenically cooled probes has substantially increased the sensitivity of NMR experiments. These probes are usually used (1) when observing nuclei of intrinsically low γ, (2) when the quantity of sample is very small or (3) when the sample has a very low solubility, or (4) in order to reduce the experimental time. They have been widely used in structure determination of small organic molecules in natural abundance, isotropically large biomolecules, smaller biomolecules in natural abundance, in metablomics, and to study ligand interactions with enzymes.

3.7.1 Structure Determination of Small Organic Molecules in Natural Abundance

Natural compounds are often obtained in low quantities that may range from a few milligrams to micrograms. It is usually necessary to acquire their ^{1}H-, and

^{13}C-NMR along with 2D- homo- and hetero-nuclear NMR spectra, which can take several days, especially when ^{13}C detection is required in natural abundance. Cryogenic probes not only increase the sensitivity in the ^{1}H detected and inverse NMR experiments but also in the ^{13}C observed experiments to a great extent. The enhanced sensitivity makes it possible to record inherently insensitive but important NMR experiments. These include the *I*ncredible *N*atural *A*bundance *D*oubl*E* *QUA*ntum *T*ransfer *E*xperiment (INADEQUATE) experiment which provides information about the carbon framework of the molecule through the observation of directly bonded [13]C–[13]C nuclei. Figure 3.17 shows

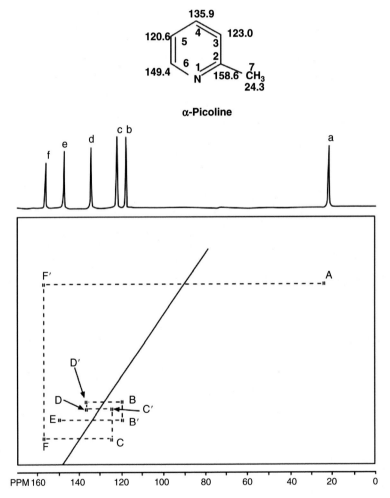

FIGURE 3.17   INADEQUATE spectrum of α-picoline on 500 MHz NMR spectrometer, recorded by using cryogenically cooled probe. Dotted lines indicate the direct [13]C–[13]C couplings in the molecule.

the INADEQUATE experiment (^{13}C–^{13}C correlations) of a natural compound, α-picoline, recorded at room temperature on a 500 MHz NMR spectrometer.

3.7.2 Ligand Binding Experiment through STD-NMR

Saturation transfer difference NMR (STD-NMR) is an important tool for molecular recognition (Section 6.9). The interaction of ligands with proteins or other biomolecules can be studied by using STD-NMR spectroscopy. During this experiment the concentration of the protein/biomolecule is very low, while the concentration of a smaller ligand molecule could be 100–200 times greater. As a result of the experiment, a specific proton in the protein is saturated and the saturation is then transferred through the protein by spin diffusion to the bound ligand. The saturation transfer effect on the ligand molecule is then observed. The difference of "on" and "off" resonance spectra shows only those protons of the ligand that interact by saturation transfer from the protein, thus indicating the binding part of the ligand to the protein. This experiment may also be combined with 2D experiments, such as NOESY, TOCSY, and HSQC. Since these experiments need higher sensitivity therefore cryogenic probes are often preferred (Cutting et al., 2007).

SOLUTIONS TO PROBLEMS

3.1 The *sensitivity* of an NMR spectrometer is a measure of the minimum number of spins detectable by the spectrometer. The greater the number of detectable spins, the more sensitive an NMR spectrometer will be.

The *signal-to-noise ratio* in an NMR experiment is, however, related to the population difference between the lower and upper spin states. The larger this difference, the larger the signal. *Since in NMR spectroscopy, the detectable spins are the ones which represent the Boltzmann distribution excess between the lower and upper spin states,* the two terms are directly related to each other, and can be used interchangeably.

3.2 1. Yes. A greater quantity of the sample means that more nuclei will be available for detection.

2. This can be understood in terms of the Avogadro number ($6.02214129 \times 10^{23}$ per mole) that is directly related to the molecular weight of a sample. One molar solution of every sample will have the same number of molecules in the solution. However, the weight of a given molar concentration of the sample will vary with its molecular weight. For example 1 M (molar) solution of a 10 kDa protein will require 10,000 g of a sample per liter, while 1 M solution of acetic acid will require a concentration of only 60 g L^{-1}.

However, it is often the solubility and not the available quantity that is the determining factor as only a limited amount of sample may be soluble in a specified volume of deuterated solvent. It is also important to note that it is the nuclei that are located opposite the detection coil which really matter since it is those nuclei that are detected. Various kinds of probeheads and NMR tubes have been designed to concentrate the detectable nuclei in the active detection area.

3.3 NMR spectroscopy was originally developed as a technique involving *absorption* spectroscopy (*CW* NMR spectroscopy), where *Rf* was absorbed by the in-resonance nuclei sequentially. This absorption of *Rf* was recorded as NMR signals, just as in other spectrophotometric techniques. However, *CW* NMR suffered from a low *S/N* ratio and a long scan time (several minutes per scan), much of which was wasted as only the base line (where no signals were present) was also being scanned.

The advent of PFT NMR was the first major development which has had a profound impact on the field of NMR spectroscopy due to following reasons.

1. In CW NMR spectroscopy, the absorption of *Rf* energy by the nuclei under study was recorded. In PFT, all nuclei are excited or perturbed simultaneously by using a short *Rf* pulse of specific power and duration. The excited spins are then allowed to *relax* back to their thermal equilibrium position through the process of relaxation. This relaxation is recorded as a free induction decay (FID). FIDs, which are time domain signals, are then converted into frequency domain signals (the so called "normal" NMR spectrum). By doing this, one can record a large number of scans in a short time. These scans are then processed through signal averaging, leading to substantial gains in sensitivity. The *S/N* ratio improves as the square root of the number of scans.

2. The process of acquiring signals in PFT NMR spectroscopy is fundamentally different from that in CW NMR spectroscopy. In PFT, the oscillating magnetization vectors (due to nuclear spins) induce a current in a nearby detector coil, creating electrical signals that oscillate at the NMR frequency, (FIDs). In CW NMR spectroscopy, however, it is the absorption of *Rf* energy by individual spins that is recorded.

3.4 If a certain *S/N* was obtained by say 400 scans (square root of 400 = 20), then to double this ratio, we will need to accumulate 1,600 scans (square root of 1600 = 40). To double the *S/N* again, we will need to record 6400 scans (square root of 6,400 = 80), etc. This indicates that the relying only on signal averaging for sensitivity enhancement is not always practical, and especially for insensitive and slow relaxing nuclei, such as ^{13}C and ^{15}N, it can take a long time.

3.5 Good solubility of various quantities of sample in the selected deuterated solvent is essential. With this condition met, one can correlate the number of scans (NS) and quantities of a sample. If the ^{13}C-NMR spectrum of 100 mg of sample, perfectly dissolved in the chosen solvent, requires

10 scans to get the same *S/N* ratio with 50 mg sample, $4 \times 10 = 40$ scans will be required. If we were to use only 25 mg, it would require 160 scans.

3.6 The number of scans (NS, or transients) that are collected depends on following factors:
1. Desired quality of the spectrum.
2. Spectrometer time available to the operator.
3. Consideration must be given to the manner in which the *S/N* ratio is improved in multiple-scan, or signal-averaging NMR experiments. In this procedure, a digitized FID is stored in the computer memory. Additional FIDs are then recorded a certain number of times, and added to the same memory locations.

   Since the *S/N* ratio increases as the square root of the number of scans in order to double the *S/N* ratio, one needs to increase the number of acquisition by four times.

3.7 This depends on the sample quantity available, the NMR spectrometer on which the measurement is being made, the molecular weight of the compound, and the type of NMR experiment to be recorded, so there is no such thing as an "appropriate number."

Following are general practices in terms of number of scans for various nuclei:

For 1D broad-band ^{13}C-NMR spectra, a concentration of 1 μM of material should be sufficient whereas for ^{1}H-NMR acquisition 30 nM can be sufficient. For a substance of molecular weight of, say, 400, ^{13}C-NMR spectra can, therefore, be recorded in a couple of hours if the sample quantity is between 0.5 and 1.0 mg, whereas for ^{1}H-NMR spectra a quantity as small as 15 μg is often sufficient to record an acceptable NMR spectrum.

3.8 For NMR measurements, the *S/N* ratio increases as the square root of the number of scans $\left(\sqrt{S/N} \right)$. In order to *double the signal-to-noise* ratio in a particular NMR spectrum, one would need to first see how many scans the original spectrum was obtained, and then quadruple that number in order to obtain a doubling of the *S/N* ratio. Thus if a spectrum was obtained after accumulating 200 scans then to double the *S/N*, one will need to record 800 additional scans. To double the *S/N* again it would need 3200 scans. It, therefore, makes no sense to keep on accumulating scans after a certain initial number, as one is just wasting instrument time without getting any significant increase in the *S/N* ratio.

3.9 The sensitivity of the NMR signal depends upon the excess of population in the lower energy state(s) as compared to the higher energy state(s). This population difference between the two (or more) states is directly related to the strength of the applied magnetic field B_0. A higher field magnet provides a stronger static magnetic field B_0, thereby increasing the population difference between the ground and excited energy states. The increased population difference results both in an improved *S/N* ratio and an increased dispersion of signals, which results in a reduction of second-order perturbations and cleaner multiplets. A cryogenically cooled probe, however, decreases the contribution of the noise to the spectrum, thus improving the *S/N* ratio but it does not increase the signal dispersion. The cooling of the radio frequency (*Rf*) coil and sections of the preamplifiers using gaseous helium in the probe results in substantial (up to four-fold) sensitivity gains.

3.10 Sensitivity enhancement by using various techniques translates into reduction of data collection time. For example, a four-fold increase in sensitivity through the use of a cryogenically cooled probe will lead to approximately 16-fold decrease in the NMR data collection time to afford a spectrum with a comparable *S/N* ratio.

3.11 *Inverse* or *indirect* selective probes (SEI) are constructed in such a manner that the *inner*-coil is tuned for ^1H, while the outer coil is tuned for decoupling on ^{13}C or ^{15}N. These probes give the highest proton (^1H) sensitivity. If inverse probes are cryogenically cooled, then the sensitivity of the ^1H-NMR spectrum is much higher.

3.12 There are two types of spectra that can be recorded for CF_3 containing compounds:
1. ^{19}F-NMR spectrum: For this purpose, a dedicated ^{19}F selective normal probe (*SEF* room temperature) probe is suitable.
2. For $^{13}C-\{^{19}F, {}^1H\}$ spectrum (^{13}C–NMR spectrum with decoupling of attached ^{19}F): In principle, a normal dual probe, such as *DUL*, should work well. The ^1H decoupling coil can also be used to decouple the ^{19}F.
   Such a spectrum can also be recorded by using special probes called triple resonance broad-band inverse (*TBI*) probes. These probes are designed to detect one particular nucleus (^{13}C in this case), while decoupling the other nuclei (^{19}F and ^1H in this case).

3.13 Such probes are designed to selectively observe a specific nucleus with high sensitivity (e.g., ^{13}C, ^{31}P, ^{11}B, ^{29}Si, ^{27}Al). The frequency of such probes can, however, be adjusted to observe ^1H, but the resulting sensitivity (for ^1H-NMR spectrum) will be much lower than that for the ^1H-selective probes (*SEL* or *SEI* probes).

3.14 Most common Shigemi microscale tubes have 5 mm outer diameter and 4.2 mm inner diameter. They have two components, an outer tube with a solid glass plug at the bottom, and a plunger insert with a solid bottom. The sample is poured into the outer tube. After the removal of gas bubbles, the plunger is inserted in the outer tube and secured with parafilm. This fully assembled Shigemi tube has a reduced volume of sample in the region of the detection coil. Ideally, the tubes are made of specialized borosilicate glass, matched with the deuterated solvent in order to afford better spectral resolution and sensitivity.

3.15 Shigemi microtubes, whose susceptibility matches D_2O, are especially suitable for protein NMR spectroscopy in structural biology studies. Often, the available quantity of protein material is very limited. A Shigemi microtube requires a reduced volume ($\sim$200 μL) of sample. The buffer containing 7% D_2O is used for the deuterium (^2H) lock.

3.16 Instruments with greater field strengths give the best results. For instance if you have the option to use 600 or 300 MHz instruments, then the 600 MHz instrument should be preferred, particularly if you have small sample quantities, as the *S/N* ratio on the 600 MHz instrument would be better than that on the 300 MHz instrument by a factor of 600/300 squared, i.e., four times better. The probe that you use is also important for optimum sensitivity. If you are employing an indirect (inverse) detection probe, then this has the ^1H observe coil closer to the sample, on the inside, and it would give a better proton signal but a weaker carbon signal (if you use it for directly detecting carbon). A standard probe, however, has the carbon coil on the inside, and it gives a better carbon signal but a weaker proton signal than the indirect detection probe.

3.17 In DNP-NMR spectroscopy, the large electron polarization of a polarizing agent is transferred to the surrounding nuclei (such as ^1H). This is *initiated and driven by microwave irradiation of the sample. During this process the large thermal polarization of the electrons spin reservoir is transferred to the nuclear spin reservoir.* The electron spin system (polarizing agent), required for DNP spectroscopy, can either be an endogenous or exogenous paramagnetic system. For solution-state NMR, the only DNP mechanism is the Overhauser effect, while for solid-state NMR spectroscopy, different DNP mechanisms can be employed, such as the solid-effect, thermal-mixing, or the cross-effect.

REFERENCES

Behnia, B., Webb, A.G., 1998. Limited-sample NMR using solenoidal microcoils perfluoucarbon plugs, and capillary spinning. Anal. Chem. 70 (24), 5326–5331.

Cai, S., Seu, C., Kovacs, Z., Sherry, A.D., Chen, Y., 2006. Sensitivity enhancement of multidimensional NMR experiments by paramagnetic relaxation effects. J. Am. Chem. Soc. 128 (41), 13474–13478.

Claridge, T.D.W., 2009. Practical aspects of high-resolution NMR. High-resolution NMR Techniques in Organic Chemistry, 27, Elsevier, Oxford, UK, pp. 55–58.

Cutting, B., Shelke, S.V., Dragic, Z., Wagner, B., Gathje, H., Kelm, S., Ernst, B., 2007. Sensitivity enhancement in saturation transfer difference (STD) experiments through optimized excitation schemes. Magn. Reson. Chem. 45 (9), 720–724.

Donovan, J.K., Frydman, 2015. HyperBIRD: a sensitivity-enhancement approach to collecting homonuclear-decoupled proton NMR spectra. Angew. Chem. Int. 54 (2), 594–598.

Fujiwara, T., Ramamoorthy, A., 2006. How far can the sensitivity of NMR be increased? Annu. Rep. NMR Spectrosc. 58, 155–157.

Granger, P., 1990. Polarization transfer INEPT, DEPT, NOE. Multinuclear Magnetic Resonance in Liquids and Solids Chemical Applications, 322, Springer, The Netherlands, pp. 63–79.

Hanabusa, M., 1971. Sensitivity enhancement of a pulsed NMR spectrometer. J. Appl. Phys. 42 (3), 1077–1079.

Hyberts, S.G., Arthanari, H., Robson, S.A., Gerhard, W., 2014. Perspectives in magnetic resonance: NMR in the post-FFT era. J. Magn. Reson. 241, 60–73.

Jarrell, H.C., 2009. Reducing the NMR sample volume using a single organic liquid: increased sensitivity for mass-limited samples with standard NMR probes. J. Magn. Reson. 198 (2), 204–208.

Kovacs, H., Moskau, D., Spraul, M., 2005. Cryogenically cooled probes: a leap in NMR technology. Prog. Nucl. Energy 46, 131–155.

Lee, H.J., Okuno, Y., Cavagnero, 2014. Sensitivity enhancement in solution NMR: emerging ideas and new frontiers. J. Magn. Reson. 241, 18–31.

Merkle, H., Wei, H., Garwood, M., Ugurbil, K., 1992. B_1-Insensitive heteronuclear adiabatic polarization transfer for signal enhancement. J. Magn. Reson. 99 (3), 480–494.

Molinski, T.F., 2010. NMR of natural products at the 'nanomole-scale'. Nat. Prod. Rep. 27 (3), 321–329.

Nanni, E.A., Barnes, A.B., Griffin, R.G., Temkin, R.J., 2011. THz dynamic nuclear polarization NMR. IEEE Trans. Terahertz Sci. Technol. 1 (1), 145–163.

Olson, D.L., Peck, T.L., Webb, A.G., Magin, R.L., Sweedler, J.V., 1995. High-resolution microcoils H-1-NMR for mass-limited, nanoliter-volume samples. Science 270 (5244), 1967–1970.

Styles, P., Soffe, N.F., Scott, C.A., Cragg, D.A., Row, F., White, D.J., White, P.C.J., 1984. A high-resolution NMR probe in which the coil and preamplifier are cooled with liquid helium. J. Magn. Reson. 60, 347–354.

Subramanian, R., Lam, M.M., Webb, A.G., 1998. Rf microcoil design for practical NMR of mass-limited samples. J. Magn. Reson. (1969) 133 (1), 227–231.

Subramanian, R., Sweedler, J.V., Webb, A.G., 1999. Rapid two-dimensional inverse detected heteronuclear correlation experiments with <100 nmol samples with solenoidal microcoil NMR probes. J. Am. Chem. Soc. 121 (10), 2333–2334.

Subramanian, R., Webb, A.G., 1998. Design of solenoidal microcoils for high-resolution C-13 NMR spectroscopy. Anal. Chem. 70 (13), 2454–2458.

Zerbe, O., Jurt, S., 2014. Acquisition and Processing. Applied NMR Spectroscopy for Chemists and Life Scientists. WILEY-VCH, Verlag GmbH & Co. KGaA, Weinheim, Germany, pp. 220–227.

Chapter 4

Spin-Echo and Polarization Transfer

Chapter Outline

In this chapter, two key concepts of NMR spectroscopy are discussed. The first is the spin echo experiment, one of the most important NMR experiments, extensively used in modern NMR and magnetic resonance imaging (MRI) for a variety of purposes. It deals with the refocussing of spin magnetization by a pulse of electromagnetic radiation. The second concept is that of polarization transfer, a remarkable development in the field of NMR spectroscopy which addresses the problem of low sensitivity.

4.1 SPIN-ECHO FORMATION IN HOMONUCLEAR AND HETERONUCLEAR SYSTEMS

Artifact signals generated due to field inhomogeneities or errors in setting pulse widths may be suppressed by *spin-echo* production. Let us consider a heteronuclear AX spin system, in which nucleus A is a proton and nucleus X is a carbon. If the behavior of nucleus X is examined, then its magnetization will be affected by nucleus A in two different ways, depending on whether nucleus A is in the

Solving Problems with NMR Spectroscopy. http://dx.doi.org/10.1016/B978-0-12-411589-7.00004-8
133

lower energy (α) state (i.e. oriented with the applied magnetic field) or in the higher energy (β) state (oriented against the applied field). The magnetization of nucleus X can, therefore, be considered to be made up of two components, $\mathbf{M}_{XA\alpha}$ and $\mathbf{M}_{XA\beta}$ (where $\mathbf{M}_{XA\alpha}$ is the magnetization vector of nucleus X when coupled to the lower energy (α) state of the neighboring nucleus A, and $\mathbf{M}_{XA\beta}$ is the magnetization vector of nucleus X when coupled to the higher energy (β) state of the nucleus A). For convenience, we will designate $\mathbf{M}_{XA\alpha}$ as $\mathbf{M}_1$, and $\mathbf{M}_{XA\beta}$ as $\mathbf{M}_2$ in the following discussion.

4.1.1 Spin-Echo Production

The basic pulse sequence for the production of a spin-echo (Hahn, 1950) is illustrated in Fig. 4.1. The behavior of ^{13}C vectors in a heteronuclear CH system is shown. The first 90°_x pulse on the ^{13}C nuclei serves to bend their z-magnetization to the y'-axis. During the subsequent delay period τ, two component vectors $\mathbf{M}_F$ and $\mathbf{M}_S$ are generated due to coupling effect with the attached proton, with $\mathbf{M}_F$ precessing a little faster than $\mathbf{M}_S$. The 180°_x pulse is then applied on the ^{13}C-nuclei. It results in the $\mathbf{M}_F$ and $\mathbf{M}_S$ vectors flipping across the x'-axis and adopting *mirror image* positions in the x'y'-plane (Fig. 4.1b (iii, iv)) so that $\mathbf{M}_F$ comes to lie as much behind the $\mathbf{M}_S$ vector as it was ahead of it prior to the application of the 180°_x pulse.

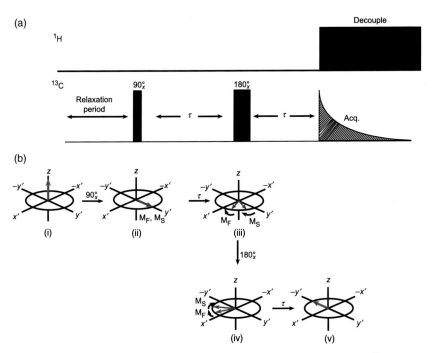

FIGURE 4.1 (a) Spin-echo pulse sequence. (b) Effect of spin-echo pulse sequence on ^{13}C magnetization vectors; (v) represents the coalescing of the two component vectors by the 180°_x refocusing pulse.

During the subsequent delay period τ, which is kept identical to the first τ delay period, M_F catches up with the M_S vector. The two magnetization vectors travel the same distance after the 180°_x pulse as they did before it and coalesce (become focused) on the y'-axis, producing a *spin-echo*.

The advantage of the 180° *refocusing pulse* is that it compensates for any errors due to field inhomogeneities. For instance, if we assume that the vector M_F was closer to the x'-axis and the vector M_S farther from it before the application of the 180°_x pulse, then the 180°_x pulse would cause a *mirror image jump* across the x'-axis, and vector M_F would adopt a position a little farther from the $-y'$-axis than vector M_S. Similarly, M_S would come to lie a little nearer the $-y'$-axis due to the 180°_x pulse. The faster vector M_F has now to cover a greater distance than the slower vector M_S in the second τ period, so that the original errors in the positioning of the vectors M_F and M_S are compensated by the "*equal and opposite error*" caused by the refocusing 180°_x pulse. As a result, refocusing occurs precisely along the $-y'$-axis at the end of the second τ period.

The 180°_x pulse applied to nucleus X under observation causes a change of its *position* but does not affect the *direction* of rotation of its magnetization vectors. This is because the direction of rotation of the vectors of nucleus X depends on the spin states of nucleus A to which nucleus X is coupled, and it is only when the spin states of nucleus A are interchanged (for instance, by irradiation of nucleus A with a 180° pulse) then the directions of rotation of the vectors of nucleus X undergo a reversal. Since refocusing at the end of the second delay period occurs along the $-y'$-axis, a *negative* signal is produced, *irrespective of the precession frequency of the nucleus*. The pulse sequence used to remove field inhomogeneities is:

$$90^\circ_x - \tau - 180^\circ_x - \tau - \text{echo}$$

$$90^\circ_x - \tau - 180_{-x}^{\;\;\circ} - \tau - \text{echo}$$

The phase alteration of the 180° pulse and co-addition of the resulting FIDs serves to cancel the pulse imperfections, thereby producing accurate spin-echoes (Fig. 4.2).

The spin-echo experiment therefore leads to the refocusing not only of the individual nuclear resonances but also of the field inhomogeneity components lying in front or behind those resonances, a maximum negative amplitude being observed at time 2τ after the initial 90° pulse. The frequency of rotation of each signal in the rotating frame will depend on its chemical shift; after the vector has been flipped by the 180° pulse, it will continue to rotate in the same direction and at the same frequency till it decays due to transverse relaxation effects. If we acquire data at the point at which each echo is produced, Fourier transformation would lead to a spectrum. By varying the delay time τ, we can collect a series of spectra from which the transverse relaxation rate R_2 constants can be measured by plotting the intensity of each signal as a function of time,

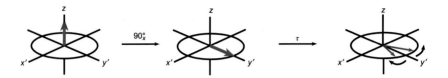

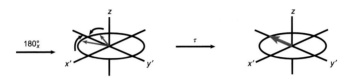

FIGURE 4.2 Effect of 180°_x pulse on phase imperfections resulting from magnetic field inhomogeneities. Spin-echo generated by 180°_x refocusing pulse removes the effects of magnetic field inhomogeneities.

provided that each signal is a singlet. In the case of multiplets, a more complex pulse sequence is used to cancel the multiplet effects and then to measure the transverse relaxation delay.

Problem 4.1

What does "magnetic inhomogeneity" mean, and how does it affects the transverse magnetization and signal intensity?

Problem 4.2

The spin-echo is used to suppress the production of spurious signals due to field inhomogeneities or to eliminate errors in the setting of pulse widths. It is also possible to use the spin-echo to follow the decay of transverse magnetization and to determine the transverse relaxation time (T_2). How might we do this in practice?

4.1.2 Spin-Echo Production in a Heteronuclear AX Spin System

In a heteronuclear AX spin system in which nucleus X is being observed, three different cases can be considered, depending on whether the 180° pulse is applied to nucleus A only, nucleus X only, or simultaneously to both nuclei A and X (Fig. 4.3).

In the first case, the 180°_x pulse is applied to nucleus X, causing the two vectors of nucleus X to flip across the x'-axis. But no change occurs in the *direction* of their rotation during the second delay period, so that at the end of this period the vectors are refocused along the $-y'$-axis producing a *negative* signal.

In the second case, the 180°_x pulse is applied only to nucleus A, causing an exchange of spin labels of the A spin states to occur, so that the direction of rotation of the X magnetization vectors is reversed during the second delay

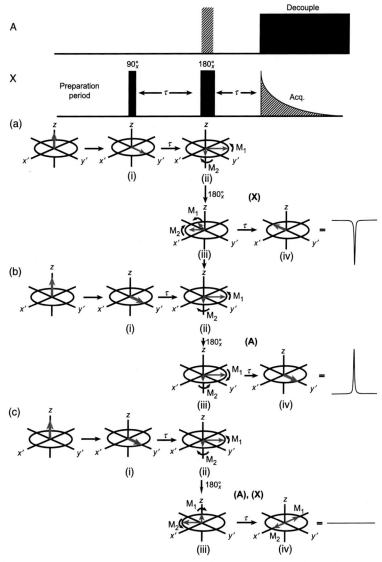

FIGURE 4.3 Spin-echo experiment. The behavior of nucleus X in an AX spin system is shown. (a) Application of the second 180°_x pulse to nucleus X in the AX heteronuclear system results in a spin-flip of the two X vectors across the x'-axis so that they adopt mirror image positions across the x'-axis. The direction of rotation of the two X vectors does not change, and they therefore refocus along the $-y'$-axis. This spin-echo at the end of the τ period along the $-y'$-axis results in a negative signal. (b) When the 180°_x pulse is applied to nucleus A in the AX heteronuclear system, the spin-flip of the X vectors across the x'-axis does not occur; only their direction of rotation changes due to relabeling of the A spin states. Spin-echo formation, therefore, takes place along the y'-axis, producing a positive signal. (c) Simultaneous application of the 180°_x to both nuclei A and X will not only cause a spin flip but also exchange the direction of their rotation so that at the end of the 2τ delay period, they will come to lie along the x'-axis resulting in the disappearance of the signal.

period, resulting in a refocusing of the X vectors to occur along the y'-axis at the end of the second delay period, thereby producing a *positive* signal.

In the third case, the 180_x° pulses are applied *simultaneously* to both nuclei A and X, so that the direction of rotation of the X vectors changes, and a flipover to mirror image positions across the x'-axis also occurs. This results in the two vectors aligning themselves along the x'-axis at the end of the second delay period. Since the detector is assumed to be aligned to detect signal components only along the y'-axis, no signal is produced. This third case is also applicable to *homonuclear spin systems*, in which the 180_x° pulse will be nonselective, affecting both the A and X nuclei.

Hence, it is clear that if the two delay periods before and after the 180_x° pulses are kept identical, then refocusing will occur only when a *selective* 180_x° pulse is applied. This can happen only in a heteronuclear spin system, since a 180_x° pulse applied at the Larmor frequency of protons, for instance, will not cause a spin flip of the ^{13}C magnetization vectors.

Problem 4.3

What is the effect of applying a spin-echo on chemical shifts?

Problem 4.4

Describe the effects of a 180_x° refocusing pulse in each of the following situations:
1. Single line dephased due to field inhomogeneity.
2. A doublet resulting from heteronuclear coupling, assuming that vectors of nucleus A are being observed and that the 180_x° pulse is applied selectively on nucleus A.

Problem 4.5

Will the vectors of a doublet be refocused at time 2τ in a spin-echo experiment?

Problem 4.6

M_α and M_β are magnetization components of the X vector of a heteronuclear AX spin system, shown here at a certain delay after the application of a 90_x° pulse. Draw the vector positions and their directions of rotation after each of the following radiofrequency pulses:

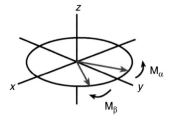

(i) 180_x° pulse applied to nucleus X
(ii) 180_x° pulse applied to nucleus A
(iii) 180_x° pulse applied to nuclei A and X

4.1.3   Measuring T_2 with a Spin-Echo Sequence

Transverse or spin–spin relaxation (T_2) results in the loss of magnetization in the xy-plane. Immediately after the 90° excitation pulse, the bulk magnetization vector possesses phase coherence and its magnitude can be measured as transverse magnetization. The bulk magnetization vector is composed of many microscopic vectors representing individual nuclei. Following the first 90°_x pulse, the transverse magnetization vectors on the y-axis start fanning out and losing the coherence. This fanning out of the individual magnetization vectors eventually leads to a complete loss of magnetization in the xy-plane. This form of relaxation is referred to as *transverse relaxation T_2**. This relaxation arises from two distinct sources, magnetic inhomogeneity $T_{2(\Delta B_0)}$, and natural spin–spin relaxation T_2. The spin–spin relaxation (T_2) can be measured by employing the spin echo pulse sequence, if the contribution of magnetic field inhomogeneity $T_{2(\Delta B_0)}$ is removed (Meiboom and Gill, 1958; Carr and Purcell, 1954).

Let us consider a sample of like spins, comprising microscopically small regions within which the magnetic field B_0 is perfectly homogeneous. The objective of this assumption is to simplify the fanning out of the transverse magnetization into millions of individual spins due to the magnetic field inhomogeneity. To measure the T_2, a simple two-pulse sequence is used (ignoring the longitudinal relaxation T_1) (Fig. 4.4a).

In the Carr–Purcell–Meiboom–Gill (CPMG) spin echo pulse sequence, during the time delay τ after the 90°_x pulse, some magnetization vectors

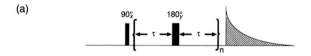

(a)

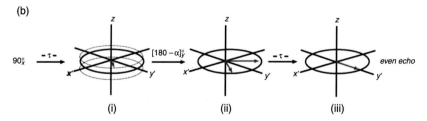

(b)

FIGURE 4.4   (a) The Carr–Purcell–Meiboom–Gill (CPMG) spin-echo sequence for the measurement of T_2. The 90°_x pulse brings the magnetization to the y'-axis. During the subsequent delay τ, some components move ahead, while some others lag behind. The 180°_y spin-echo pulse rotates the vectors around $+y$. Following a further delay τ the faster moving vectors coincide with the slower vectors along the $+y$-axis. T_2 is measured by repeating the $-\tau-180^\circ-\tau-$ sequence with increasing $2\tau n$ (by increasing n) and by acquiring the data, following the last even echo peak in each case, (b) the vector representation resulting from CPMG pulse sequence.

(isochromats) move ahead while some other lag behind, in a process called "dephasing" (Fig. 4.4b(ii)). The 180°_y pulse rotates the magnetization vectors along the $+y$-axis and serves to rephase the magnetization vectors in the transverse plane. This reversing of the phase order of the transverse magnetization vectors results in the slower moving vectors being repositioned ahead of the faster moving vectors. During the subsequent time delay τ, the faster moving magnetization vectors coincide with the slower moving vectors on the $+y$-axis, and thus the effect of field inhomogeneity in the xy-plane is eliminated. The $-\tau-180^\circ-\tau-$ sequence is repeated "n" times (Fig. 4.4b(iii)). In the CPMG pulse sequence, a series of experiments with increasing $2\tau n$ (by increasing n) are performed. This is followed by acquiring of data following the last even echo peak in each case. However, some loss of phase coherence by genuine transverse relaxation T_2 is also taking place during the 2τ time delay, which is not refocused by the spin echo sequence, since there is no phase memory associated with this. The magnetization intensity will therefore slowly fade away due to the natural T_2 relaxation, independent of the field homogeneity. With the increasing n, the fading of transverse magnetization due to T_2 relaxation (independent of magnetic inhomogeneity) is measured, providing the approximate value of spin–spin relaxation time T_2. So the CPMG method involves the application of a series of 180° pulses at intervals of τ, 3τ, 5τ, 7τ, etc., after the 90° pulse. This results in the formation of echoes at times 2τ, 4τ, 6τ, 8τ, etc. The spin echo train will decay due to the T_2 relaxation processes resulting from molecular interactions. The height of the multiple echoes will decrease because of T_2 dephasing (Fig. 4.5). In principle, the amplitude of spin echo should be proportional to $\exp(-2\tau/T_2)$. As the delay τ increases, the diffusion loss becomes more prominent, making the relaxation data less reliable.

The mapping of T_2 is not straightforward as a number of factors can complicate this process. These include the splitting of the coherent transverse magnetization due to homonuclear couplings (J-coupling) and magnetic inhomogeneity. Longitudinal or spin-lattice relaxation T_1 also acts to destroy the transverse magnetization.

The significance of measuring the T_2 is to properly design the multi-pulse experiments. The T_2 value effectively defines how long a multi-pulse NMR experiment can be conducted before the transverse magnetization (magnetization in the xy-plane) has decayed significantly.

4.1.4 Attached Proton Test (APT), Gated Spin-Echo (GASPE), and Spin-Echo Fourier Transform (SEFT)

When a 90°_x pulse is applied to the ^{13}C nuclei, then during the subsequent delay interval τ after the pulse, the magnetization vectors of the ^{13}C nuclei of CH_3, CH_2, CH, and quaternary carbons do not rotate synchronously with one another but rotate with characteristically different angular velocities. If the value of the delay time is kept at $1/J$ seconds, then CH_2 and quaternary carbons give signals with positive amplitudes while CH_3 and CH carbons give

(a)

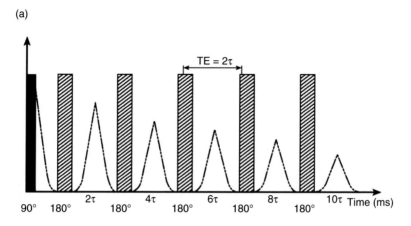

(b)

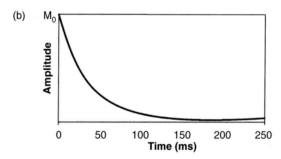

FIGURE 4.5 (a) Carr–Purcell–Meiboom–Gill (CPMG) spin echo pulse sequence and (b) exponential decaying curve of spin echo with T_2.

signals with negative amplitudes. This provides a method, variously known as *A*ttached *P*roton *T*est (APT), *GA*ted *S*pin-*E*cho (GASPE), or *S*pin-*E*cho *F*ourier *T*ransform (SEFT) for distinguishing between them. The heteronuclear spin echo experiment can, therefore, be used to differentiate between the quaternary and other carbons (CH_3, CH_2, and CH). However, in modern NMR spectroscopy, the polarization transfer techniques have superseded the APT, or SEFT experiments. These methods are described here because of the importance of underlying principles.

The pulse sequence used in the APT experiment is shown in Fig. 4.6; the movement of the magnetization vectors at various delay intervals τ after the 90°_x ^{13}C pulse is shown in Fig. 4.7. The initial 90°_x ^{13}C pulse bends the magnetization vectors of the CH_3, CH_2, CH, and quaternary carbons so that they come to lie together along the y'-axis in the $x'y'$-plane. During the subsequent delay interval, these vectors separate from each other and rotate in the $x'y'$-plane at their characteristic Larmor frequencies. The CH_3, CH_2, CH carbon magnetization

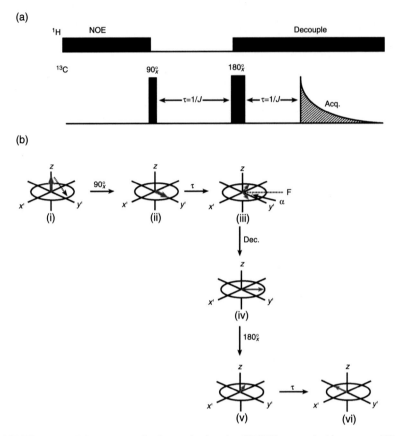

FIGURE 4.6   (a) Pulse sequence for the gated spin-echo (GASPE) or attached proton test (APT) experiment. (b) Effect of the pulse sequence on the ^{13}C magnetization vectors of a CH group.

vectors also split into 4-, 3-, or 2-vector components, respectively, due to coupling with their attached protons (Patt and Shoolery, 1982; Brown et al., 1981). The detector, which by convention is regarded as being located along the y'-axis, detects only the sum of these individual vector components, with the signal amplitude increasing to a maximum value when the vector sum approaches the $+y'$-axis and decreasing as they move away from it toward the $-y'$-axis. Since this cosinusoidal modulation in signal strength with delay time τ depends on the positions (or the respective angular velocities) of the split magnetization components, it is termed *J-modulation* (Bildsøe et al., 1983; Sørensen and Ernst, 1983).

The decoupler is off during the first $1/J$ delay period, so it is during this period that the effect of *J*-splitting comes into play and the coupling information is provided. This information is contained in the phase and magnitude of the signal, which in turn are dependent on the positions, angular velocities,

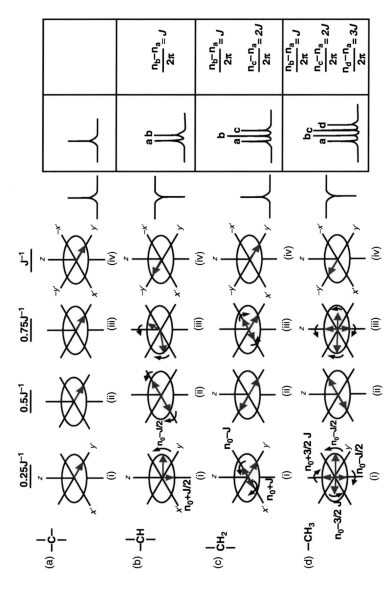

FIGURE 4.7 Evolution of magnetization vectors of C, CH, CH_2, and CH_3 carbons in the $x'y'$-plane after the 90_x° pulse. The magnetization vectors of C, CH, CH_2, and CH_3 carbons evolve to different extents during the delays set at $\tau = 0.25\ J^{-1}$, $0.5\ J^{-1}$, $0.75 J^{-1}$, and J^{-1}. At J^{-1}, the quaternary and CH_2 carbons appear along the $+y'$-axis, while the CH_3 and CH carbons appear along the $-y'$-axis, giving negative signals. This forms the basic of GASPE (or APT) experiments.

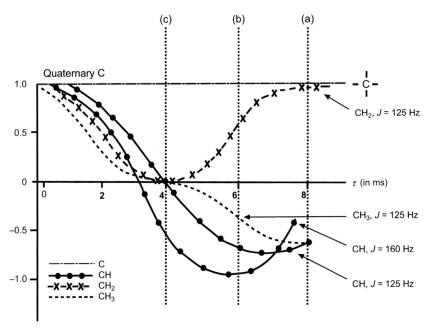

FIGURE 4.8   Signal intensities in the APT experiment depend on delay time τ. The signal intensities of CH_3, CH_2, and CH carbons are shown as various curves as a function of τ.

and number of individual split magnetization components. The decoupler is switched on at the beginning of the second delay interval that causes the split magnetization vectors to collapse into resultant singlets. The 180°_x pulse causes these vectors to adopt mirror image positions across the x'-axis (Fig. 4.6). At the end of the second delay interval, the CH_2 and quaternary carbons appear with positive phases, while the CH_3 and CH carbons appear with negative phases, thus allowing the CH_3 and CH carbons to be distinguished from the CH_2 and quaternary carbons (Fig. 4.7).

   The dependence of signal phases and intensities on delay time τ in the APT experiment is shown in Fig. 4.8. If J_{CH} is assumed to be 125 Hz and the delay time is accordingly set at $1/J = {}^1/_{125} = 8$ ms, then, as is apparent from Fig. 4.8, the quaternary and CH_2 carbons appear with maximum positive amplitudes while the CH_3 and CH carbons afford maximum negative amplitudes [see vertical line at (a)]. If the delay time is adjusted to 6 ms [see vertical line at (b)], then quaternary carbons still appear with similar positive amplitudes, the CH_2 carbons have weaker positive amplitudes, and the CH_3 and CH carbons have weak negative amplitudes.

   The signal magnitudes at the end of the 2τ period are given by the following equations:

   For quaternary carbons (C): $M_{2\tau} = M_0 e^{-2\tau/T_2}$
   For methine carbons (CH): $M_{2\tau} = M_0 e^{-2\tau/T_2} \cos \pi \tau J$

For methylene carbons (CH_2): $M_{2\tau} = \frac{1}{2\tau} M_0 e^{-2\tau/T_2} (1 + \cos 2\pi\tau J)$

For methyl carbons (CH_3): $M_{2\tau} = \frac{3}{4} M_0 e^{-2\tau/T_2} \left(\cos \pi\tau J + \frac{1}{3}\cos 3\pi\tau J \right)$

One disadvantage of the APT experiment is that it does not readily allow us to distinguish between carbon signals with the same phases, i.e., between CH_3 and CH carbons or between CH_2 and quaternary carbons, although the chemical shifts may provide some discriminatory information. The signal strengths also provide some useful information, since CH_3 carbons tend to be more intense than CH carbons, and the CH_2 carbons are usually more intense than quaternary carbons due to the greater nuclear Overhauser enhancements on account of the attached protons.

In interpreting APT spectra, we have to be careful about erroneous phasing of the signals that may occur if the actual J_{CH} values differ significantly from those sets for the experiment. For instance, J_{CH} may be about 125 Hz for alkyl groups, so the optimum $1/J$ will be 8 ms, but J_{CH} in alkenes (olefinic protons) may be about 160 Hz, so $1/J$ for such functional groups should be set at 6.25 ms (Fig. 4.8). Such alkyl CH carbons will, therefore, afford the most intense negative signals at $2\tau = 16$ ms, while in alkene CH carbons the most intense negative signals will appear at $2\tau = 12.5$ ms. If 2τ is set at 16 ms, then the alkene CH carbons could appear with weak negative intensities; if the J values differ significantly, we may even obtain signals with the "wrong" phasing, leading to misinterpretation of the results. The 2τ values are thus of practical importance since if $\tau = \frac{1}{2} J$, it will lead to the fanning out of all ^{1}H-coupled ^{13}C resonances, but if $\tau = 1/J$ it will lead to positive signals for C and CH_2 and negative signals for CH and CH_3. With τ set at $\frac{1}{2} J$, the quaternary carbons generally appear with greater intensity than the other carbons, which will be of near-zero intensities, thereby allowing them to be distinguished, particularly from the CH_2 carbons, as compared to the "normal" APT spectrum, in which both CH_2 and quaternary carbons appear with positive amplitudes. A *difference* APT spectrum, in which an APT spectrum recorded with τ set at 2/5 J is subtracted from another APT spectrum recorded with τ set at 3/5 J, can provide useful information. The methyl carbons will then appear with reduced intensities in the difference spectrum as compared to the methine carbons, allowing us to distinguish between them.

The errors in the APT spectra may also arise due to (1) variation in the T_2 relaxation times of different carbons, (2) wide variation in J_{CH} values, and (3) modulations caused by long-range ^{13}C–^{1}H couplings. A procedure known as Error Self COmpensation Research by Tao scrambling (ESCORT) has been developed to yield cleaner subspectra with reduced J "cross-talk" (Madsen et al., 1986). This involves replacing the normal APT spectral editing procedure, which is based on using an average $^1J_{CH}$ coupling constant, by a linear combination of experiments that is partly independent of the error in setting J values.

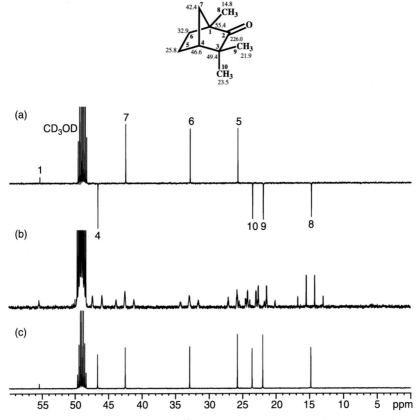

FIGURE 4.9   Signal assignment for the ^{13}C NMR spectrum of fenchone isolated from *Illicium religiosum* Sieb.; (a) SEFT experiment with 1/*J*. (b) Off-resonance spectrum; (c) spectrum with ^{1}H broad-band decoupling.

The variation in the T_2 relaxation times of the individual carbons can be minimized by keeping the time between excitation and detection constant (Radeglia and Porzel, 1984).

Differences between SEFT, off resonance, and broad-band decoupled spectra are shown in Fig. 4.9.

Problem 4.7

How can you distinguish between C, CH, CH$_2$, and CH$_3$ carbons through the APT experiment?

Problem 4.8

The GASPE (or APT) spectrum of 2-methylsuccinic acid isolated from *Ziziphora tenuior* L. is shown here. Assign the signals to the various carbons.

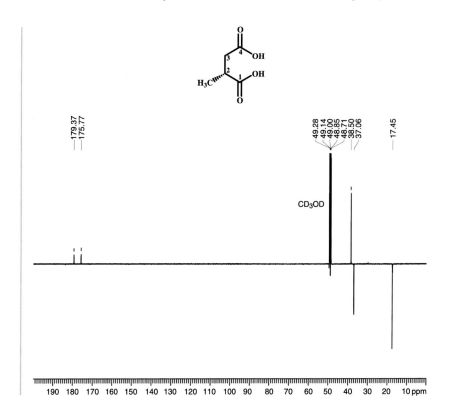

I am missing some peaks in my APT spectrum. Why is that?

4.1.5 Nonobservable Magnetization by 90° Pulses: *Mixing Spin States*

In the preceding discussion, we have been concerned with the generation and detection of single-quantum coherence (i.e., $\Delta M = +1$) magnetization. This magnetization corresponds to the creation of a vector in the $x'y'$-plane, which is detected. There are many methods to generate magnetization with transitions other than $\Delta M = +1$. Such *multiple- or zero-quantum coherences* are "invisible," since they do not induce any fluctuation in a receiver directly, and they do not follow the selection rules. However, through the application of a suitable pulse or a set of pulses, these coherences can be converted into single-quantum transitions before being detected. Such magnetization can be generated simply by using two 90° pulses separated by a time period. The two 90° pulses together serve to prepare the spin system; and after this preparation period, the coherences are allowed to evolve in the time period τ, before some of them are converted into a single-quantum coherence by a third 90° pulse before detection. The second 90° pulse thus serves to mix the orders of coherences, converting some of the single-quantum coherence generated by the first 90° pulse into other orders of coherence, whereas the third 90° pulse does the reverse—it converts nonsingle-quantum coherence into single-quantum coherence.

Problem 4.10

What is the difference between single-quantum coherence and zero- or multiple-quantum coherences?

Problem 4.11

Can the simple vector presentation be used to display the effects of a 90° pulse on zero- or multiple-quantum magnetizations?

4.2   CROSS-POLARIZATION

The strength of an NMR signal depends on the magnetogyric ratio γ of a nucleus, which in turn determines the population difference ΔE between the upper and lower energy states. In a given applied magnetic field, B_0: $\Delta E = \gamma h\, B_0 / 2\pi$ where h is Planck's constant, certain nuclei, such as ^{13}C, occur in low natural abundance and have unfavorable magnetogyric ratios, leading to poor sensitivity. Thus, the magnetogyric ratio γ of ^{13}C is about a quarter that of ^1H, and since the signal obtainable from a nucleus is proportional to γ^3 of that nucleus, the lower γ of ^{13}C results in weakening of its signal intensity by a factor of $(1/4)^3 \cong 64$. Moreover, since ^{13}C has a natural abundance of 1.1%, this leads to a further hundred-fold reduction in the sensitivity of the signal, so that ^{13}C signals are about 5800-fold weaker than ^1H signals. Some increase in sensitivity is achievable through nuclear Overhauser enhancement. Alternatively, we can transfer magnetization from ^1H to ^{13}C nuclei to intensify the ^{13}C signals, a process known as *cross-polarization*. Such population-transfer or polarization-transfer procedures are now used extensively in many NMR experiments, such as INEPT and DEPT (Granger et al. 1990).

Some of the most important 2D experiments involve chemical shift correlations between either the same type of nuclei (e.g., $^1H/^1H$ homonuclear shift correlation) or between nuclei of different types (e.g., $^1H/^{13}C$ heteronuclear shift correlation). Such experiments depend on the modulation of the nucleus under observation by the chemical shift frequency of other nuclei. Thus, if ^1H nuclei are being observed and they are being modulated by the chemical shifts of other ^1H nuclei in the molecule, then homonuclear shift correlation spectra are obtained. In contrast, if ^{13}C nuclei are being modulated by ^1H chemical shift frequencies, then heteronuclear shift correlation spectra result. One way to accomplish such modulation is by transfer of polarization from one nucleus to the other nucleus. Thus, the magnitude and sign of the polarization of one nucleus are modulated at its chemical shift frequency, and its polarization transferred to another nucleus, before being recorded in the form of a 2D spectrum. Such *polarization transfer* between nuclei can be accomplished by the simultaneous application of a pair of 90° pulses on the two nuclei involved in the exchange of polarization, provided that the spin components are properly aligned.

Let us consider a simple heteronuclear ^{1}H/^{13}C spin system in which the ^{1}H polarization is transferred to ^{13}C nuclei. The first $90°_x$ ^{1}H pulse bends the z-magnetization to the $+y'$-axis. During the subsequent time interval t_1 the ^{1}H magnetization is split into counter rotating components and, at the end of $t_1 = \frac{1}{2}$ J period, are aligned along the x'-axis of the rotating frame. The *Rf* field is then phase shifted in order to allow a 90° pulse to be applied along the y'-axis. This pulse causes the two magnetization component vectors lying along the y'-axis to rotate in the $x'z$-plane so that they come to lie along the z-axis. The two vectors α_H and β_H are now directed in opposite directions along the z-axis (see Fig. 4.10a(v)), whereas in the equilibrium state (see Fig. 4.10a(i)) they were both pointing toward the z-axis. What we have therefore succeeded in doing is to invert the population of *one* of the two ^{1}H spin states. Since the ^{13}C nuclei, which are coupled to ^{1}H nuclei, have their spin states determined by the spin labels of the ^{1}H spins, the inversion (or "relabeling") of one of the ^{1}H spins causes a corresponding relabeling of one of the ^{13}C spins (shown as $^\alpha$C and $^\beta$C in Fig. 4.10b(iv), with $^\beta$C having undergone the inversion in the illustration). A $90°_x$ pulse (along the x'-axis) rotates these two ^{13}C vectors so that they come to lie along the y'-axis (b(vii)). This antiphase magnetization can then be detected immediately, or it can be allowed to evolve for a certain time before detection (b(ix)). Alternatively, if decoupling is applied to the ^{1}H nuclei during detection, it can be made to focus onto a single component before detection.

Clearly, the extent to which the ^{1}H nuclei line up along the x'-axis (a(iv)) will depend on the duration of t_1 which in turn will determine the extent of the antiphase z-magnetization created by the second 90° ^{1}H pulse. The extent of divergence of the two magnetization components of A between the two 90° ^{1}H pulses depends on J_{HC}, while the positions of the *resultant* of these two vectors will depend on the extent of precession due to the chemical shift of ^{1}H. Hence, the positioning of the component vectors (and consequently the extent of antiphase z-magnetization created by the second 90° pulse) will depend both on the magnitude of the coupling constant J_{CH} and on the chemical shift δ_H. The antiphase polarization of A along the x'-axis (in a(iv)) before the application of the second 90° pulse is modulated by both the chemical shift and the coupling constant frequencies, and it is this information that is transferred to the ^{13}C nuclei by polarization transfer before detection.

Problem 4.12

Which of the following terms is technically correct:

(a) Population transfer

(b) Polarization transfer

The experiment just discussed presents a heteronuclear spin system (Nolis and Parella, 2007). In a homonuclear case, the separate 90° ^{13}C pulse necessary

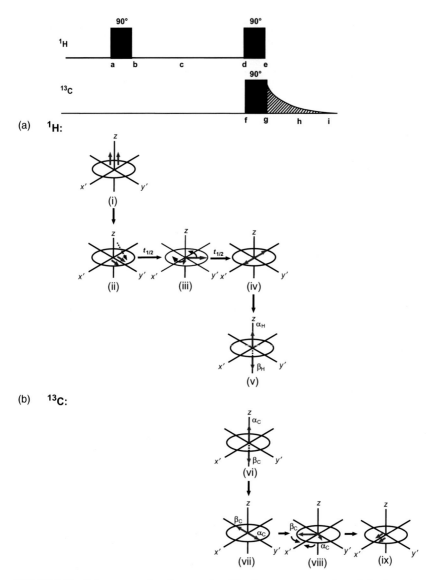

FIGURE 4.10 Pulse sequence for a heteronuclear AX spin system representing polarization transfer from ^{1}H to ^{13}C nuclei.

in the heteronuclear system is not required, since the second 90° ^{1}H pulse affects the coupled partner ^{1}H nucleus as well. The nucleus detected therefore has its two transitions antiphase with respect to each other, corresponding to the states represented in Fig. 4.10b(vii), b(viii), etc. at detection.

The ^{1}H/^{13}C energy level diagram in Fig. 4.11 will help to clarify how polarization transfer, i.e., inversion of one of the ^{1}H spin states, intensifies the

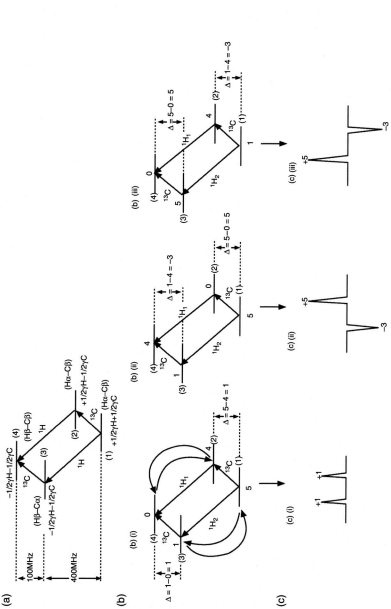

FIGURE 4.11 (a) Energy levels, populations, and single-quantum transitions for a CH system. (b) (i) Populations at thermal equilibrium (Boltzmann distribution) responsible for the normal ^{13}C intensities, for example, of the doublet of CHCl$_3$. The amplitudes and intensities corresponding to this population difference are shown in (c) (i). (b) (ii) Population inversions of protons through the $^1H_1$ transition result in an increased negative difference (-3) between levels 3 and 4, and an increased positive difference ($+5$) between levels 1 and 2. The corresponding amplitudes and intensities of CH doublet are shown in (c) (ii). (b) (iii) The alternative exchange of proton populations achieved through $^1H_2$ transition (b) (i) gives rise to a reverse situation in (b) (iii) to that seen in (b) (ii). The population differences are: level 3 − level 4 = 5 − 0 = +5, and level 1 − level 2 = 1 − 4 = −3. Fig. (c) (iii) shows corresponding amplitudes and intensities of the CH doublet.

two antiphase ^{13}C lines. The coupled ^{1}H/^{13}C nuclei can be represented by four energy levels, corresponding to (1) ^{1}H: α, ^{13}C: α, (2) ^{1}H: α, ^{13}C: β, (3) ^{1}H: β, ^{13}C: α, and (4) ^{1}H: β, ^{13}C: β (i.e., $\alpha\alpha$, $\alpha\beta$, $\beta\alpha$, $\beta\beta$) states. In the lowest energy ($\alpha\alpha$) orientation, both ^{1}H and ^{13}C nuclei are aligned with the applied magnetic field; in the highest energy ($\beta\beta$) orientation, both are aligned against the applied magnetic field. In the two other energy levels ($\alpha\beta$ and $\beta\alpha$), the two nuclei are not parallel to each other. The population of the lowest energy ($\alpha\alpha$) state is $\left(\frac{1}{2}\right)\gamma_H + \left(\frac{1}{2}\right)\gamma_C$, while in the highest energy ($\beta\beta$) state the population is $-\left(\frac{1}{2}\right)\gamma_H - \left(\frac{1}{2}\right)\gamma_C$. In the two intermediate $\alpha\beta$ and $\beta\alpha$ states, the respective populations will be $\left(\frac{1}{2}\right)\gamma_H - \left(\frac{1}{2}\right)\gamma_C$ and $-\left(\frac{1}{2}\right)\gamma_H + \left(\frac{1}{2}\right)\gamma_C$. This is shown in Fig. 4.11a.

Suppose, for clarity, we add a common factor $\left[\left(\frac{1}{2}\right)\gamma_H + \left(\frac{1}{2}\right)\gamma_C\right]$ to all four energy levels so that the energy difference between them does not change. The values of these energy levels will then become γ, γ_C, γ_H, and $\gamma_H + \gamma_C$. Since γ_H is about four times γ_C, let us also assume that $\gamma_H = 4$ and $\gamma_C = 1$. The population of the four energy levels will then be 0, 1, 4, and 5. The population difference between the two ^{13}C spin states before the application of the ^{1}H polarization transfer pulse corresponds to the lower energy state minus the upper energy state, i.e., $1 - 0 = 1$ or $5 - 4 = 1$.

Applying the polarization transfer pulse inverts the population of one of the two ^{1}H states, producing a corresponding change in the populations of the ^{13}C states. This occurs because the two ^{13}C transitions share a common energy level with this proton transition. Thus, if the populations 0 and 4 of one of the ^{1}H transitions are exchanged, then the ^{13}C population differences become $5 - 0 = 5$ and $1 - 4 = 3$. The original ^{13}C population differences were $5 - 4 = 1$ and $1 - 0 = 1$, so we see that an intensification of the ^{13}C signal amplitudes has occurred. The *net* population difference was $(5 - 4) + (1 - 0) = 2$; after the polarization transfer pulse the *net* population difference $(5 - 3)$ is still 2. There is, therefore, no net transfer of magnetization, the intensification of the positive contribution being exactly balanced by a corresponding change in the negative contribution. However, since the original ^{13}C line intensities *before* the polarization transfer pulse were +1 and +1 and *after* the pulse they are +5 and −3, we see a significant increase in the intensity of the individual antiphase ^{13}C signals has occurred (Fig. 4.11c, ii). Figure 4.11c, iii, shows that the alternative exchange of proton populations (5 with 1) through the other ^{1}H transition leads to the ^{13}C population differences of $5 - 0 = 5$ and $1 - 4 = -3$, i.e., the first peak of the ^{13}C doublet has an intensity of +5, and the second peak has an intensity of −3. The spectra therefore show an enhancement of the order of $\gamma_H/\gamma_C = {\sim}4$. The two ^{13}C transitions would, therefore, have asymmetric intensities with an intensification of the individual peaks by factors of approximately +5 and −3.

Problem 4.13

How does polarization transfer make it possible to record $^{13}C-^{13}C$ INADEQUATE experiments?

Population transfer experiments may be selective or nonselective. Selective population transfer experiments have found only limited use for signal multiplicity assignments (Sørensen et al., 1974) or for determining signs of coupling constants (Chalmers et al., 1974; Pachler and Wessels, 1973), since this is better done by employing distortionless enhancement by polarization transfer (DEPT) or correlated spectroscopy (COSY) experiments. However, nonselective population transfer experiments, such as INEPT or DEPT (presented later), have found wide application.

Problem 4.14

How does magnetization transfer from ^1H to a coupled ^{13}C nucleus (polarization transfer or population transfer) affect the signal intensity of the ^{13}C nucleus?

Problem 4.15

What is the difference between nonselective and selective polarization transfer?

Problem 4.16

Among ^{13}C and ^{15}N, which of these elements get the maximum benefit of polarization transfer from an attached ^1H nucleus?

4.2.1 INEPT

The polarization transfer experiment just described involved inversion of one of the two vector components so that it is pointed toward the $-z$-axis, in contrast to the equilibrium situation, in which both components were pointing toward the $+z$-axis. It is this inversion of the ^1H spins that caused a corresponding change in the population difference of the coupled ^{13}C nuclei, which was detected. _I_ntensive _N_uclei _E_nhancement by _P_opulation _T_ransfer (INEPT) is a nonselective polarization transfer experiment (Baishya and Khetrapal, 2014; Morris et al. 2012; Pegg et al., 1981a,b; Avent and Freeman, 1980; Bolton, 1980; Burum and Ernst, 1980; Doddrell et al., 1980; Morris, 1980a, 1980b; Morris and Freeman, 1979), in which polarization is transferred from all protons to all nuclei having the appropriate $^1H/^{13}C$ coupling.

The pulse sequence employed is basically similar to that already described, except that (1) a 180° ^1H pulse is applied in the middle of the evolution period to refocus the effect due to proton chemical shift evolution, and (2) a 180° ^{13}C pulse is applied simultaneously with the 180° ^1H pulse, which serves to invert the carbon spin labels so that the ^{13}C vectors continue to diverge from one

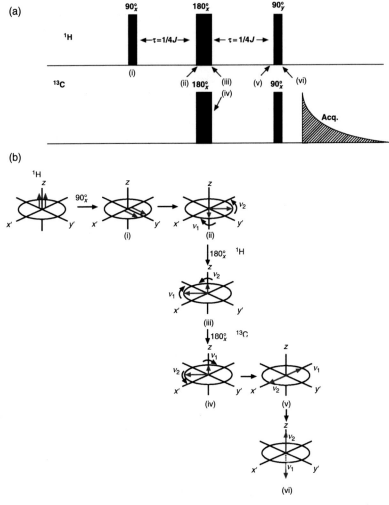

FIGURE 4.12 (a) Pulse sequence for the INEPT experiment. (b) Effect of pulses on ^1H magnetization. Application of the pulse sequence shown results in population inversion of one of the two proton vectors of the CH doublet and therefore causes an intensification of the corresponding ^{13}C lines.

another. The basic INEPT pulse sequence applied to CH doublets is shown in Fig. 4.12. Application of the 90° ^1H pulse aligns the vectors along the y'-axis. During the subsequent $\tau = \frac{1}{4} J$ period, the two vectors of the ^1H nucleus fan out in the $x'y'$-plane. Applying the 180° ^1H pulse causes them to jump across the x'-axis and adopt mirror image positions in the $x'y'$-plane. The simultaneous application of the 180° ^{13}C pulse causes an interchange of the spin labels of the two vectors, resulting in an exchange of their "identities" and it is accompanied by a corresponding change in the direction of rotation, so that in the subsequent $\tau = \frac{1}{4} J$ period they move away from each other, thereby aligning them along the x'-axis at the end of the evolution period. As in the previously described polarization transfer experiment, a $90°_y$ ^{13}C pulse results in their rotation in the $x'z$-plane so they point in *opposite* directions along the z-axis. Thus, while at equilibrium both component vectors of the ^1H doublet pointed toward the $+z$-axis, by applying the INEPT sequence one of these two vector components becomes inverted so that it now points toward the $-z$-axis. This $-z$-magnetization created in one of the ^1H vectors causes a corresponding change in the population difference of the ^{13}C vectors. In order to read this magnetization, it must be first brought to the $x'y'$-plane as it is only the magnitude of the component on the y'-axis that is detected. This is read by applying a $90°_x$ ^{13}C pulse that converts the z-magnetization (*longitudinal* magnetization) of the ^{13}C nuclei to *transverse* magnetization in the $x'y'$-plane. The ^{13}C-NMR spectrum now shows signals that are enhanced by a factor $K = \gamma_H/\gamma_C$ due to the greater population differences between the upper and lower energy states of the ^{13}C nuclei.

Problem 4.17

In what respect is the INEPT experiment superior to the APT or GASPE experiment?

Problem 4.18

Describe the fundamental properties of an INEPT experiment.

4.2.1.1 Signal Intensification with INEPT

In the case of a CH group, the two lines of the ^{13}C doublet have intensities of $+5$ and -3 (see the previous discussion), and since they have different phases, they will partially cancel each other if they were to become focused. A suitable delay is, therefore, introduced to bring them into phase, the time delay differing for CH_3, CH_2, and CH groups. Once both lines are in phase then each ^{13}C line exhibits a four-fold increase in intensity. This is significantly superior to the three-fold enhancement in sensitivity obtainable by nuclear Overhauser enhancement when recording proton-decoupled ^{13}C-NMR spectra. In practice, five-fold to six-fold enhancement of sensitivity of INEPT spectra can be obtained, since the repetition rate of the entire INEPT sequence is governed by the relatively shorter *proton* spin-lattice relaxation times, whereas in proton-decoupled ^{13}C spectra the repetition rate of successive scans is determined by the longer ^{13}C spin-lattice relaxation times.

In polarization transfer experiments such as INEPT, the recorded multiplets do not have the standard binomial intensity distributions of 1:1 for doublets, 1:2:1 for triplets, and 1:3:3:1 for quartets, etc. Instead, they exhibit

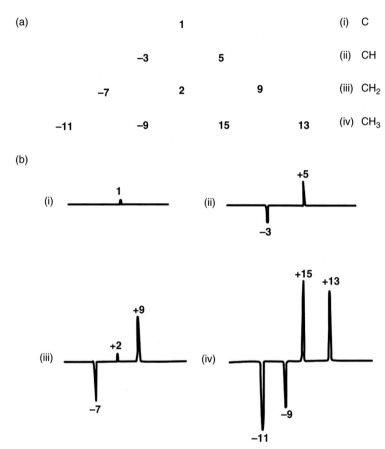

FIGURE 4.13 (a) Relative intensities in an INEPT spectrum presented as a variation of the Pascal triangle. (b) Signal phase and amplitudes of (i) quaternary, (ii) CH, (iii) CH_2, and (iv) CH_3 carbons in a normal INEPT experiment.

intensities – as indicated in the Pascal triangle in Fig. 4.13 – of −3:5 for doublets, −7:2:9 for triplets, −11, −9, 15, and 13 for quartets, etc.

The sensitivity enhancement in the INEPT experiment is particularly marked when nuclei of low magnetogyric ratios are being detected. A comparison of the signal intensities obtained by polarization transfer against those obtained by full NOE for various nuclei is presented in Table 4.1.

4.2.1.2 Spectral Editing with INEPT

INEPT can be used for determining the multiplicities of ^{13}C nuclei coupled to protons. This is done by deleting the 180° ^1H and ^{13}C refocusing pulses and by including a delay Δ the value of which is adjusted to give a particular angle θ,

TABLE 4.1 A Comparison of Signal Intensities Obtained by Polarization Transfer against Those Obtained by Full nOe from Protons to the Heteronucleus*

Nucleus	Polarization Transfer	Maximum nOe
^{13}C	3.98	2.99
^{15}N	9.87	−3.94
^{31}P	2.47	2.24
^{29}Si	5.03	−1.52
^{57}Fe	30.95	16.48
^{103}Rh	31.78	−14.98

The intensities given are relative to those observed without nOe.

TABLE 4.2 Relative Signal Intensities in INEPT

θ^*	CH_3	CH_2	CH
$\pi/4$	$\frac{3}{2}\sqrt{2}$	1	$1/\sqrt{2}$
$\pi/2$	0	0	1
$3\pi/4$	$\frac{3}{2}\sqrt{2}$	−1	$1/\sqrt{2}$
$\theta^* = \Delta$			

before data acquisition. The CH_3, CH_2, and CH carbon intensities depend on the delay Δ (i.e., on the angle θ), as shown in Table 4.2.

Problem 4.19

Based on the equation $K = \gamma_H/\gamma_C$, we would expect about a three-fold enhancement in the signal intensities of proton-bearing ^{13}C nuclei in the broad-band decoupled INEPT experiment. In practice, the intensification is more than three-fold. Why?

The precession of the ^{13}C multiplet components during the delay Δ by angle θ (where $\theta = \pi J \Delta$) is related to the signal intensities of CH, CH_2, and CH_3 carbons as follows:

CH: $I \alpha \sin \theta$

CH_2: $I \alpha \sin 2\theta$

CH_3: $I \alpha (3/4) (\sin \theta + \sin 3\theta)$.

Since in the rotating frame the doublet vectors of the CH carbons precess at a precessional frequency of $+J/2$ and $-J/2$, if the delay Δ is set at $\frac{1}{2} J$ (i.e., $\theta = 90°$) then only CH carbons will appear. With Δ set at $\frac{1}{4} J$, all

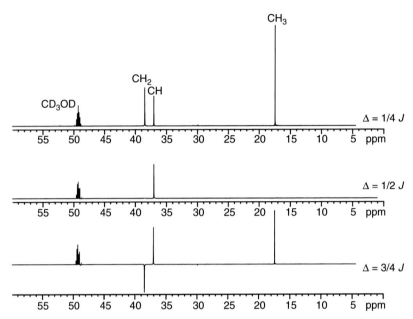

FIGURE 4.14   The behavior of CH, CH_2, and CH_3 groups during spectral editing in INEPT experiments.

protonated carbons will appear with positive amplitudes; with Δ set at ¾ J, CH and CH_3 carbons will produce positively phased signals and CH_2 carbons will give negatively phased signals. The quaternary carbons do not appear, since there are no attached protons from which polarization transfer can occur. Comparing the spectra at various Δ values with the broad-band decoupled spectrum yields a procedure for determining carbon multiplicities (*spectral editing*). This is illustrated in Fig. 4.14. To distinguish between the three kinds of carbon atoms (CH_3, CH_2, and CH), we need to record three different spectra with the delays Δ adjusted to make θ equal $\pi/4$, $\pi/2$, and $3\pi/4$. Suitable combinations of these spectra allow us to generate subspectra containing CH, CH_2, or CH_3 carbons.

As mentioned already, the INEPT spectra are characterized by the antiphase character of the individual multiplets. The INEPT ^{13}C-NMR spectrum of isobutylbromide is shown in Fig. 4.15. Doublets show one peak with positive phase and the other with negative phase. Triplets show the outer two peaks with positive and negative amplitudes and the central peak with a weak positive amplitude. Quartets have the first two peaks with positive amplitudes and the remaining two peaks with negative amplitudes.

Since the natural magnetization has been removed by phase alternation, we cannot apply ^{1}H decoupling during data acquisition, because there is no *net* magnetization contributing to any of the multiplets. Decoupling would therefore result in the mutual cancellation of the antiphase peaks within each

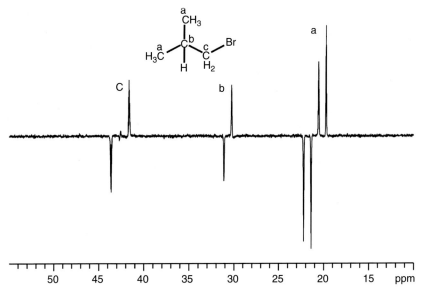

FIGURE 4.15 INEPT spectrum of isobutyl bromide.

multiplet when they collapse together, leading to the disappearance of the signals. It is therefore necessary to collect coupled spectra with antiphase peaks within multiplets, and if decoupled spectra are required, we need to record a refocused INEPT experiment or a variant of it.

Problem 4.20

How can spectral editing in INEPT be used to generate separate spectra for CH, CH_2, and CH_3 groups, and how is it superior to the traditional off-resonance decoupled spectra?

4.2.2 Refocused INEPT

The INEPT experiment can be modified to allow the antiphase magnetization to precess for a further time period so that it comes into phase before data acquisition. The pulse sequence for the refocused INEPT experiment (Pegg et al., 1981b; Burum and Ernst, 1980) is shown in Fig. 4.16. Another delay D_3 is introduced and 180° pulses applied at the center of this delay simultaneously to both the ^1H and the ^{13}C nuclei. Decoupling during data acquisition allows the carbons to be recorded as singlets. The value of D_3 is adjusted to enable the desired type of carbon atoms to be recorded. Thus, with D_3 set at 1/4 J, the CH carbons are recorded; at 1/8 J, the CH_2 carbons are recorded; and at 1/6 J, all protonated carbons are recorded. With D_3 at 3/8 J, the CH and CH_3 carbons appear out of phase from the CH_2 carbons.

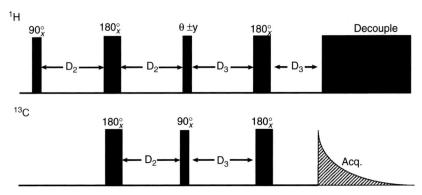

FIGURE 4.16   Refocused INEPT pulse sequence.

The additional delay causes a decrease in signal strength due to loss of magnetization from transverse relaxation. Moreover, severe phase distortions, particularly for CH_3 carbons, can produce anomalous results. A further modification of INEPT known as INEPT$^+$ incorporates an additional 90° purging-pulse on the ^1H nuclei before detection (Sørensen and Ernst, 1983). This serves to remove the unwanted phase and multiplet anomalies. Many other variants of the INEPT experiment are known (Mohebbi and Gonen, 1996; Bax and Drobny, 1985; Neuhaus et al., 1985; King and Wright, 1983; Wagner, 1983; Eich et al., 1982; Bodenhausen and Ruben, 1980), but their discussion is beyond the scope of this book.

> **Problem 4.21**
>
> What are the characteristic differences between basic INEPT and refocused INEPT experiment?

4.2.3   DEPT

The most widely used method for determining multiplicities of carbon atoms is _D_istortionless _E_nhancement by _P_olarization _T_ransfer (DEPT). This has replaced the classical method of recording off-resonance ^{13}C spectra with reduced CH couplings from which the multiplicity can be read directly.

The DEPT experiment (Doddrell et al., 1982) involves a similar polarization transfer as the INEPT experiment, except that it has the advantage that all the ^{13}C signals are in phase at the start of acquisition so that there is no need for an extra refocusing delay as in the refocused INEPT experiment. Coupled DEPT spectra, if recorded, would therefore retain the familiar phasing and multiplet structures (1:1 for doublets, 1:2:1 for triplets, etc.). Moreover, DEPT experiments do not require as accurate a setting of delays between pulses as do INEPT experiments.

The pulse sequence used in DEPT is shown in Fig. 4.17 with reference to a CH group. A $90°_x$ ^1H pulse bends the equilibrium magnetization from the z-axis to the

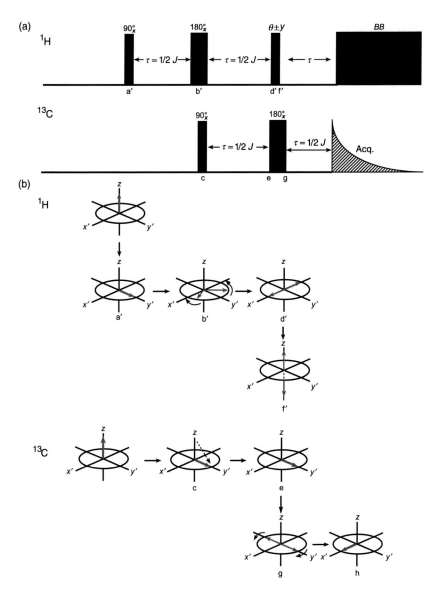

FIGURE 4.17   (a) Pulse sequence for the DEPT experiment. (b) Effect of the pulse sequence on
^1H and ^{13}C magnetization vectors. ^{13}C magnetization can be recorded either as multiplets or, if
broad-band decoupling is applied during the acquisition period, as singlets.

y'-axis. During the subsequent evolution period, the transverse ^1H magnetization
is modulated by the CH coupling so that it splits into two counter-rotating vectors;
at the end of the first ½ J delay, the vectors point along the $+x'$- and $-x'$-axis.
Next, a 180° ^1H pulse is applied to remove any field inhomogeneities; simultaneously, a 90°_x ^{13}C pulse serves to bend the ^{13}C magnetization from its equilibrium

position along the z-axis to the y'-axis. Since there is no z magnetization for either the ^1H or the ^{13}C nuclei, they are effectively decoupled. Therefore during the subsequent ½ J delay period, they remain stationary at their original positions, with the delay serving only to remove the field inhomogeneity effects by the preceding 180°_x ^1H pulse. At the end of the second ½ J delay period, a proton pulse of angle θ is applied. The value of the angle θ is adjusted according to the type of spectrum desired. Thus, if we wish to record only CH carbons, then the value of θ is kept at 90° with the result that there is maximum polarization for CH carbons and negligible x'-magnetization for CH_3 and CH_2 carbons.

With the 90°_x pulse we have rotated one of the two ^1H spin vectors to the $+z$-axis (i.e., to its original equilibrium position) and the other to the $-z$-axis, i.e., inverted its population. As in the APT and INEPT experiments, this inversion of population of one of the two spin vectors enhances the population difference between the two ^{13}C spin states with which the ^1H is coupled (see Section 4.2 discussion on polarization transfer for details on this phenomenon) resulting in an intensification of the ^{13}C signal. Simultaneously, a 180° pulse is applied to the ^{13}C nuclei to remove any field inhomogeneities during the third ½ J delay period. It is during this last delay that the ^{13}C signals are modulated by $^{13}C–^1H$ coupling, since z-magnetization now exists for the ^1H nuclei. The ^{13}C vectors rotate in the $x'y'$-plane, and at the end of the third ½ J delay, the FID is acquired with or without ^1H decoupling. With θ set at 90°, only CH carbons are recorded. With θ at 135°, the CH and CH_3 carbons are modulated differently from CH_2 carbons so that CH_3 and CH carbons give positively phased signals and CH_2 carbons give negatively phased signals. The two spectra on comparison with the broad-band decoupled spectrum suffice to identify CH_3, CH_2, CH, and quaternary carbons. With θ at 45°, all protonated carbons (CH_3, CH_2, and CH) appear with positive phases. The dependence of signal intensities of CH_3, CH_2, and CH carbons in the DEPT experiment on the angle θ_y of the last polarization pulse is shown in Fig. 4.18. The

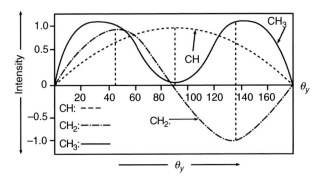

FIGURE 4.18   In DEPT experiments, signal intensities of CH_3, CH_2, and CH carbons depend on the angle θ_y of the last polarization pulse. For instance, at θ_y = 90°, only CH carbons can be seen, while at 135°, CH_3 and CH carbons will appear with one phase and CH_2 carbons will appear with the opposite phase.

type of DEPT spectra obtained depends on the angle θ_y of the polarization pulse. Moreover, DEPT spectra are less dependent than INEPT spectra on the delay times between pulses, an error of ±20% in the setting of these delays still giving acceptable DEPT spectra.

Many modifications of DEPT spectra such as DEPT$^+$, DEPT^{++} (Sørensen and Ernst, 1983), DEPT GL (Sørensen et al., 1983), modified DEPT (MODEPT) (Sørensen et al., 1984), universal polarization transfer (UPT) (Bendall et al., 1983), phase oscillations to maximum editing (POMMIE) (Bulsing et al., 1984; Bendall and Pegg, 1983), and detection of quaternary carbons (DEPTQ) (Burger and Bigler, 1998) have been reported.

Problem 4.22

DEPT (θ=135° and 90°) and broad-band decoupled ^{13}C-NMR spectra of 2-methylsuccinic acid isolated from *Ziziphora tenuior* L. are shown. Assign the signals to various carbons of the molecule.

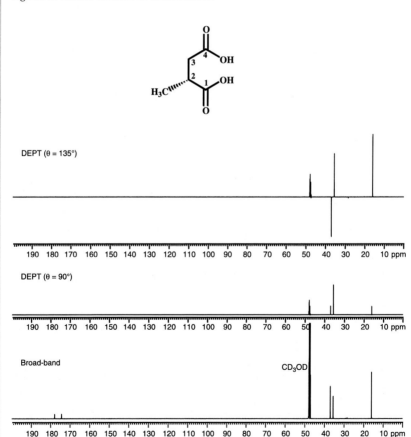

> **Problem 4.23**
>
> What are the differences between INEPT and DEPT experiments? Why is DEPT considered a *distortionless* experiment?

> **Problem 4.24**
>
> Why do quaternary carbons not appear in polarization transfer experiments?

4.3 POLARIZATION TRANSFER IN REVERSE

In the preceding experiments, polarization was being transferred from a nucleus with a high magnetogyric ratio, such as ^{1}H, to a nucleus with a low magnetogyric ratio, such as ^{13}C. It is possible to transfer the polarization from a nucleus with low magnetogyric ratio, e.g., ^{13}C, to the one with a higher magnetogyric ratio, e.g., ^{1}H, before detection of the transferred magnetization. Such a *reverse transfer of polarization* has advantages of both selectivity and sensitivity (Section 3.5.3).

Consider the selective polarization transfer in the normal (forward) sense, i.e., from a ^{1}H nucleus to a ^{13}C nucleus to which ^{1}H is attached. To carry out this transfer from a given ^{1}H nucleus, we have to identify and irradiate it selectively. This may not always be possible, particularly in larger molecules such as lipids or proteins, due to extensive overlap, or because the ^{1}H signals may be obscured by a large water signal. In the corresponding ^{13}C-NMR spectrum, however, it is easier to identify and irradiate individual carbons selectively because of the greater dispersion of the signals, particularly if enrichment of the particular carbon has been carried out. This is especially useful in studies of metabolic pathways. For instance, if we feed a ^{13}C-labeled drug to a living system and we wish to study its metabolism, we can use the ^{13}C label as a handle for detecting its attached protons selectively. The ^{13}C label helps us to identify the carbon unambiguously because of its more intense signal as compared to other unenriched carbons in the same molecule, and polarization transfer from such a carbon to the corresponding protons would therefore allow us to follow readily the fate of such protons in the corresponding metabolite.

The inverse INEPT (Bodenhausen and Ruben, 1980) and inverse DEPT (Brooks et al., 1984) experiments utilize such an approach. In the inverse INEPT experiment, successive 90° pulses are applied to the ^{13}C nucleus, followed by a ^{1}H read pulse. Protons not coupled to the ^{13}C nucleus are suppressed by presaturation of the entire ^{1}H-NMR spectrum before the polarization transfer, so only those signals will be detected that are generated by polarization transfer from the ^{13}C nucleus.

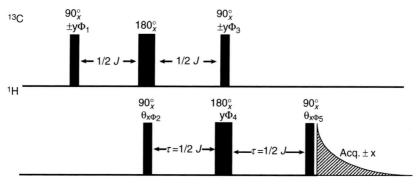

FIGURE 4.19 Pulse sequence for the inverse (reverse) DEPT experiment.

Transfer of polarization from ^{13}C nuclei to ^{1}H nuclei and their subsequent detection leads to a 16-fold increase in sensitivity because the ^{13}C magnetization is being measured indirectly through detecting it *via* the nucleus with the higher magnetogyric ratio (i.e., ^{1}H). Irradiation of the protons between the scans causes a further three-fold increase in the population of the ^{13}C nuclei due to nOe, so an overall 50-fold increase in sensitivity is achievable in contrast to direct ^{13}C measurements. However, because of the longer relaxation times of the ^{13}C nuclei, the scan rate has to be reduced, and since ^{1}H signals are split, a 15-fold to 20-fold increase in sensitivity may actually be achieved.

The pulse sequence used in the reverse DEPT experiment is shown in Fig. 4.19. Presaturation of the protons removes all ^{1}H magnetization and pumps up the ^{13}C population difference due to nOe. Broad-band decoupling of the ^{13}C nuclei may be carried out. The final spectrum obtained is a 1D ^{1}H-NMR plot that contains only the ^{1}H signals to which polarization has been transferred—for instance, from the enriched ^{13}C nucleus.

The application of the reverse DEPT pulse sequence to monitor ^{13}C-labeled glucose by mouse liver-cell extract is shown in Fig. 4.20. The α-and β-anomeric proton resonances are shown in the starting material; these are transformed to CH_2 proton resonances in the metabolite.

Problem 4.25

What are the principles governing reverse polarization transfer experiments? Why are reverse experiments more sensitive than "normal" experiments?

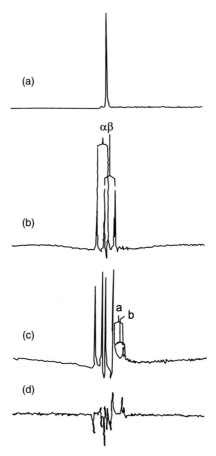

FIGURE 4.20   Application of the reverse DEPT pulse sequence to monitor ^{13}C-labeled glucose by mouse liver-cell extract. (a) Normal FT spectrum. (b) Reverse DEPT spectrum showing the α- and β-anomeric proton resonances. (c) Two different CH_2 proton resonances, a and b, appear after 1.5 h of metabolism. (d) Edited ^1H spectrum confirming that the CH_2 resonances arise from metabolic products. *(Reprinted from Brooks et al. (1984). J. Magn. Reson., 56, 521, copyright ©, Academic Press.)*

13C-NMR Chemical shift Assignments by DEPT

The broad-band decoupled and DEPT spectra of podophyllotoxin, $C_{22}H_{22}O_8$, isolated from *Podophyllum hexandrum* Royle., are shown. Although the molecule contains 22 carbons, only 18 signals are visible in the broad-band ^{13}C-NMR spectrum. This would obviously mean that some of these signals represent more than one carbon. Pairs of C-10/C-14, C-11/C-13, and 11-OCH₃/13-OCH₃ are magnetically equivalent, while C-6 and C-7 could show accidental overlap. With this information, assign chemical shifts to various carbon atoms in the molecule.

(a) DEPT (0 = 135°), (b) DEPT (0 = 90°), and (c) broad-band decoupled ^{13}C-NMR spectra of podophyllotoxin.

The DEPT and broad-band decoupled ^{13}C-NMR spectra of tschimganin $(1R,2S,4R)$-1,7,7-trimethylbicyclo[2.2.1]heptan-2-yl-4-hydroxy-3-methoxybenzoate, $C_{18}H_{24}O_4$, isolated from a medicinal plant *Ferula ovina* (Bioss.), are shown. Spectrum (a) displays CH and CH_3 signals with positive phase and CH_2 signals with negative phase; (b) shows only CH signals; and (c) shows the broad-band decoupled spectrum with all carbons (C, CH, CH_2, and CH_3). (c) Assign the ^{13}C-NMR chemical shifts to the various carbons in the molecule.

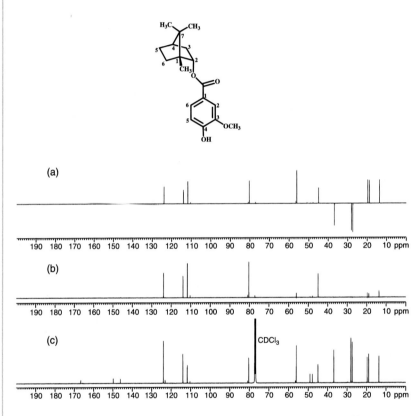

(a) DEPT ($\theta = 135°$), (b) DEPT ($\theta = 90°$) and (c) broad-band decoupled ^{13}C-NMR spectra of $(1R,2S,4R)$-1,7,7-trimethylbicyclo[2.2.1]heptan-2-yl-4-hydroxy-3-methoxybenzoate.

How can DEPT spectra help in ^{13}C-NMR chemical shift assignments? The DEPT spectra (a and b) and the broad-band decoupled (c) ^{13}C-NMR spectra of 7-deoxyloganic acid, $C_{16}H_{24}O_9$, isolated from *Osyris wightiana* Wall. ex Wight are shown below. Assign the resonances to the various carbons.

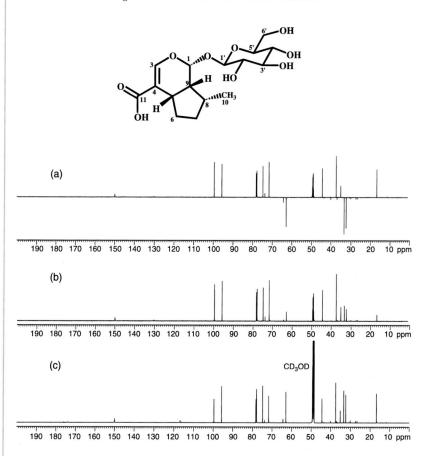

DEPT ($\theta = 135°$), (b) DEPT ($\theta = 90°$), and (c) broad-band decoupled ^{13}C-NMR spectra of 7-deoxyloganic acid.

^{13}C-NMR Chemical Shift Assignments by DEPT

The DEPT and broad-band decoupled ^{13}C-NMR spectra of a natural product, citroside B, $C_{19}H_{30}O_8$, isolated from *Osyris wightiana* Wall. ex Wight are presented here. Assign the signals to the various carbons in the molecule.

(a) DEPT ($\theta = 135°$), (b) DEPT ($\theta = 90°$), and (c) broad-band decoupled ^{13}C-NMR spectra of citroside B.

^{13}C-NMR Chemical Shift Assignments by GASPE

How are ^{13}C-NMR chemical shift assignments made by GASPE? The GASPE spectrum of a natural lignin, podophyllotoxin, is shown. The signals at δ 56.2, 108.3, and 152.6 represent two carbons each in identical magnetic environments, while the signal at δ 147.7 also represents two carbons that accidentally appear at about the same chemical shift. Assign chemical shift values to various protonated and quaternary carbons in the molecule.

GASPE ^{13}C-NMR Spectrum of podophyllotoxin.

The GASPE spectrum of tschimganin (1R,2S,4R)-1,7,7-trimethylbicyclo[2.2.1] heptan-2-yl-4-hydroxy-3-methoxybenzoate, $C_{18}H_{24}O_4$, is shown. Assign the chemical shift values to various protonated and quaternary carbons in the structure.

GASPE ^{13}C-NMR spectrum of tschimganin (1R,2S,4R)-1,7,7-trimethylbicyclo[2.2.1]heptan-2-yl-4-hydroxy- 3-methoxybenzoate.

SOLUTIONS TO PROBLEMS

4.1 Despite major developments in the design of high field super-conducting magnets and shimming technologies, the applied magnetic field B_0 still has some level of inhomogeneity. This slight variation in field strength can grossly affect the quality of the NMR signals, if it is not corrected. Even the most modern and stable NMR magnets have a field B_0 which can vary over the entire width of the sample holding area.

The Larmor frequency of an NMR active nucleus is directly proportional to the magnetic field strength (B_0). This means that, for example, if the CH_3 group of ethyl alcohol is exposed to variable B_0, then it will resonate not with one but several Larmor frequencies, resulting in the broadening of its signals or, in the extreme case, it can lead to the appearance of several signals. The spin echo experiment is capable of correcting the influence of this magnetic inhomogeneity by refocusing the spin vector.

4.2 The spin-echo is an elegant method for the measurement of transverse relaxation time T_2. In practice, this is done by repeating the spin-echo experiment many times, with different delay intervals ($2\tau, 4\tau, 6\tau$, etc.). The following sequence is used:

$$90^\circ_x - \tau - 180^\circ_x - \tau(\text{1st echo}) - \tau - 180^\circ_x - \tau_x(\text{2nd echo})\ldots$$

The intensities of the echoes (signal intensities) are than plotted as a function of delays. This yields a straight line whose gradient is $-1/T_2$. The decay in the intensities of echoes (i.e., of the succession of signals which follow the FID) is determined solely by T_2.

4.3 Nuclei resonating at different chemical shifts will also experience similar refocusing effects. This is illustrated by the accompanying diagram of a two-vector system (acetone and water), the nuclei of which have different chemical shifts but are refocused together by the spin-echo pulse (M_A = magnetization vector of acetone methyl protons, M_W = magnetization vector of water protons).

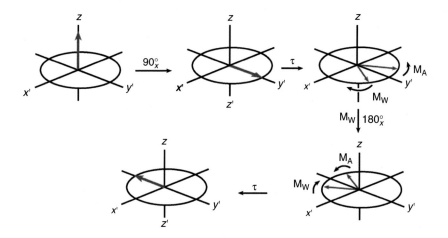

4.4 1. The 180°_x pulse will refocus the magnetization vectors, dephased due to field inhomogeneities and other instrumental errors. Following is the vector representation of this refocusing effect:

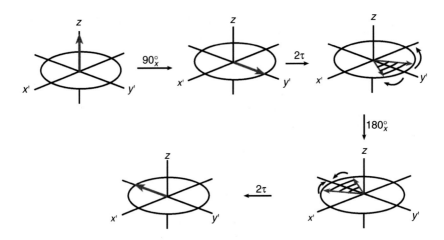

2. Assuming that we are observing the magnetization vector of nucleus A, which is coupled to nucleus X in an AX heteronuclear spin system, on application of a 90_x° pulse, the magnetization vector of A will come to lie on the y'-axis. Coupling with the two spin states of nucleus X will cause the A vector to split into two components. One of the A vectors (associated with the α state of nucleus X) will rotate at a frequency $\delta + (J_{AX}/2)$, while the other A vector (associated with the β state of X) will rotate at a frequency $\delta - (J_{AX}/2)$. Applying a 180_x° selective pulse on A at the end of the first τ period will rotate these vectors onto their mirror image positions. During the second τ period, they will continue to rotate toward the $-y'$-axis so that at the end of this delay they will realign together along the $-y'$-axis. This can be readily visualized by the following vector diagram.

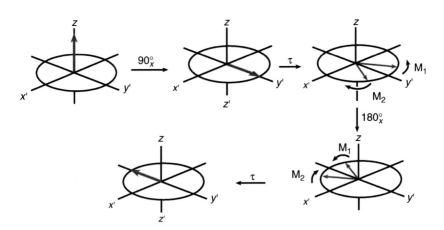

4.5 No. Since the direction of precession of the vectors is reversed during the second half of the sequence, they continue to diverge from each other. The effect of a spin-echo experiment on spin–spin coupling in a first-order homonuclear spin system is shown in the following vector representation:

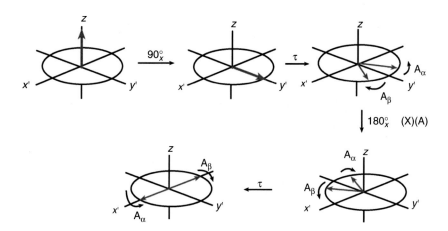

The initial 90°_x pulse will bring vector A of a homonuclear AX system to lie along the $+y'$-axis. After a delay τ, vector A will split into A_α and A_β due to its coupling with nucleus X. Labels A_α and A_β represent the effect of α and β spin states of the coupled nucleus X. The two components A_α and A_β will move away from each other, creating a relative phase difference of $2\pi J\tau$ by the end of the first delay. Applying a 180°_x pulse to nuclei A and X at this point will have two distinct effects. First, it will rotate A_α and A_β onto mirror image positions. Second, it will also flip the spin labels of the coupled nuclei, i.e., interchange A_α with A_β. Since these are the labels that define the direction of rotation of the individual components, the two components A_α and A_β will now move in opposite directions to their previous directions of rotation, i.e., away from each other, building up a phase difference of $4\pi J\tau$ at the time of the echo. The inversion of both coupled spins is the key to echo modulation in several J-resolved experiments.

4.6

(i)

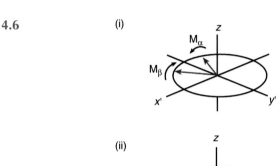

(ii)

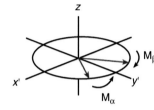

(iii)

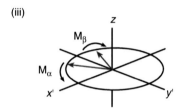

4.7 After the 90°pulse, the transverse magnetization vectors of ^{13}C nuclei of C, CH, CH_2, and CH_3 do not rotate synchronously with one another but rotate with characteristically different angular velocities during the same delay interval. This results in their appearing with differing (positive or negative) amplitudes. This forms the basis of the APT experiment.

4.8 As explained in the text, the signals for CH_3 and CH carbons will appear with one phase, whereas CH_2 and quaternary carbons will appear with opposite phase. In the GASPE spectrum shown, the negative phased signals represent the CH_3 and CH carbons in the molecule, and CH_2 and C carbons appear with positive amplitude. The signal at δ 17.4 is, therefore, due to the methyl carbon, while the signal at δ 37.0 is assigned to the CH carbon. The signals with positive amplitude at δ 38.5 are assigned to the CH_2 carbon. The signals at δ 179.3 and 175.7 with positive amplitude can be assigned to the acid carbonyl carbon.

4.9 The spectrum in the APT experiment is dependent on both the magnitude of the $^1J_{CH}$ coupling and the number of attached protons. In setting up the experiment, the delays are matched to the inverse of $^1J_{CH}$ (usually set around 140 Hz, which is the average $^1J_{CH}$ for the sp^3 and sp^2 carbons). If the size of $^1J_{CH}$ is much larger than this default $^1J_{CH}$ value, then many peaks will disappear or appear with an incorrect phase. This is often seen in carbons, such as terminal ethynyl groups ($^1J_{CH}$ = 250 Hz approx.), epoxide carbons ($^1J_{CH}$ = 175 Hz), and many aromatic carbons, such as those in furans, pyrones, isoflavones ($^1J_{CH}$ = 200 Hz), and others.

4.10 Single-quantum coherence is the type of magnetization that induces a voltage in a receiver coil (i.e., *Rf* signal) when oriented in the *xy*-plane. This signal is observable, since it can be amplified and Fourier-transformed into a frequency-domain signal. Zero- or multiple-quantum coherences do not obey the normal selection rules and do not induce any voltage in the receiver coil. These nonobservable magnetizations can only be detected indirectly after conversion to single-quantum coherence.

4.11 No. The vector presentation is suitable for depicting single-quantum magnetizations but is not appropriate when considering zero-, double-, and higher-order quantum coherences. Quantum mechanical treatment can be employed when such magnetizations are considered.

4.12 The pulse sequence used in the polarization transfer experiments, selectively or nonselectively, transfers the magnetization of the sensitive nuclei (^1H or ^{19}F) to the insensitive nuclei (such as ^{15}N and ^{13}C). This leads to sensitivity enhancement. It is therefore correct to term this phenomenon as population transfer or polarization transfer.

4.13 INADEQUATE is an inherently insensitive, though a powerful, technique. It provides the ^{13}C–^{13}C coupling framework directly. However, the probability of having two ^{13}C nuclei adjacent to one another is very small (1.1% $\times$ 1.1% = about 1 in 8,300). Moreover, the low magnetogyric ratio (γ) of the ^{13}C nuclei further reduces their sensitivity, making their detection difficult. Therefore, in the INADEQUATE experiment, large quantities (500 mg or more) of samples (of molecular weights below 1,000) are required to obtain a reasonable result within 24 hours on a 400 MHz instrument. This is why despite the utility of the INADEQUATE experiment, it is only seldom used. Incorporating a polarization transfer sequence before the INADEQUATE pulse can increase the sensitivity to a substantial level. This allows the INADEQUATE experiment to be performed with enhanced sensitivity

4.14 The signal intensity of the ^{13}C nucleus depends on the population difference between its two energy states, α and β. At thermal equilibrium, the ^1H population difference is about four times higher than that of ^{13}C nuclei, so there are four times as many protons in Boltzmann excess population of the lower energy state as there are ^{13}C nuclei (based on their respective magnetogyric ratios). Inversion of one of the ^1H spin states causes a corresponding intensification of the population of the ^{13}C nuclei, since the ^{13}C transitions share a common energy level with this proton transition.

The following diagram explains the phenomenon in a rather simplified fashion.

1. Populations at thermal equilibrium: ^1H population difference = (1) − (3) or (2) − (4); ^{13}C population difference = (3) − (4) or (1) − (2):

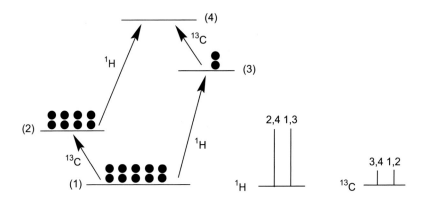

2. Populations after inverting one of the ^1H transitions:

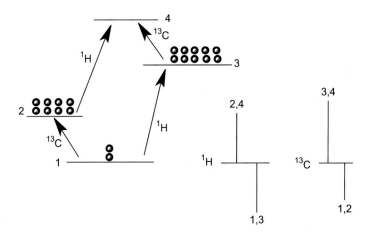

4.15 Nonselective polarization transfer, as implied by the term, represents a process that allows simultaneous polarization transfer from *all* protons to *all* X nuclei. In selective polarizations transfer, however, the population of just one nucleus is inverted at any *one* time. The selective polarization transfer sequence therefore cannot be used to generate a proton decoupled ^{13}C-spectrum containing all sensitivity-enhanced ^{13}C resonances.

4.16 The magnetogyric ratio $(-2.712 \times 10^7$ rad $T^{-1}s^{-1})$ and natural abundance of ^{15}N (0.038%) are much lower than those of ^{13}C $(6.7283 \times 10^7$ rad $T^{-1}s^{-1}$ and natural abundance 1.1%). The sensitivity enhancement for ^{15}N due to polarization transfer is about 10-fold, as compared to ^{13}C where the sensitivity enhancement is about four-fold. It is therefore correct to say that the sensitivity enhancement in the case of ^{15}N is more significant.

4.17 The APT pulse sequence provides limited information about the number of hydrogens bonded to the carbons in a molecule, since it does not readily allow us to distinguish between the CH_3 and CH carbons or between CH_2 and quaternary carbons. The INEPT spectrum can not only yield information about the multiplicity of all the carbons, but also afford sensitivity-enhanced ^{13}C signals due to polarization transfer.

4.18 The INEPT experiment has the following important characteristics:
   1. It enhances the sensitivity of an insensitive nucleus X, such as ^{13}C or ^{15}N, by a factor of γ_H/γ_x due to population transfer from an attached sensitive nucleus such as ^1H, and by a factor of $1 + \gamma_H/\gamma_x$ due to NOE generated by broad-band decoupling.
   2. It reduces the dependence of signal strength on γ.
   3. The INEPT experiment, after incorporation of a variable delay Δ, also provides a means for determining the number of protons attached to the ^{13}C nuclei in the molecule (i.e., establishing the multiplicity of the carbon signals). It is thus possible to generate separate sub-spectra for CH, CH_2, and CH_3 groups.

4.19 The nuclear Overhauser effect resulting from the broad-band decoupling during the decoupled INEPT experiment also contributes to the signal enhancement of the ^{13}C signals.

4.20 The precession of CH, CH_2, and CH_3 vectors during Δ occurs at different angular velocities. In order to distinguish between the three kinds of carbon resonances, it is essential to record three spectra with variable delays Δ. The delay should be adjusted to make θ (where $\theta = \pi J \Delta$) equal to $\pi/4$, $\pi/2$, and $3\pi/4$. The $\theta = \pi/4$ delay spectrum will contain all the resonances, the $\theta = \pi/2$ delay spectrum will give only CH signals, and the spectrum obtained with the delay $\theta = 3\pi/4$ will exhibit all CH_2 signals inverted and CH and CH_3 signals in positive phase. Careful examination of these three spectra allows generation of CH, CH_2, and CH_3 subspectra.

4.21 The INEPT spectrum cannot be recorded with broad-band proton decoupling, since the components of multiplets have antiphase disposition. With an appropriate increase in delay time, the antiphase components of the multiplets appear in phase. In the refocused INEPT experiment, a suitable refocusing delay is therefore introduced that allows the ^{13}C spin multiplet components to get back into phase. The pulse sequences and the resulting spectra of fenchone from the two experiments are given below:

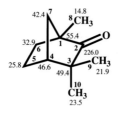

1. **INEPT**

$$^1\text{H} \quad 90^\circ_x - \tau - 180^\circ_x - \tau - 90^\circ_y$$

$$^{13}\text{C} \quad 180^\circ_x - 90^\circ_x - Acquisition$$

INEPT 135
CH, CH₃ positive, CH₂ negative

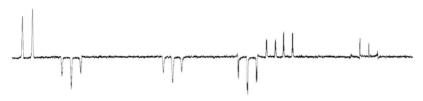

INEPT 45
CH, CH₂, and CH₃ positive

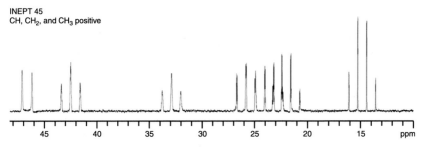

2. **Refocussed INEPT**

$$^1\text{H} \quad 90^\circ_x - (1/4)J - 180^\circ_x - (1/4)J - 90^\circ \pm y - \Delta/2 - 180^\circ - \Delta/2 - Decoupling$$

$$^{13}\text{C} \quad 180^\circ_x - 90^\circ_x - 180^\circ_x - Acquisition$$

Refocussed INEPT 135
CH, CH₃ positive, CH₂ negative

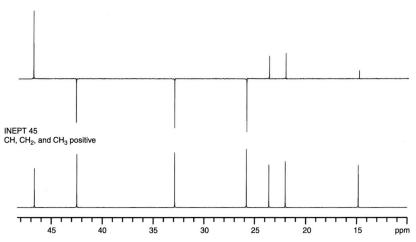

INEPT 45
CH, CH₂, and CH₃ positive

3. Broad-band decoupled

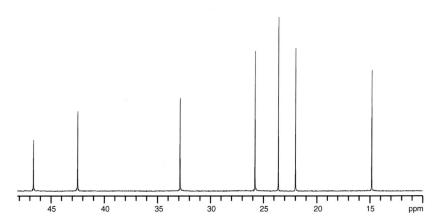

4.22 The broad-band decoupled ^{13}C-NMR spectrum of 2-methylsuccinic acid shows five carbon resonances; the DEPT ($\theta = 135°$) spectrum displays only three signals; i.e., only the protonated carbons appear, since the quaternary carbonyl carbon signals do not appear in the DEPT spectrum. The CH and CH₃ carbons appear with positive amplitudes, while CH₂ carbon appeared with negative amplitude. The DEPT ($\theta = 90°$) spectrum displays only the methine carbons. It is therefore possible to distinguish between CH₃ carbons from CH carbons. Since the broad-band decoupled ^{13}C spectrum contains all carbons (including quaternary carbons), whereas the DEPT spectra do not show the quaternary carbons, it is possible to differentiate between quaternary carbons from CH, CH_2, and CH_3 carbons by examining the additional peaks in the broad-band spectrum in comparison to the DEPT spectra. The chemical shifts assigned to the various carbons are presented around the structure.

O
‖
179.3
37.0 $\overset{4}{3}$ OH
38.5 $\overset{}{2}$
$H_3C^{\text{\tiny IIII}}$ $\overset{1}{}$ OH
17.4 ‖ 175.7
O

4.23 Both experiments are based on polarization transfer from sensitive nuclei to insensitive nuclei, and therefore the major portions of their pulse sequences are common. The INEPT experiment, without refocusing and decoupling, however, yields spectra with "*distorted*" multiplets. For instance, the two lines of a doublet appear in antiphase with respect to one another. Similarly, the central line of a triple may be too small to be visible, while the outer two lines of the triplet will be antiphase to one another. Introducing a variable refocusing delay Δ and broad-band decoupling in the INEPT sequence can convert this experiment into a more useful one.

In the DEPT experiment, all the signals of the insensitive nuclei are in phase at the start of acquisition so that no refocusing period Δ (with accompanying loss in sensitivity) is required. Since the multiplets appear in-phase, it is called a *distortionless* experiment. Moreover, DEPT spectra depend on the angle θ of the last polarization transfer pulse, and are less dependent on the delay times $\frac{1}{2} J$ between the pulses. An error of $\pm 20\%$ in the estimation of J values still affords acceptable DEPT spectra. This offers a distinct advantage over INEPT, which requires a more accurate setting of delays between the pulses.

4.24 In these experiments, the signals of insensitive nuclei appear through polarization transfer from the more sensitive nuclei. Since quaternary carbons lack any attached hydrogen atoms, they cannot benefit from the polarization transfer and therefore do not appear in the spectra resulting from these experiments.

4.25 Reverse polarization transfer utilizes the fact that protons, due to their greater magnetogyric ratio, have greater *sensitivity*, while carbons or other low-abundant nuclei, because of their greater dispersion, have a higher *selectivity*. By transferring polarization from a less sensitive nucleus (such as ^{13}C) to a more sensitive nucleus (e.g., ^1H) and then recording the carbon nuclei through their effects on proton magnetization, we transfer the desired selectivity to the resulting spectrum, which has the character of a carbon spectrum. As compared to direct ^{13}C detection, the reverse polarization transfer signals are about 16 times stronger than with "normal" detection.

4.26 An examination of the structure shows that the aromatic ring at C-1 contains three pairs of identical carbons (the two –OCH$_3$ carbons at δ 56.2, the two carbons bearing the OCH$_3$ groups at δ 152.6, and the two carbons *ortho* to the methoxy group at δ 108.3). This accounts for three out of the four "missing" signals. The fourth "missing" carbon could be the one at δ 147.7, due to the accidental overlap of the two oxygen-bearing carbons in ring A to which the methylenedioxy group is attached. Only 11 resonances are visible in the DEPT ($\theta = 135°$) spectrum. This allows the quaternary carbon signals to be distinguished in the broad-band spectrum, which are at δ 131.2, 133.0, 135.3, 137.2, 147.7, 152.6, and 174.3. Quaternary signals between δ 135 and 155 generally represent oxygen-bearing aromatic carbons. For instance, the signal at δ 152.6 can be assigned to the methoxy-bearing magnetically equivalent C-11 and C-13, while the signal at δ 137.2 may be assigned to C-12. The upfield chemical shift of C-12 is due to the electron-donating mesomeric effect of the methoxy groups at C-11 and C-13. The quaternary signal at δ 147.7 may be due to the accidental overlap of the oxygen-bearing C-6 and C-7. The signal for the C-16 carbonyl is readily distinguishable at δ 174.3, while the assignments for C-4a, C-8a, and C-9 at δ 133.0, 131.2, and 135.3 are interchangeable and need additional heteroCOSY experiments [such as heteronuclear multiple bond connectivity (HMBC), see later] to be distinguished from one another. Two signals appear with negative phase in the DEPT 135° spectrum at δ 71.2 and 101.4, representing the two methylene carbons. The low-field methylene signal at δ 101.4 is characteristic for the methylenedioxy carbon containing two electron-withdrawing oxygen atoms, while the signal at δ 71.2 is assigned to C-15. The methyl carbons of the –OCH$_3$ groups appear at δ 56.2 and 60.7. Differentiation between CH$_3$ and CH carbons is possible from DEPT 90°, which shows only CH signals. The chemical shift assignments to the various carbons of podophyllotoxin are as follows.

4.27 The broad-band ^{13}C-NMR spectrum of tschimganin (spectrum c) shows 18 resonances. The DEPT spectra ($\theta = 135°, 90°$) display signals of only protonated carbons. By comparison with the broad-band decoupled spectrum, it is easy to identify the peaks for quaternary carbons (δ 47.8, 49.0, 123.0, 146.1, 149.8, and 166.6). The DEPT spectrum ($\theta = 90°$) (b) shows five signals for methine carbons resonating at δ 44.9, 80.3, 111.6, 113.9, and 123.9. Signals with negative amplitude in the DEPT 135° spectrum (a) are those of the methylene carbons resonating at δ 27.4, 28.0, and 36.9; signals with positive amplitude are of methyl and methine carbons. Comparison with the DEPT 90° spectrum indicates that the signals at δ 13.6, 18.9, 19.7, and 56.0 are for CH$_3$ groups. The downfield quaternary signal is assigned to the ester carbonyl carbon at δ 166.6 (COO). The other two downfield quaternary carbon signals at δ 149.8 and 146.1 may be due to the methoxy- and hydroxyl-substituted aromatic carbons (C-4′ and C-3′, respectively). The quaternary carbon signal resonating at δ 123.0 may be assigned to the aromatic C-1′. The upfield quaternary carbon signals at δ 49.0 and 47.8 may be assigned to the bridge head carbons C-1 and C-7, respectively. The CH signals at δ 123.0, 113.9, and 111.6 may be assigned to the aromatic methine carbons, i.e., C-6′, C-5′, and C-2′, respectively. Another methine signal at δ 80.3 can be assigned to C-2. The downfield chemical shift of this carbon is due to its direct connectivity with the ester oxygen. The remaining methine carbon signal at δ 44.9 can be assigned to C-4. The signals for the CH$_2$ carbons at δ 36.9, 28.0, and 27.4 may be assigned to C-3, C-6, and C-5, respectively. The methoxy carbon resonates at 56.0. The remaining three methyl carbon signals at δ 18.9, 19.7, and 13.6 are due to the two methyl substituents at C-7, and the bridge head methyl at C-1.

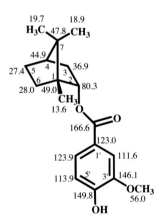

4.28 The resonances for all 16 carbons of 7-deoxyloganic acid are visible in the broad-band decoupled ^{13}C-NMR spectrum, whereas the DEPT 135°

spectrum (a) (CH_3, CH upright, and CH_2 inverted) exhibits only 14 signals. A careful comparison of these three spectra helps in identifying one CH_3, three CH_2, ten CH, and two quaternary carbons. Two quaternary signals appear at δ 116.8 and 173.9 in the broad-band decoupled spectrum. The latter signal (δ 173.9) is due to an α, β-unsaturated carboxylic acid carbonyl (C-11 in this case). The former signal (δ 116.8) can be readily assigned to the olefinic carbon (C-4) attached to the $-COOH$. Ten methine carbon resonances appear in the DEPT 90° spectrum at δ 35.1, 37.4, 44.4, 71.7, 74.8, 77.9, 78.3, 95.7, 99.6, and 149.9. The most downfield signal δ 149.9 may be assigned to the olefinic C-3. The signals at δ 99.6 and 95.7 are characteristic of anomeric C-1', and acetal C-1 carbons, respectively. The characteristic signals for the sugar moiety resonating at δ 71.7, 74.8, 77.9, and 78.3 can be assigned to C-4', C-2', C-3', and C-5', respectively. The remaining three methine signals at δ 35.1, 37.4, and 44.4 may be due to C-5, C-8, and C-9, respectively. The DEPT spectra (a and b) show one CH_3 signal at δ 16.8 which can easily be assigned to the C-10 CH_3, substituted at C-8. The two methylene signals, resonating at δ 32.4 and 33.4, may be assigned to C-6 and C-7, respectively. Another methylene signal at δ 62.9 may be assigned to the C-6' hydroxy methylene of the sugar moiety.

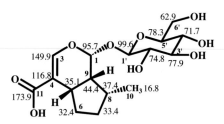

4.29 The broad-band decoupled ^{13}C-NMR spectrum of citroside B, shows signals for all 19 carbon atoms, including four methyl, three methylene, seven methine, and five quaternary carbons. The DEPT spectrum (θ = 135°) exhibits 14 signals for the protonated carbons. A comparison of broad-band spectrum (c) with the DEPT-135° spectrum indicates the presence of five quaternary carbons resonating at δ 37.0, 78.5, 118.9, 200, and 212.9. The most downfield carbon at δ 212.9 is characteristic of the ketonic carbonyl carbon (C-9), conjugated with the allene group. The other downfield quaternary carbon signal at δ 200.7 can be assigned to the allene carbon (C-7). The signal resonating at δ 118.9, indicated the presence of an olefinic carbon and hence was assigned to C-6. The remaining two quaternary carbons at δ 78.5, and 37.0 can be assigned to C-5 and C-1, respectively. The downfield shift of C-5 indicates the presence of an oxygen containing moiety. The three resonances with negative amplitudes at δ 48.0, 49.8, and 62.8 are due to the methylene carbons. The downfield shift of the methylene carbon at δ 62.8 indicated the presence of a hydroxyl substituent and was therefore assigned to the C-6' hydroxyl methylene carbon. The other two resonances at δ 49.8 (C-2) and 48.0 (C-4) are due to methylene carbons of the cyclohexane ring. A comparison between DEPT 90° and 135° spectra helps to identify the methyl and methine carbons. The four methyl signals at δ 26.6, 26.7, 30.0, and 32.5 can thus be assigned to the C-10, C-11, C-12, and C-13, respectively. The DEPT spectrum ($\theta = 90°$) shows seven signals for the methine carbons at δ 63.8, 71.6, 75.2, 77.8, 78.5, 98.6, and 101.3. The most downfield methine carbon signal at δ 101.3 may be assigned to the C-8 of the allene moiety. The signal at δ 98.6 is due to the anomeric carbon of a sugar moiety. The characteristic resonances for the sugar moiety were at δ 71.6 (C-4'), 75.2 (C-2'), 77.8 (C-3'), and 78.5 (C-5'). The ^{13}C-NMR assignments to the various carbons are presented around the following structure.

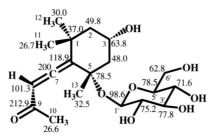

4.30 The GASPE ^{13}C-NMR spectrum of podophyllotoxin shows 18 peaks having positive and negative amplitudes. Nine signals with positive phase (pointing upward) are due to the quaternary and CH$_2$ carbons. The remaining nine resonances, having negative phase, are due to the CH$_3$ and CH carbons.

Let us first consider the overlapping signals. Since the signals at δ 56.2 (CH$_3$), 108.3 (CH), and 152.6 (−C−) each represent two carbons in identical magnetic environments, they can be assigned to the two OCH$_3$ carbons (substituted at C-11 and C-13) to the C-10 and C-14 methine carbons and to the C-11 and C-13 methoxy-bearing quaternary carbons, respectively. The quaternary signal at δ 147.7 represents the overlapping resonances of C-6 and C-7, which accidentally appear at the same chemical shift. The signals with positive phase, representing CH$_2$ and quaternary carbons, will be considered next. The peaks at δ 101.4 and 71.2 may be readily assigned to the CH$_2$ of the methylenedioxy group and C-15 (geminal to oxygen), respectively, based on their characteristic chemical shifts. The remaining seven signals with this phase represent the nine quaternary carbons. Two of these (δ 152.6 and 147.7) represent two-carbon signals that have already been assigned to C-11/C-13 and C-6/C-7.

The downfield quaternary carbon signal at δ 174.3 is readily assigned to the lactone carbonyl carbon. The remaining four signals (δ 137.2, 135.3, 133.0, and 131.2) could be assigned to C-12, C-9, C-4a, and C-8a, with interchangeable assignments.

Signals with negative phase (pointing downward) are due to CH_3 and methine carbons. The peaks at δ 56.2 and 60.7 may be assigned to the methoxy carbons, while the remaining signals are due to the methine carbons. As is apparent, most of the assignments are based on chemical shift values, and there is always an element of doubt in the differentiation of the quaternary carbons from the CH_2 carbons and of the CH carbons from the CH_3 carbons. Polarization transfer experiments (DEPT, etc.) are therefore superior to the GASPE or APT experiments, in both sensitivity and multiplicity assignments.

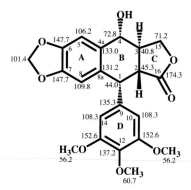

4.31 The GASPE (APT) ^{13}C-NMR spectrum of tschimganin displays signals for all 18 carbons. The peaks having positive amplitude in the GASPE spectrum are due to the quaternary and CH_2 carbons, while signals with negative amplitude represent CH and CH_3 carbons. Unlike the DEPT spectrum, the GASPE spectrum contains signals for the quaternary carbons.

Among the positive peaks, the most downfield signal at δ 166.6 may be assigned to the ester carbonyl carbon. The signals at δ 149.8, and 146.1 are due to the substituted (C-4') and the hydroxyl substituted (C-3') carbons. Another downfield quaternary signal at δ 123.0 may be assigned to the C-1' quaternary carbon. The upfield shift of the signal may be due to the substitution of electron withdrawing ester carbonyl carbon. The upfield quaternary carbon resonances at δ 49.0, 47.8, and 44.9 may be assigned to C-1 and C-4 and the bridge carbon C-7, respectively. The signals resonating at δ 36.9, 28.0, and 27.4 may be assigned to the C-3, C-6, and C-5 methylenes, respectively.

The remaining nine signals having negative phase (pointing downward) are due to methyl and methine carbons. The downfield signal at

δ 56.0 is typical of the methoxy carbon at C-3'. The methyl carbons, resonating at δ 18.9 and 19.7, are due to the two methyl substituents at C-7, while the signal at δ 13.6 may be assigned to the methyl substituted at C-1. The downfield methine carbons at δ 123.9, 113.9, and 111.6 may be assigned to the aromatic methines C-6', C-5', and C-2,' respectively. The signal at δ 80.3 can be assigned to C-2, which resonates downfield due to its direct connectivity with the electronegative ester oxygen atom. The methine carbon at δ 44.9 can be assigned to C-4. The chemical shift values are presented on the structure of tschimganin.

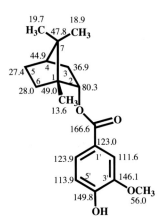

REFERENCES

Avent, A.G., Freeman, R., 1980. NMR spin-lattice relaxation studied by magnetization transfer. J. Magn. Reson. (1969) 39 (1), 169–174.

Baishya, B., Khetrapal, C.L., 2014. Perfect echo INEPT: more efficient heteronuclear polarization transfer by refocusing homonuclear J-coupling interaction. J. Magn. Reson. 242, 143–154.

Bax, A., Drobny, G., 1985. Optimization of two-dimensional homonuclear relayed coherence transfer NMR spectroscopy. J. Magn. Reson. (1969) 61 (2), 306–320.

Bendall, M.R., Pegg, D.T., 1983. EPT with two variable pulse angles, a universal polarization transfer sequence. J. Magn. Reson. (1969) 52 (1), 164–168.

Bendall, M.R., Pegg, D.T., Tyburn, G.M., Brevard, C., 1983. Polarization transfer from and between quadrupolar nuclei by the UPT sequence. J. Magn. Reson. (1969) 55 (2), 322–328.

Bildsøe, H., Donstrup, S., Jakobsen, H.J., Sørensen, O.E., 1983. Subspectral editing using a multiple quantum trap: Analysis of J cross-talk. J. Magn. Reson. 53 (1), 154–162.

Bodenhausen, G., Ruben, D.J., 1980. Natural abundance nitrogen-15 NMR by enhanced heteronuclear spectroscopy. Chem. Phys. Lett. 69 (1), 185–189.

Bolton, P.H., 1980. Enhancement of signals from insensitive nuclei by magnetization transfer *via* J couplings. J. Magn. Reson. (1969) 41 (2), 287–292.

Brooks, W.M., Irving, M.G., Simpson, S.J., Doddrell, D.M., 1984. Application of the inverse DEPT polarization-transfer pulse sequence as a powerful water suppression method to follow metabolic processes in H_2O solution by ^{1}H NMR spectroscopy. J. Magn. Reson. (1969) 56 (3), 521–526.

Brown, D.W., Nakashima, T.T., Rabenstein, D.L., 1981. Simplification and assignment of carbon-13 NMR spectra with spin-echo Fourier transform techniques. J. Magn. Reson. (1969) 45 (2), 302–314.

Bulsing, J.M., Brooks, W.M., Field, J., Doddrell, D.M., 1984. Polarization transfer *via* an intermediate multiple-quantum state of maximum order. J. Magn. Reson. (1969) 56 (1), 167–173.

Burger, R., Bigler, P., 1998. DEPTQ: distorsionless enhancement by polarization transfer including the detection of quaternary nuclei. J. Magn. Reson. 135 (2), 529–534.

Burum, D.P., Ernst, R.R., 1980. Net polarization transfer *via* a J-ordered state for signal enhancement of low-sensitivity nuclei. J. Magn. Reson. (1969) 39 (1), 163–168.

Carr, H.Y., Purcell, E.M., 1954. Effects of diffusion on free precession in nuclear magnetic resonance experiments. Phys. Rev. 94 (3), 630–638.

Chalmers, A.A., Pachler, K.G.R., Wessels, P.L., 1974. Assignment of carbon-13 NMR spectra and sign determination of long range carbon-13 − proton coupling constants with SPI. Org. Magn. Reson. 6 (8), 445–447.

Doddrell, D.M., Pegg, D.T., Bendall, M.R., 1982. Distortionless enhancement of NMR signals by polarization transfer. J. Magn. Reson. 48 (2), 323–327.

Doddrell, D.M., Pegg, D.T., Bendall, M.R., Brooks, W.M., Thomas, D.M., 1980. Enhancement of ^{14}N nuclear magnetic resonance signals in ammonium nitrate using proton polarization transfer. J. Magn. Reson. (1969) 41 (3), 492–495.

Eich, G., Bodenhausen, G., Ernst, R.R., 1982. Exploring nuclear spin systems by relayed magnetization transfer. J. Am. Chem. Soc. 104 (13), 3731–3732.

Granger, P., N.A.T.O., Series, A.S.I., Series, C., 1990. Polarization transfer INEPT, DEPT, NOE, Mathematical and Physical Sciences. Multinucl. Magn. Reson. Liq. Solids: Chem. Appl. 322, 63–79.

Hahn, E.L., 1950. Spin echoes. Phys. Rev. 80 (4), 580–594.

King, G., Wright, P.E., 1983. Application of two-dimensional relayed coherence transfer experiments to ^{1}H NMR studies of macromolecules. J. Magn. Reson. (1969) 54 (2), 328–332.

Madsen, J.C., Bildsøe, H., Jacobsen, H.J., Sørensen, O.W., 1986. ESCORT editing: an update of the APT experiment. J. Magn. Reson. (1969) 67 (2), 243–257.

Meiboom, S., Gill, D., 1958. Modified spin-echo method for measuring nuclear relaxation times. Rev. Sci. Instrum. 29 (8), 688–691.

Mohebbi, A., Gonen, O., 1996. Recovery of heteronuclear coherence-transfer efficiency losses due to ^{1}H–^{1}H J coupling in proton to phosphorus RINEPT. J. Magn. Reson. (A) 123 (2), 237–241.

Morris, G.A., 1980a. Sensitivity enhancement in nitrogen-15 NMR: polarization transfer using the INEPT pulse sequence. J. Am. Chem. Soc. 102 (1), 428–429.

Morris, G.A., 1980b. Indirect measurement of proton relaxation rates by "INEPT" polarization transfer to carbon-13: Proton spin-lattice relaxation in cholesteryl acetate solutions. J. Magn. Reson. (1969) 41 (1), 185–188.

Morris, G.A. In: Harris, R.K., Wasylishen, R.E. (Eds.). INEPT, Encyclopedia of NMR (2012).

Morris, G.A., Freeman, R., 1979. Enhancement of nuclear magnetic resonance signals by polarization transfer. J. Am. Chem. Soc. 101 (3), 760–762.

Neuhaus, D., Wagner, G., Vasak, M., Kaegi, J.H.R., Wüthrich, K., 1985. Systematic application of high-resolution, phase-sensitive two dimensional H-NMR techniques for the identification of the amino acid-proton spin systems in proteins: rabbit metallothioein-2. Eur. J. Biochem. 151 (2), 257–273.

Nolis, P., Parella, T., 2007. Solution-state NMR experiments based on heteronuclear cross-polarization. Curr. Anal. Chem. 3 (1), 47–68.

Pachler, K.G.R., Wessels, P.L., 1973. Selective population inversion (SPI). A pulsed double resonance method in FT NMR spectroscopy equivalent to INDOR. J. Magn. Reson. (1969) 12 (3), 337–339.

Patt, S.L., Shoolery, J.N., 1982. Attached proton test for carbon-13 NMR. J. Magn. Reson. (1969) 46 (3), 535–539.

Pegg, D.T., Bendall, M.R., Doddrell, D.M., 1981a. Heisenberg vector model for precession via heteronuclear scalar coupling. J. Magn. Reson. (1969) 44 (2), 238–249.

Pegg, D.T., Doddrell, D.M., Brooks, W.M., Bendall, M.R., 1981b. Proton polarization transfer enhancement for a nucleus with arbitrary spin quantum number from n scalar coupled protons for arbitrary preparation times. J. Magn. Reson. 44 (1), 32–40.

Radeglia, R., Porzel, A., 1984. Two modified pulse sequences for the ^{13}C NMR spectroscopic characterization of CH_n building groups. An effective method for the determination of n and for the estimation of $^1J(^{13}C\text{-}^1H)$. J. Prakt. Chem. 326 (3), 524–528.

Sørensen, O.W., Dønstrup, S., Bildsøe, H., Jakobsen, H.J., 1983. Suppression of J cross-talk in subspectral editing. The SEMUT GL pulse sequence. J. Magn. Reson. (1969) 55 (2), 347–354.

Sørensen, O.W., Ernst, R.R., 1983. Elimination of spectral distortion in polarization transfer experiments. Improvements and comparison of techniques. J. Magn. Reson. (1969) 51 (3), 477–489.

Sørensen, S., Hansen, R.S., Jakobsen, H.J., 1974. Assignments and relative signs of ^{13}C-X coupling constants in ^{13}C FT NMR from selective population transfer (SPT). J. Magn. Reson. (1969) 14 (2), 243–245.

Sørensen, U.B., Bildsøe, H., Jakobsen, H.J., 1984. Purging sandwiches and purging pulses in polarization transfer experiments. The equivalence of DEPT and refocused INEPT with a purging sandwich. J. Magn. Reson. (1969) 58 (3), 517–525.

Wagner, W., 1983. Two-dimensional relayed coherence transfer spectroscopy of a protein. J. Magn. Reson. (1969) 55 (1), 151–156.

Chapter 5

The Second Dimension

Chapter Outline

5.1 CONCEPT AND GENESIS

The development of two-dimensional (2D) spectroscopy was a major milestone in the history of NMR spectroscopy. It has formed the basis of hundreds of powerful experiments which have found wide applications in various fields of science. The advent of 2D NMR spectroscopy has greatly facilitated the interpretation of NMR spectra and allowed the structures of complex macromolecules to be studied in depth.

In the one-dimensional (1D) NMR experiments discussed earlier, the FID was recorded immediately after the first pulse, and the only time domain involved (t_2) was the one in which the FID was obtained. If, however, the signal is not recorded immediately after the pulse but a certain time interval (t_1) is allowed to elapse before detection, then during this time interval (the *evolution period*) the nuclei can be made to interact with each other in various ways, depending on the pulse sequences applied. Introduction of this second dimension in NMR spectroscopy, triggered by Jeener's original experiment, has resulted in

Solving Problems with NMR Spectroscopy. http://dx.doi.org/10.1016/B978-0-12-411589-7.00005-X

191

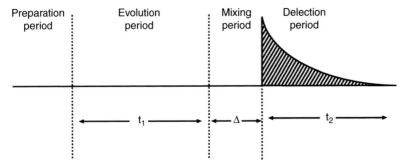

FIGURE 5.1 The various time periods in a 2D NMR experiment. Nuclei are allowed to approach a state of thermal equilibrium during the preparation period before the first pulse is applied. This pulse disturbs the equilibrium polarization state established during the preparation period, and during the subsequent evolution period t_1 the nuclei may be subjected to the influence of other, neighboring spins. If the amplitudes of the nuclei are modulated by the chemical shifts of the nuclei to which they are coupled, 2D shift-correlated spectra are obtained. If, however, their amplitudes are modulated by the coupling frequencies, then 2D J-resolved spectra result. The evolution period may be followed by a mixing period Δ, as in Nuclear Overhauser Effect Spectroscopy (NOESY) or 2D exchange spectra. The mixing period is followed by the second evolution (detection period) t_2.

tremendous advances in NMR spectroscopy and in the development of a multitude of powerful NMR techniques for structure elucidation of complex organic and biomolecules (Aue et al., 1976; Jeener, 1971).

The multidimensional NMR spectroscopic techniques have greatly facilitated the interpretation of NMR data by providing the much needed separation of crowded and overlapped signals into two or more dimensions. The resulting ability to separate the chemical shifts and couplings of nuclei into two (or more) different dimensions allowed ready correlation between them (Mevers et al., 2014; Mohamed et al., 2014; Luo et al., 2014a; Luo et al., 2014b). This has expanded the applications of NMR spectroscopy to fields, such as structural biology, proteomics, and metabolomics. Once the concept of 2D NMR came into general use, higher dimensional NMR techniques (3D, 4D, etc.) were soon developed, opening up the applications of NMR spectroscopy to new fields involving the study of large complex biomolecules.

2D NMR spectroscopy is a spectral method in which the data are collected in two different time domains: acquisition of the FID (t_2), and a successively incremented delay (t_1). The resulting FID (data matrix) is accordingly subjected to two successive sets of Fourier transformations to furnish a 2D NMR spectrum along the two frequency axes. The time sequence of a typical 2D NMR experiment is given in Fig. 5.1. The major difference between one- and 2D NMR methods is, therefore, the insertion of an evolution time, t_1, that is systematically incremented within a sequence of pulse cycles. Many experiments are generally performed with variable t_1, which is incremented by a constant Δt_1. The resulting signals (FIDs) from this experiment depend on the two time variables (t_1, t_2). Two sets of successive Fourier transformations are, therefore, required to convert the signals obtained into the two frequency axes, v_2 and v_1.

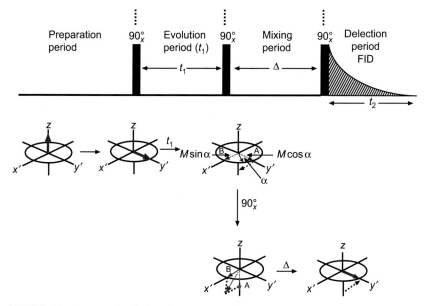

FIGURE 5.2 An example of a 2D NMR pulse experiment. The time axis may be divided into a *preparation period* (during which the nuclei are allowed to reach an equilibrium state) and an *evolution period* t_1 during which the nuclei interact with each other. In some experiments (e.g., NOESY), this may be followed by a mixing period. Finally, the magnetization is detected during the *detection period* t_2. The experiment is repeated with incremented values of t_1, and a second set of Fourier transformations (after transposition of data) gives the 2D plot. The second 90_x° pulse serves to remove the y' component **A** of the main magnetization (having magnitude $M \cos \alpha$) by bending it away toward the $-z$-axis, leaving component **B** along the x'-axis (with magnitude $M \sin \alpha$). After the *mixing period* Δ, the magnetization along the z-axis is returned to the y'-axis by the 90_x° pulse for detection. It is only the signal component along the y'-axis that is detected. The signal intensity, therefore, varies sinusoidally as a function of the angle by which the main vector has diverged from the y'-axis.

A convenient way to understand the modern 2D NMR experiment is in terms of magnetization vectors. Figure 5.2 presents a pulse sequence and the corresponding vector diagram of a 2D NMR experiment of a single-line ^{13}C spectrum (e.g., the deuterium decoupled ^{13}C-NMR spectrum of $CDCl_3$).

The preparation period allows the spin system to relax back to its equilibrium state by T_1 and T_2 relaxation processes. The first 90_x° pulse bends the magnetization of the sample (M_z) so that it becomes aligned with the y'-axis in the $x'y'$-plane. During the subsequent evolution period (t_2), the magnetization vector will rotate in the $x'y'$-plane. By the end of a certain time interval, t_2, the vector would have moved through an angle α, *which is proportional to the evolution time t_1 and the frequency v* (angle $\alpha = 2\pi v t_1$ radians).

The vector may be considered to have two components, component **A**, aligned along the y'-axis with magnitude $M \cos (2\pi v t_1)$, and component **B**, aligned along the x'-axis with magnitude $M \sin (2\pi v t_1)$. The second 90_x° pulse at the end of the evolution period rotates component **A** of the magnetization, aligned with the y'-axis onto the $-z'$-axis, leaving behind component **B**, located along the x'-axis, with magnitude $M \sin(2\pi \upsilon t_1)$ in the $x'y'$-plane. The magnitude of both components **A** and **B** will also be reduced to a certain extent by the inherent relaxation that occurs during the evolution period t_1. Component **B**, which remains in the $x'y'$-plane after the second 90°_x pulse, is detected as transverse magnetization in the correlated spectroscopy (COSY) experiments. If the FID is recorded immediately after the second 90°_x pulse, the detector will "catch" the transverse magnetization at its maximum amplitude. The transverse magnetization would, however, decrease with time as the magnetization vector moves away from the y'-axis in the $x'y'$-plane. It is, therefore, possible to perform the experiment with discrete changes in the evolution time t_1 and to record many FIDs as a function of t_1. Fourier transformation of each of these FIDs will yield the corresponding "normal" 1D NMR spectrum, with the peaks appearing at frequency υ_1, except that the amplitude of the peaks will oscillate sinusoidally according to $\sin(2\pi \upsilon t_1)$ in the various spectra thus obtained.

It is, therefore, apparent that the first series of Fourier transformations of the data (FIDs) with respect to t_2 yields a series of so-called "1D" spectra, in which the amplitude of the peak oscillates (with frequency υ_2) as a function of t_1 (Fig. 5.3):

$$S(t_1, t_2) \xrightarrow{\text{F.T. over } t_2} S(t_1, \upsilon_2)$$

If the resulting "1D" spectra are arranged in rows one behind the other (a process known as *data transposition*), and the peak is viewed along a *column* (i.e., at 90° to the rows), then the sinusoidally oscillating peak gives the appearance of a *pseudo*-FID (or *interferogram*, Fig. 5.4), with the sinusoidal variation of the peak intensity being evident with respect to the vertical axis (i.e., along t_1). The second Fourier transformation of this interferogram with respect to t_1 then generates a 2D NMR spectrum:

$$S(t_1, \upsilon_2) \xrightarrow{\text{F.T. over } t_2} S(\upsilon_1, \upsilon_2)$$

Problem 5.1

The 1D ^{1}H-NMR spectrum is a plot of intensity *versus* frequency; chemical shifts (δ, ppm) are scaled on the x-axis, while the integration/intensity is presented on the y-axis. Then why is this not considered a 2D-NMR spectrum?

In the case of a nuclear Overhauser enhancement spectroscopy (NOESY) experiment, a third *mixing period* is inserted after the second 90°_x pulse, and a third 90°_{-x} pulse is then applied at the end of the mixing period (Fig. 5.2). The final 90°_{-x} pulse does not affect component **B** lying along the x'-axis (proportional to $M \sin \alpha$) so that this component is not converted into detectable magnetization. Component **A** lying along the $-z$-axis is, however, returned to the y'-axis by the third 90°_{-x} pulse for detection. Component **B** along the x'-axis

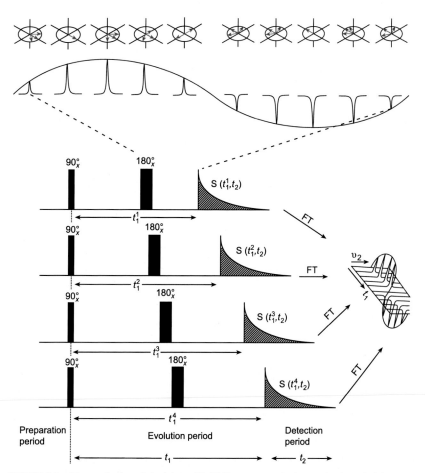

FIGURE 5.3 The mechanics of obtaining a 2D NMR spectrum. As the t_1 value is varied the magnetization vectors are "caught" during detection at their various positions on the $x'y'$-plane. The value of the detection time t_2 is kept constant. The first set of Fourier transformations across t_2 is followed by transposition of the data, which aligns the peak behind one another, and a second set of Fourier transformations across t_1 then affords the 2D plot.

is, therefore, successively removed at various values of t_1, and component **A** is returned to the y'-axis and detected. The change in signal strength (ignoring loss due to relaxation) depends on the magnitude of component B removed, which in turn depends on the sine of angle α [i.e., $M \sin(2\pi v t_1)$]. As different experiments are performed at various values of t_1, the NMR signal is accordingly seen to oscillate sinusoidally as a function of t_1. Data transposition (i.e., arrangement in rows), followed by Fourier transformation of the resulting *columns* of oscillating peaks, will afford the 2D NOESY spectrum.

The first Fourier transformation of the FID yields a complex function of frequency with real (cosine) and imaginary (sine) coefficients. Each FID,

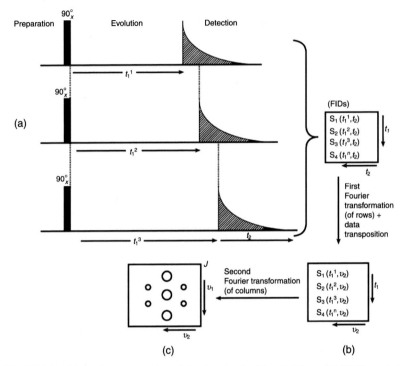

FIGURE 5.4 Schematic representation of the steps involved in obtaining a 2D NMR spectrum. (a) Many FIDs are recoded with incremented values of the evolution time t_1 and stored. (b) Each of the FIDs is subjected to Fourier transformation to give a corresponding number of spectra. The data are transposed in such a manner that the spectra are arranged behind one another so that each peak is seen to undergo a sinusoidal modulation with respect to the changing values of t_1. A second series of Fourier transformations is carried out across these columns of peaks to produce the 2D plot shown in the form of contours, as shown in (c).

therefore, has a real half and an imaginary half, and when subjected to the first Fourier transformation the resulting spectrum will also have real and imaginary data points. When these real and imaginary data points are arranged behind one another, vertical columns result. This transposed data matrix when subject to a second Fourier transformation will furnish a 2D NMR spectrum with both real (absorptive) and imaginary (dispersive) components. The resulting spectrum can be displayed in the absolute-value mode, which contains both these contributions, or in the pure-absorption mode, which contains real contributions only (States et al., 1982; Keeler and Nauhaus, 1985).

It is immaterial which Fourier transformation is carried out first and which second (i.e., whether the function is first transformed with respect to the t_1 or the t_2 variable). Usually, it is more economical to transform individual FIDs (transformation with respect to t_2) and then, after transposition, to transform the resulting interferograms with respect to t_1 to produce the 2D spectra (or, more precisely, into sections of the 2D NMR spectrum).

Problem 5.2

What was so exciting about Jeener's original experiment? What is the major difference between 1D and 2D NMR experiments?

Problem 5.3

Explain how the *second dimension* is generated in NMR spectroscopy?

Problem 5.4

Is it correct to call F_1 as indirect dimension?

There are basically four main types of 2D NMR experiments: *J*-resolved, shift correlation through bonds (e.g., COSY), shift correlations through space (e.g., NOESY) and chemical and conformation exchanges (e.g., EXSY). These spectra may be of homonuclear or heteronuclear type involving interactions between similar nuclei (e.g., protons) or between different nuclear species (e.g., ^1H with ^{13}C). Recently, some interesting "2D NMR" experiments have been developed, "diffusion order spectroscopy" (DOSY) with chemical shifts on one axis and diffusion coefficients on the other axis. The DOSY experiments are extensively used for mixture analysis (Section 9.3.1).

In 2D *J*-resolved experiments, the precession due to chemical shift frequencies is suppressed by refocusing or reversal in the second half of the evolution period. Precession due to couplings, however, remains present in at least one of the two halves of the evolution period. The modulation of the nucleus being observed by the *J*-coupling frequencies of the nuclei with which it is coupled leads to 2D *J*-resolved spectra, which contain chemical shifts of the observed nuclei along one axis (F_2-axis) and their splitting by *J*-coupling along the other axis (F_1-axis).

In shift-correlated spectra (e.g., COSY), the signals of the observed nuclei are modulated by the chemical shift frequencies of the partner nuclei to which they are coupled, and often by *J*-coupling frequencies as well. The result is that after the second Fourier transformation, a 2D plot is obtained in which the unmodulated signals lie along the diagonal axis while the modulated signals appear as cross-peaks lying on either side of the diagonal at the chemical shifts of the coupled nuclei, i.e., centered at the coordinates $F_2 = \delta_A$ and $F_2 = \delta_B$, where A and B are the coupled nuclei. In NOESY spectra, the chemical shifts of nuclei lying close to one another in space are correlated.

Problem 5.5

How is it possible to obtain information about the behavior of a spin system during the evolution time t_1?

Problem 5.6

When there is only one time variable during a 2D experiment, i.e., t_1, why do we need to process the data through *two* Fourier transformation operations?

Problem 5.7

What are the major benefits of 2D NMR spectra?

Problem 5.8

A number of 2D NMR experiments, such as NOESY, have a mixing period incorporated in their pulse sequence. In principle, precession of *xy*-magnetization also occurs during the mixing period. Why do we not need to have a third Fourier transformation to monitor the precession frequencies that occur during the mixing period?

Problem 5.9

How is peak amplitude modulated during the preparation, evolution, mixing, and detection phases of a COSY experiment?

5.2   DATA ACQUISITION IN 2D NMR

A number of parameters have to be chosen when recording 2D NMR spectra: (1) the pulse sequence to be used, which depends on the experiment required to be conducted, (2) the pulse lengths and the delays in the pulse sequence, (3) the spectral widths SW_1 and SW_2 to be used for F_1 and F_2, (4) the number of data points or time increments that define t_1 and t_2, (5) the number of transients for each value of t_1, (6) the relaxation delays between each set of pulses that allows an equilibrium state to be reached, and (7) the number of preparatory dummy transients (DS) per FID required for the establishment of the steady state for each FID. Table 5.1 summarizes some important acquisition parameters for 2D NMR experiments.

5.2.1   Pulse Sequence

The choice of the pulse sequence to use is of fundamental importance. We must decide carefully what information is required, and choose the right experiment to obtain it. Although hundreds of 2D pulse sequences are now available for various experiments, only some have proven themselves to be of general utility. Only such proven techniques should be chosen to solve structural problems.

5.2.2   Setting Pulse Lengths and Delays

Accurate calibration of pulse lengths is essential for the success of most 2D NMR experiments. Wide variations ($>20\%$) in the setting of pulse lengths may significantly reduce sensitivity and lead to the appearance of artifact signals. In some experiments, such as inverse NMR experiments, accurately set pulse lengths are even more critical for successful outcomes.

TABLE 5.1 Some Important 2D Acquisition Parameters*

F_1 Domain	F_2 Domain
$N_1 = TD_1$ = number of FIDs, t_1 domain point pairs	$N_2 = TD_2$ = number of total t_2 domain points (quardrature channels A and B)
SI_1 = total number of data points (size) for F_1 transform; normally this = $2TD_1$	SI_2 = total number of data points (size) for F_2 transform; normally, this = TD_2
$\pm SW_1$ = frequency domain, in Hz	SW_2 = frequency domain, in Hz (quadrature)
Hz/PT_1= $2SW_1/SI_1$ or $4SW_1/SI_1$ (phase sensitive)	Hz/PT_2 = $2SW_2/SI_2$
$DW_1 = 0.5/SW_1 = t_1$ increment for a single evolution period	$DW_2 = 0.5/SW_2$ (sequential sampling) or $1/SW_2$ (simultaneous sampling)
$DW_{1'} = DW_{1/2} = t_1$ increment for each half of a split evolution (echo) or for the TPPI method in general	t_2 = increment (dwell time) for quadrature
$t_1 = t_1^\circ + n(DW_1)$; n = 0, 1, 2,. . ., $N_1 - 1$ (evolution period)	$t_2 = t_2^\circ + n(DW_2)$; n = 0, 1, 2,. . .,$N_2 - 1$ (detection period)
$t_1' = t_1 / 2$	$AQ = t_2^{max}$ (acquisition time)
D_1 = relaxation delay (preparation period)	
t_m, t_d, t_c, etc. = fixed mixing periods NS = number of transients per FID DS = number of preparatory dummy transients per FID	

Based on the nomenclature and conventions of Bruker software.

Problem 5.10

Why is accurate calibration of pulse widths and delays essential for the success of a 2D NMR experiment?

5.2.3   Spectral Widths (SW_1 and SW_2)

The spectral width SW_2 relates to the frequency domain. With the variation of the evolution period t_1, the intensity and phase of the signals undergo changes that are determined by the behavior of the nuclear spins during the evolution and mixing periods. During the detection period t_2, FIDs are accumulated, each at a discrete value of t_1, and Fourier transformation with respect to t_2 produces a set of spectra each with the frequency domain F_2. Arranging these spectra in *rows*, corresponding to rows of the data (matrix), and Fourier transformation at different points across *columns* of this data matrix creates the second frequency domain F_2. In quadrature detection, the F_2 domain covers the frequency range $SW_2/2$.

Bruker instruments use quadrature detection, with channels **A** and **B** being sampled alternately so that the dwell time is given by:

$$DW_2 = \frac{1}{SW_2} \text{ and } t_2 = t_2^0 + n(DW_2)$$

where $n = 0, 1, 2,. . ., N_2 - 1$, t_2^0 is the initial value of t_2, and N_2 is the total number of A and B data points.

In Varian instruments, simultaneous sampling occurs – i.e., points **A** and **B** are taken together as pairs – so that the dwell time is given by

$$DW_2 = \frac{1}{SW_2} \text{ and } t_2 = t_2^0 + n(DW_2)$$

where $n = 0, 1, 2,. . ., (N_2/2) - 1$.

In quadrature detection, the transmitter offset frequency is positioned at the center of the F_2 domain (i.e., at $F_2 = 0$; in single-channel detection it is positioned at the left edge). Frequencies to the left (or downfield) of the transmitter offset frequency are positive; those to the right (or upfield) of it are negative (Turner and Hill, 1986).

The spectral width SW_1 associated with the F_1 frequency domain may be defined as $F_1 = SW_1$. The time increment for the t_1 domain, which is the effective dwell time, DW_1 for this period is related to SW_1 as follows: $DW_1 = (\frac{1}{2}) SW_1$. The time increments during t_1 are kept equal. In successive FIDs, the time t_1 is incremented systematically: $t_1^0 = t_1 + n(DW_1)$, where $n = 0, 1, 2, 3,. . . ., N_1 - 1$, N_1 is the number of FIDs, t_1^0 is the value of t_1 at the beginning of the experiment, and DW_1 is the time increment for the t_2 domain.

In homonuclear-shift-correlated experiments, the F_1 domain corresponds to the nucleus under observation; in heteronuclear-shift-correlated experiments, F_1 relates to the "unobserved" or decoupled nucleus. It is, therefore, necessary to set spectral width SW_1 after considering the 1D spectrum of the nucleus corresponding to the F_1 domain. In 2D J-resolved spectra, the value of SW_1 depends on the magnitude of the coupling constants and the type of experiment. In both homonuclear and heteronuclear experiments, the size of the largest multiplet structure, in hertz, determines SW_1, which in turn is related to the homonuclear or heteronuclear coupling constants. In homonuclear 2D spectra, the transmitter offset frequency is kept at the center of F_1 (i.e., at $F_1 = 0$) and F_2 domains. In heteronuclear-shift-correlated spectra, the decoupler offset frequency is kept at the center ($F_1 = 0$) of the F_1 domain, with the F_1 domain corresponding to the "invisible" or decoupled nucleus.

Problem 5.11

Why are we much more likely to have signals outside the spectral width (SW) in a 2D NMR experiment than in a 1D NMR experiment? Why do spectral widths in a 2D NMR need to be defined very carefully, and what effects will this have on the spectrum?

5.2.4 Number of Data Points in t_1 and t_2

The effective resolution is determined by the number of data points in each domain, which in turn determines the length of t_1 and t_2. Thus, though digital resolution can be improved by zero-filling (Bartholdi and Ernst, 1973), the basic resolution, which determines the separation of close-lying multiplets and line widths of individual signals, will not be altered by zero-filling.

The effective resolution R for N time-domain data points is given by the reciprocal of the acquisition time, AQ: $R = 1/ AQ = 2SW/N$. Since there must be at least two data points to define each signal, the digital resolution DR (in hertz per point, or, in other words, the spacing between adjoining data points) must be DR $= R/2 = 2SW/2N$. The real part of the spectrum is, therefore, obtained by the transform of $2N$ data points, of which N are time-domain points and another N are zeros. This level of zero-filling allows the resolution contained in the time domain to be recorded. Additional zero-filling will lead only to the appearance of a smoother spectrum, but will not cause any genuine improvement in resolution.

The number of data points to be chosen in the F_1 and F_2 domains is dictated not only by the desired resolution but by other external considerations, such as the available storage space in the computer and the time that can be allocated for data acquisition, transformation, and other instrument operations. Clearly, to avoid any unnecessary waste of time, we should choose the *minimum* resolution that would yield the desired information. Thus, if the peaks are separated by at least 1 Hz, then the desired digital resolution should be $R/2 = \frac{1}{2}$ Hz to allow for signal separation in the F_2 domain. The resolution considerations in the F_2 and F_1 domains may be different, depending on what information is required from each domain in the experiment. Since the acquisition time AQ is small as compared to the total length of the pulse sequence with its various relaxation delays, it is preferable to choose N_2 (the number of data points in F_2) based on the final digital resolution required *without zero-filling*. The sensitivity and resolution in the F_2 domain will be improved as long as t_2 is equal to or less than the effective relaxation time T_2.

In the case of the t_1 domain, since it is only the number N of data points that determines the resolution, and not the time involved in the pulse sequence with various delays, it is advisable to acquire only half the theoretical number of FIDs and to obtain the required digital resolution by zero-filling. Thus, the resolution R_1 in the F_1 domain will be given by $R_1 = 2SW_1/ N_1$; that in the F_2 domain is given by $R_2 = 1/AQ = 2SW_2/N_2$.

Problem 5.12

What factors dictate the choice of the number of data points (SI_1 and SI_2) in 2D NMR spectroscopy?

5.2.5 Number of Transients

The number of transients (NS) required varies depending on the experiment to be performed. For most experiments, four transients are required to provide quadrature detection and suppress the axial peaks at $F_1 = 0$. Additional phase cycles may be necessary involving 16–32 transients if artifacts arising from imperfect 180° pulses are to be suppressed or if quadrature images appear in the F_2 dimension. The number of transients will, therefore, be the number of cycles in the phase cycling routine multiplied by the minimum number N_1 of FIDs to be acquired. The signal-to-noise ratio in the 2D plot will be defined by the number of transients (Levitt et al., 1984), provided that the time domains are not significantly greater than the relaxation times T_2 or T_2^*. Normally, only the minimum number of transients required should be used, unless the sample quantities are very small or the experiment is insensitive. Table 5.2 summarizes some of the data requirements for the common NMR experiments.

> **Problem 5.13**
> What factors govern the minimum number of transients (NS)?

5.2.6 Dummy Scans and Relaxation Delays between Successive Pulses

To reach a steady state before data acquisition, a certain number of "dummy" scans are usually required. If the relaxation delay between the repetition time of the pulse sequence is sufficiently long ($5T_1$ of the slowest-relaxing nuclei), then dummy scans may not be necessary. In practice, it is usual to have the relaxation delay set at about $2T_1$ or $3T_1$ and to use two dummy scans if the number of transients (NS) is 4 or 8, or use four dummy scans if NS is greater than 16.

> **Problem 5.14**
> What are "dummy" scans, and why are a number of these scans acquired before actual data acquisition?

5.3 DATA PROCESSING IN 2D NMR

At the end of the 2D experiment, we will have acquired a set of N_1 FIDs composed of N_2 quadrature data points, with $N_2/2$ points from channel A and $N_2/2$ points from channel B, acquired with sequential (alternate) sampling. How the data are processed is critical for a successful outcome. The data processing involves (1) dc (direct current) correction (performed automatically by the instrument software), (2) apodization (window multiplication) of the t_2 time-domain data, (3) F_2 Fourier transformation and phase correction, (4) window multiplication of the t_1 domain data and phase correction (unless it is a magnitude or a power-mode spectrum, in which case phase correction is not required), (5) complex Fourier transformation in F_1, (6) co-addition of real and imaginary data (if phase-sensitive representation is required) to give a magnitude (M) or a power-mode (P) spectrum. Additional steps may be tilting, symmetrization, and calculation of projections.

TABLE 5.2 Summary of Data Requirements for 2D NMR Experiments

Experiments	^{1}H freq. (MHz)	$TD_2 = SI_2$ (words)	Hz/PT_2	$TD_1 = SI_1/2$ (words)	Hz/PT_1	NS	DS	D1 (s)	Time (h)	FIDs
^{1}H J-resolved (4.5 ppm, ±18 Hz)	400	8K	0.439	32	1.125	32	8	1.5	1H, 11 M	32
^{1}H COSY, NOESY (4.5 ppm)	400	2K	1.757	256	1.757	16	8	1.5	3H	256
						32	8	2	5H, 48 M	256
^{1}H COSY-TPPI (4.5 ppm)	400	2K	1.757	256	1.757	16	8	1.5		256
^{13}C-^{1}H J-res. (gated dec.) (142 ppm, ±120 MHz)	400	8K	3.02	256	0.93	32	4	1.5	4H	256
		32 K	0.756	256	0.93	32	4	1.5	6H, 15 M	
^{13}C-^{1}H correlated (94 × 4.5 ppm)	400	1K	18.35	256	7.03	32	2	1.5	4H, 6M	256
^{13}C INADEQUATE (142 ppm)	400	32 K	0.866	256	55.46	128	16	5	2D, 2H	256

TD_2, t_2 time domain data points; TD_1, t_1 time domain pairs of data points; SI_2 and SI_1, total data points; Hz/PT_2 and Hz/PT_1, digital resolution in F_2 and F_1 (real) domains, respectively; NS, number of acquired transients; DS, number of dummy transients; D1, recycle time.

The first set of Fourier transformations in 2D NMR experiments, such as COSY, produces signals in the F_2 dimension. These signals have real (R) and imaginary (I) components. The second set of Fourier transformation across t_1 gives signals in the F_1 domain, which also have real (R) and imaginary (I) components, leading to four different quadrants arising from the four possible combinations of the real and imaginary components (RR, RI, IR, and II). The pure absorption -mode signals correspond to the real (RR) quadrant (Fig. 5.5). Each of these data-processing procedures will be considered next, briefly.

5.3.1 DC Correction

Since there is some contribution of the receiver dc to the signal, this needs to be removed. In 1D experiments, the FIDs decay substantially, and the last portion of the FID gives a reasonably good estimate of the dc level. The computer works out the average dc level automatically, and subtracts it from each of the two quadrature data sets.

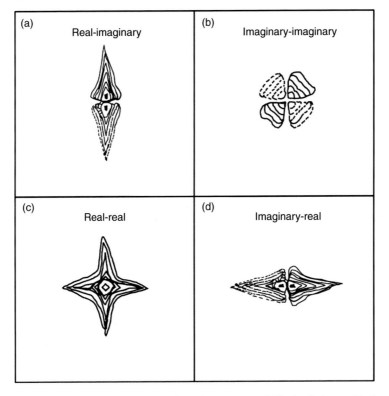

FIGURE 5.5 The first set of Fourier transformations across t_2 yields signals in v_2, with absorption and dispersion components corresponding to real and imaginary parts. The second FT across t_1 yields signals in v_1, with absorption (i.e., real) and dispersion (i.e., imaginary) components; quadrants (a), (b), (c), and (d) represent four different combinations of real and imaginary components and four different line shapes. These line shapes are normally visible in phase-sensitive 2D plots.

In 2D NMR experiments, the FIDs are relatively short and with fewer data points, so dc correction is more difficult to carry out accurately. Phase cycling procedures are recommended whenever required to remove dc offsets before dc correction, which is carried out before the first (F_2) Fourier transform. Since the data points in the F_1 transform arise from frequency-domain spectra, no dc correction is normally required (we expect to see unchanging dc components at $F_1 = 0$).

Problem 5.15

What is dc correction?

Problem 5.16

Why is it not necessary to perform dc correction before second (F_1) Fourier transformation?

5.3.2 Peak Shapes in 2D NMR Spectra

The apodization functions mentioned earlier (Section 2.1.12) have been applied extensively in 1D NMR spectra, and many of them have also proved useful in 2D NMR spectra. Before discussing the apodization functions as employed in 2D NMR spectra, we shall consider the kind of peak shapes we are dealing with.

The free induction decay may be considered as a complex function of frequency, having real and imaginary coefficients. There are two coefficients in each point of the frequency spectrum, and these real and imaginary coefficients may be described as the cosine and sine terms, respectively. Each set of coefficients occupies half the memory in the time-domain spectrum. Thus, if the memory size is N, then there will be $N/2$ real and $N/2$ imaginary data points in the spectrum. Through the process of phasing, we can display signals in the absorption mode. Such *pure* 2D *absorption peaks* are characterized by the following property: if we take sections across them parallel to the F_1 or F_2 axes, they produce pure 1D absorption Lorentzian lines (Fig. 5.6a and b). However, they lack cylindrical or elliptical symmetry, and display protruding ridges running down the lines parallel to F_1 and F_2, so they give star-shaped cross-sections (Fig. 5.6c). This distortion can be suppressed by Lorentz–Gaussian transformation.

The *pure negative 2D dispersion peak shapes* are shown in Fig. 5.7. The peaks have negative and positive lines, with vanishing signal contributions as they pass through the baseline, and they are broadened at the base, so they decay slowly, leading to a poor signal-to-noise ratio. Horizontal cross-sections of such peaks give the appearance of butterfly wings. To improve the signal-to-noise ratio as well as the appearance of 2D spectra, it is desirable to remove the dispersive components.

The *phase-twisted peak shapes* (or mixed absorption–dispersion peak shape) are shown in Fig. 5.8. Such peak shapes arise by the overlapping of the absorptive and dispersive contributions in the peak. The center of the peak contains

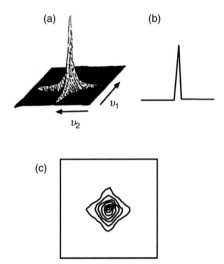

FIGURE 5.6 Different views of pure 2D absorption peak shapes: (a) a 3D view, (b) a vertical cross-section, i.e., a vertical 2D view, and (c) a horizontal 2D view.

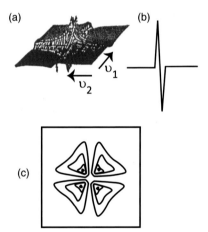

FIGURE 5.7 Different views of pure negative 2D dispersion peak shapes: (a) a 3D view, (b) a vertical cross-section, and (c) a horizontal cross-section.

mainly the absorptive component, while as we move away from the center there is an increasing dispersive component. Such mixed phases in peaks reduce the signal-to-noise ratio; complicated interference effects can arise when such lines lie close to one another. Overlap between positive regions of two different peaks can mutually reinforce the lines (constructive interference), while overlap between positive and negative lobes can mutually cancel the signals in the region of overlap (destructive interference).

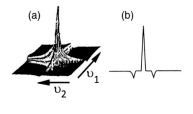

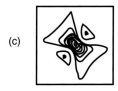

FIGURE 5.8 Different views of an absorption–dispersion peak: (a) a 3D view, (b) a 2D vertical cross-section, and (c) a 2D horizontal cross-section.

2D spectra are often recorded in the absolute-value mode. The absolute value A is square root of the sum of the squares of the real (R) and imaginary (I) coefficients:

$$A = \sqrt{R^2 + I^2} \tag{5.1}$$

Although this eliminates negative contributions, since the imaginary part of the spectrum is also incorporated in the absolute-value mode, it produces broad dispersive components. This leads to the broadening of the base of the peaks ("tailing"), so that the lines recorded in the absolute-value mode are usually broader and show more tailing than those recorded in the pure absorption mode.

Problem 5.17

What are the common peak shapes, and why is it necessary to know the peak shapes before applying apodization?

5.3.2.1 Apodization in 2D NMR Spectra

Some of the apodization functions described in Section 2.1.12 can also be adopted for use in 2D spectra. The types of functions used will vary according to the experiment and the information wanted.

The majority of 2D experiments produce lines with phase-twisted line shapes that cannot be satisfactorily phased. An absolute-value display is the most convenient, since phasing is not required, the lines having positive components only. However, the lines have long dispersive tails, so we need to apply strong weighting functions to suppress such tailing, which would otherwise cause undesirable interference effects between neighboring signals. A price has to be paid for such cosmetic improvement of spectra in terms of sensitivity, since multiplication by weighting functions leads to rejection of some of the signals present in the early part of the FID.

The weighting functions used to improve line shapes for such absolute-value-mode spectra are *sine-bell, sine-bell squared, phase-shifted sine-bell, phase-shifted sine-bell squared*, and a *Lorentz–Gauss transformation* function. The effects of various window functions on COSY data (absolute-value mode) are presented in Fig. 5.9. One advantage of multiplying the time domain $S(t_1)$ or $S(t_2)$ by such functions is to enhance the intensities of the cross-peaks relative to the noncorrelation peaks lying on the diagonal.

Another resolution-enhancement procedure used is *convolution difference* (Campbell et al., 1973). This suppresses the ridges from the cross-peaks and weakens the peaks on the diagonal. Alternatively, we can use a shaping function that involves production of *pseudoechoes*. This makes the envelope of the time-domain signal symmetrical about its midpoint, so that the dispersion-mode contributions in both halves are equal and opposite in sign (Bax et al., 1981, 1979). Fourier transformation of the pseudoecho produces signals in the pure-absorption mode. A Gaussian shaping function may be employed with the pseudoecho to give it a symmetrical envelope.

The sine-bell functions are attractive because, having only one adjustable parameter, they are simple to use. Moreover, they go to zero at the end of the time domain, which is important when zero-filling to avoid artifacts. Generally, the sine-bell squared and the pseudoecho window functions are the most suitable for eliminating dispersive tails in COSY spectra.

With COSY and 2D *J*-resolved spectra, it is normally necessary to apply weighting functions in both dimensions. Multiplication with a sine-bell squared function is recommended.

Phase-sensitive spectra, such as phase-sensitive COSY, do not show phase-twisted line shapes that arise due to the mixing of the absorptive and dispersive components. The peaks lying on or near the diagonal tend to differ in phase by 90° from the correlation cross-peaks, which appear on either side of the diagonal. This means that if we adjust the phasing of the cross-peaks to the absorption mode, then the diagonal peaks will appear in the dispersive mode, and *vice versa*. Since it is the cross-peaks that are of interest, we normally adjust the phasing so that they appear in the absorptive mode. Improvements in peak shape can be achieved by multiplying with a cosine-bell (sine-bell shifted by 90°) function.

Heteronuclear-shift-correlation spectra, which are usually presented in the absolute-value mode, normally contain long dispersive tails that are suppressed by applying a Gaussian or sine-bell function in the F_1 domain. In the F_2 dimension, the choice of a weighting function is less critical. If a better signal-to-noise ratio is wanted, then an exponential broadening multiplication may be employed. If better resolution is needed, then a resolution-enhancing function can be used.

Problem 5.18

What are the effects of various window functions on the shapes of peaks in a COSY data set?

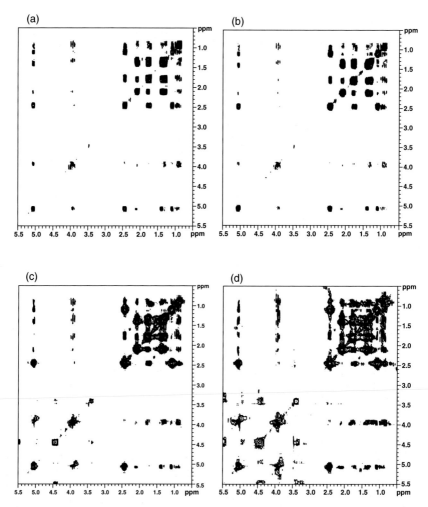

FIGURE 5.9 Effect of different window functions (apodization functions) on the appearance of COSY plot (magnitude mode). The spectra are recorded on 500 MHz in CDCl₃ at 298 K of (1R,2S,4R)-1,7,7-trimethylbicyclo[2.2.1]heptan-2-yl 4-hydroxy-3-methoxybenzoate. (a) Sine-bell squared and (b) sine-bell. (c) Shifted sine-bell squared with 2. (d) Shifted sine-bell squared with 4. (a) and (b) are virtually identical in the case of delayed COSY, whereas sine-bell squared multiplication gives noticeably better suppression of the stronger dispersion-mode components observed when no delay is used. A difference in the effective resolution in the two axes is apparent, with F_2 having better resolution than F_1. The spectrum in (c) has a significant amount of dispersion-mode line shape.

5.3.2.2 F_2 Fourier Transformation

The next step after apodization of the t_2 time-domain the data is subjected to F_2 Fourier transformation and phase correction. As a result of the Fourier transformations of the t_2 time domain, a number of different spectra are generated. Each spectrum corresponds to the behavior of the nuclear spins during the corresponding evolution period, with one spectrum resulting from each t_1 value. A set of spectra is thus obtained, with the rows of the matrix now containing N real and N imaginary data points. These real and imaginary parts are rearranged so that the individual spectra are in rows one behind the other in a matrix constituted by the F_2 and t_1 domains.

5.3.2.3 *F_1 Fourier Transformation and Window Multiplication*

The matrix obtained after the F_2 Fourier transformation and rearrangement of the data set contains a number of spectra. If we look down the columns of these spectra parallel to t_1, we can see the variation of signal intensities with different evolution periods. Subdivision of the data matrix parallel to t_1 gives columns of data containing both the real and the imaginary parts of each spectrum. An equal number of zeros is now added and the data sets subjected to Fourier transformation along t_1. This Fourier transformation may be either a Redfield transform, if the t_2 data are acquired alternately (as on the Bruker instruments), or a complex Fourier transform, if the t_2 data are collected as simultaneous A and B quadrature pairs (as on the Varian instruments). Window multiplication for t_1 may be with the same function as that employed for t_2 (e.g. in COSY), or it may be with a different function (e.g. in 2D *J*-resolved or heteronuclear-shift correlation experiments).

Problem 5.19

The 2D data set S(t_1, t_2) requires two Fourier transformation operations. Explain why the time variable t_2 is almost always Fourier transformed before t_1.

5.3.2.4 *Magnitude-Mode Spectra*

If phase-sensitive spectra are not required, then magnitude-mode $P(\omega)$ (or "absolutes-mode") spectra may be recorded by combining the real and imaginary data points. These produce only positive signals and do not require phase correction. Since this procedure gives the best signal-to-noise ratio, it has found wide use. In heteronuclear experiments, in which the dynamic range tends to be low, the power-mode spectrum may be preferred, since the *S/N* ratio is squared and a better line shape is obtained so that wider window functions can be applied.

Problem 5.20

How are magnitude-mode spectra computed from the FID?

5.3.2.5 *Tilting, Symmetrization, and Projections*

Finally, certain other procedures — such as tilting, symmetrization, or plotting of projections — may be required, depending on the type of spectra being recorded. In homonuclear 2D *J*-resolved experiments, *J*-coupling information is present in both t_1 and t_2 domains, so multiplets appear at an angle of 45°. A tilt-correction procedure (Baumann et al., 1981a) is, therefore, applied mathematically to produce orthogonal J and δ axes, making the multiplets more readable. However, this tilt correction leads to a 45° tilt in the F_1 line shape. This can be suppressed by using a window function for F_1, such as multiplication by sine-bell squared, to reduce the tailing of signals in F_1. Symmetrization procedures can also be employed for cosmetic improvement of the spectrum when the peaks occurring in a spectrum are symmetrically arranged· such as the multiplets in a 2D J-resolved spectrum, or when the cross-peaks lie symmetrically on either sides of the diagonal in COSY and NOESY spectra. Symmetrization can be carried out by *triangular multiplication* (Baumann et al., 1981b) in which each pair (a, b) of symmetry-related points is replaced by the geometric mean $\sqrt{ab}$. A simpler symmetrization procedure is to replace each pair of symmetry-related points by the smaller of the two values a and b.

We can take slices at various points of the 2D spectrum, along either the F_1 or the F_2 axes or, alternatively, we can record projections. Such subspectra can provide useful information. There are two ways such projection can be produced. The first method involves co-addition through a row or column to create a point in the projection spectrum. In this so-called "sum" mode, weak multiplets in the 2D spectrum (e.g., the 2D J-resolved spectrum) will appear much stronger than the noise in the projection spectrum. One disadvantage is a broadening of signals due to the addition of "tails" from neighboring peaks. The other "maximum point" method relies on the selection of the maximum data point from each column or row. This improves line shapes but produces weak signals, and some multiplets may be too weak to be recognizable in the projection spectrum. *In homonuclear experiments, it is therefore advisable to record F_2 projections rather than F_1 projections because of the resulting higher resolution and sensitivity.*

Problem 5.21

What could be the possible reasons for noise in 2D NMR spectra, and how can symmetrization be used to improve the quality and readability of the plot?

Problem 5.22

What are different projection spectra, and how are they different from normal 1D *NMR spectra?*

5.4 PLOTTING OF 2D SPECTRA

2D NMR spectra are normally presented as contour plots (Fig. 5.10a), in which the peaks appear as contours. Although the peaks can be readily visualized by such an "overhead" view, the relative intensities of the signals and the structures of the multiplets are less readily perceived. Such information can be easily obtained by plotting slices (cross-sections) across rows or columns at different points along the F_1 or F_2 axes. Stacked plots (Fig. 5.10c) are pleasing esthetically, since they provide a pseudo-3D representation of the spectrum (Fig. 5.10b). However, except for providing information about noise and artifacts, they offer no advantage over contour plots. Finally, the projection spectra mentioned in the previous section may also be recorded.

The recording of contour plots involves selecting the optimum contour level. At too low a contour level, noise signals may make the genuine cross-peaks difficult to recognize; if the contour level is too high, then some of the weaker signals may be eliminated along with the noise (Fig. 5.11).

In the discussions that follow, the theory behind the common 2D NMR experiments is presented briefly, the main emphasis being on how newcomers can solve practical problems utilizing each type of experiment.

Problem 5.23

How many types of plots are generally used in 2D NMR spectroscopy?

SOLUTIONS TO PROBLEMS

5.1   In NMR spectroscopy, only the frequencies (relating to chemical shifts and coupling constants (Hz) are counted as dimensions, and not the intensity or the integration. A 2D NMR spectrum contains two frequency axes. The intensities are present on the third axis and these are, therefore, usually displayed as contour plots.

5.2   Jeener's idea was to introduce an incremented time t_1 into the basic 1D NMR pulse sequence and to record a series of experiments at different values of t_1, thereby adding a second dimension to NMR spectroscopy. Jeener described a novel experiment in which a coupled spin system is excited by a sequence of *two* pulses separated by a variable time interval t_1. During these variable intervals, the spin system is allowed to evolve to different extents. This variable time t_1 is, therefore, termed the *evolution time*. The insertion of a variable time period between two pulses represents the prime feature that distinguishes 2D NMR experiments from 1D NMR experiments.

5.3   In principle, all 2D experiments are designed according to the same principle: they consist of a series of 1D experiments in which a single delay (evolution time t_1) is altered in length before further manipulation. During the evolution and detection time periods, called t_1 and t_2, the chemical shift and scalar couplings evolve. Therefore, signal intensities and phases vary in these time periods. The key difference between one- and 2D spectra is the introduction of the evolution time t_1, which is systematically incremented during which the amplitude of the NMR signal varies as well. The FIDs are then recorded for each time period and Fourier transformed. This gives a series of 1D signals that vary in their intensities from short to long t_1 values. The modulation of intensities in these 1D spectra gives birth to the *second dimension* after processing.

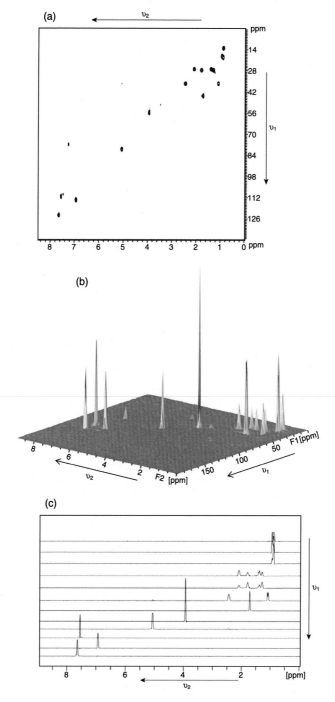

FIGURE 5.10 (a) Contour plot, (b) *pseudo* 3D presentation, and (c) stacked plot.

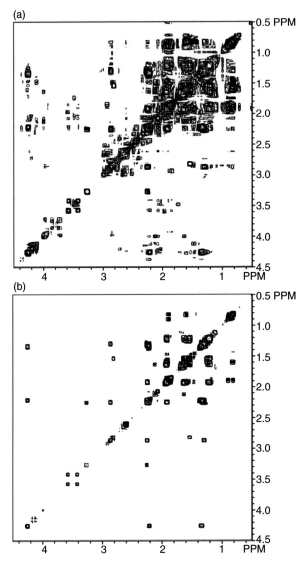

FIGURE 5.11 2D NMR plots recorded at different contour levels. (a) 2D spectra recorded at low contour level usually have noise lines across the plot. (b) With a higher contour level, many of the noise peaks are eliminated and the peaks become clearer.

5.4 Frequencies are derived from the amplitude modulation of the signals indirectly; the first Fourier transformation of the FIDs produces F_2, which is called the *direct dimension*. The F_1 is generated through the second Fourier transformation of the *pseudo* FID (interferogram) and it is, therefore, called the *indirect dimension.*

5.5 Information about the behavior of the spin system during the evolution time can be obtained indirectly by observing its influence on a set of FIDs. Since the FIDs obtained portray the positions and status of the magnetization vectors, we can map the behavior of the nuclei during the evolution time from the FIDs.

5.6 There are actually two independent time periods involved, t_1 and t_2. The time period t_1 after application of the first pulse is incremented systematically, and separate FIDs are obtained at each value of t_1. The second time period, t_2, represents the detection period and it is kept constant. The first set of Fourier transformations (of rows) yields frequency-domain spectra, as in the 1D experiment. When these frequency-domain spectra are stacked together (data transposition), a new data matrix, or "*pseudo-FID*," is obtained, $S(t_1, F_2)$, in which absorption-mode signals are modulated in amplitude as a function of t_1. It is, therefore, necessary to carry out a second set of Fourier transformations to convert this "pseudo-FID" to frequency domain spectra. The second set of Fourier transformations (across columns) on $S(t_1, F_2)$ produces a 2D spectrum $S(F_1, F_2)$. This represents a general procedure for obtaining 2D spectra.

5.7 Some important benefits of 2D NMR spectroscopy are:
1. *J*-Modulated 2D NMR experiments (2D *J*-resolved) provide an excellent method to resolve highly overlapping resonances into readily interpretable and recognizable multiplets.
2. Homonuclear-shift-correlation 2D NMR techniques (COSY, NOESY, etc.) offer a way of identifying spin couplings of nuclei. By interpretation of homonuclear 2D correlated spectra, we can obtain information about the spin networks. Similarly, heteronuclear-shift-correlation 2D NMR techniques help to identify one-bond or long-range heteronuclear connectivities.
3. 2D NMR techniques, such as homonuclear Hartmann–Hahn spectroscopy (HOHAHA) and heteronulear multiple bond connectivity (HMBC), are excellent ways to identify nuclei separated by several bonds, thereby allowing chemists to build complex structures from substructural fragments.
4. 2D INADEQUATE spectra can allow the direct deduction of the carbon framework of the molecules.

5.8 Precession of xy-magnetization may also occur during the mixing period. This is normally a constant effect that does not vary from one FID to the next, since the mixing period is a constant time period and changes taking place during it contribute equally to every FID.

5.9 The first 90° pulse rotates the magnetization to the transverse plane. During the incremented evolution period (t_1), the magnetization rotates ("evolve") in the transverse plane as a function of the chemical shift frequency. This rotation is time dependent (we see the magnetization positioned at various places depending on the evolution time). During the mixing period, the second 90° pulse transfers the magnetization from one spin to another scalarly coupled spin. The non-transferred component of the magnetization eventually appears as a diagonal peak. During the detection period, a series of FIDs are recorded, each differing from the other because of the incremented/variable t_1. Amplitude modulation during the 2D COSY experiment pulse sequence is shown as follows:

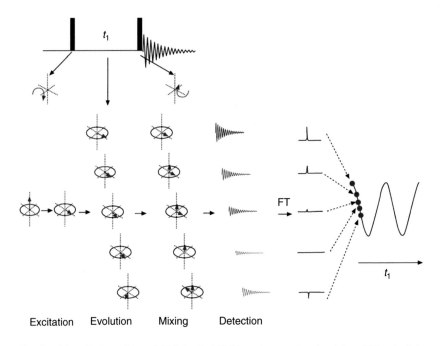

Excitation   Evolution   Mixing   Detection

(Reprinted from Zerbe and Jurt, (2014) Applied NMR spectroscopy for chemists and life scientists, pp. 355, with permission from Wiley-VCH Verlag GmbH & Co. KGaA, Germany).

5.10 Although accurate calibration of pulse lengths and delays is desirable for every type of NMR experiment, it becomes essential in 2D NMR experiments. Errors in pulse length greater than 10–20% can significantly lower the *S/N* ratio and increase the amplitude of artifact peaks. When such errors approach ±40–50%, the spectra become severely distorted. This is easily understood, since a 180° pulse with 50% error becomes a 90° pulse and the pulse sequence is transformed into a totally different pulse sequence. Similarly, serious errors in mixing delays, expressed in terms of scalar couplings (i.e., $1/nJ$) are generally catastrophic for the experiment, and they may not only lower the *S/N* ratio, but may cause complete loss of certain signals. This is illustrated by the following example.

1. COSY 45° spectrum with accurate pulse lengths

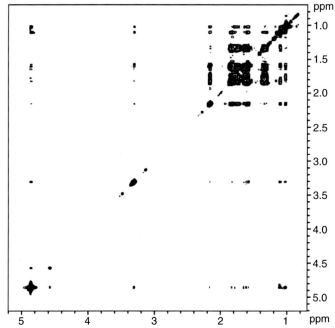

2. COSY 45° with poor pulse lengths (45% error) affords cross-peaks with weaker intensities. Some cross-peaks were found missing.

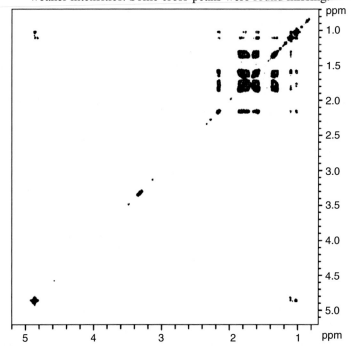

5.11 In a 2D NMR experiment it is essential to minimize the size of the data set by cutting down the spectral width so that valuable instrumentation time is not wasted in accumulating noise. However, if the spectral width chosen is too small, then the signals falling outside the spectral width give rise to erroneously positioned peaks, either on the diagonal line or on one of the axes of the 2D spectra. The spectral widths (SW_1 and SW_2) in 2D NMR experiments, therefore, need to be defined carefully, since they depend not only on the type of the nuclei being detected but also on the kind of experiment. The spectral width SW_2 (for the F_2 domain) is generally defined by the nature of the nucleus being observed during the detection period and the spectral regions to be acquired, while the spectral width SW_1 (for the F_1 domain) depends mostly on the type of experiment. For instance, SW_1 in homonuclear-shift-correlation experiments depends on the observed nucleus, whereas in heteronuclear-shift-correlation experiments it corresponds to the "unobserved" or decoupled nucleus. For J-resolved 2D NMR experiments, SW_1 corresponds to the coupling constants. The complexities in defining SW are the prime reason for peak folding problems encountered in 2D NMR spectroscopy.

5.12 The following factors are considered when determining the number of data points, SI, in 2D NMR experiments:
1. Desired resolution
2. Time available for data acquisition
3. Time required for FT and spectral manipulation

5.13 The requirement of the minimum number of transients (NS) depends on the following factors:
1. The insensitivity of the 2D experiment; for example, 32 transients are required in the inherently insensitive 2D INADEQUATE experiment, in contrast to 16 transients for the more sensitive 2D J-resolved spectrum.
2. To attain the required sensitivity, for example, to detect very small couplings, it may become necessary to accumulate more transients.
3. Sample quantities: large quantities of the sample require a smaller number of transients.
4. Appearance of artifacts and quadrature images in the spectrum necessitate more transients to be recorded.

5.14 "Dummy" scans are preparatory scans within the complete time course of the experiment (pulses, evolution, delays, acquisition time). A certain number of these "dummy" scans are generally acquired before each FID in order to attain a stable steady state. Though time-consuming, they are useful for suppressing artifact peaks.

5.15 DC correction is a process by which the contribution of the receiver *dc* is omitted from the FID. The *dc* level is generally determined by examining the last (one-fourth) portion of the FID (tail), which is more likely to have the maximum dc contribution of the receiver. The level is then subtracted automatically from each FID of the data set before F_2 Fourier transformation.

5.16 Since data points in F_1 transformation are taken directly from the frequency-domain spectra (resulting from F_2 transformations), there is no need for a second *dc* correction.

5.17 There are generally three types of peaks: pure 2D absorption peaks, pure negative 2D dispersion peaks, and phase-twisted absorption–dispersion peaks. Since the prime purpose of apodization is to enhance resolution and optimize sensitivity, it is necessary to know the peak shape on which apodization is planned. For example, absorption-mode lines, which display protruding ridges from top to bottom, can be dealt with by applying Lorentz–Gauss window functions, while phase-twisted absorption–dispersion peaks will need some special apodization operations, such as multiplication by sine-bell or phase-shifted sine-bell functions.

5.18 The sine-bell, sine-bell squared, phase-shifted sine-bell and phase-shifted sine-bell squared window functions are generally used in 2D NMR spectroscopy. Each of these has a different effect on the appearance of the peak shape. For all these functions, a certain price may have to be paid in terms of the signal-to-noise ratio, since they remove the dispersive components of the magnitude spectrum. This is illustrated in the following COSY spectra:

1. COSY 45° spectrum with sine-bell multiplication

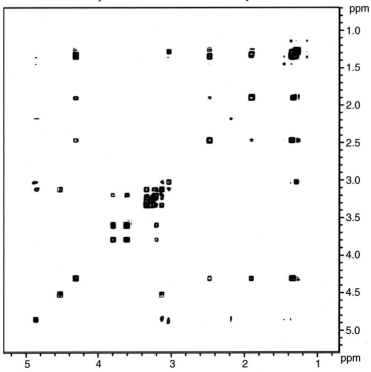

2. COSY 45° spectrum with sine-bell squared multiplication

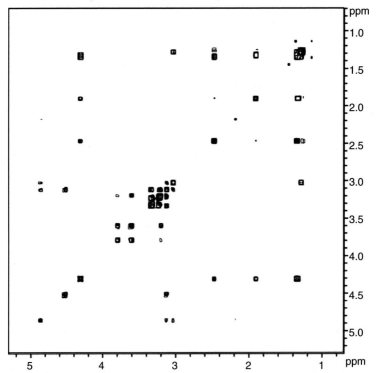

3. COSY 45° spectrum with phase-shifted sine-bell multiplication

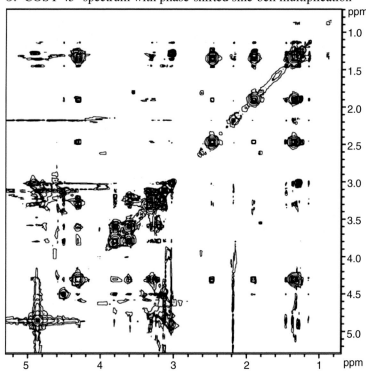

4. COSY 45° spectrum with phase-shifted sine-bell squared multiplication

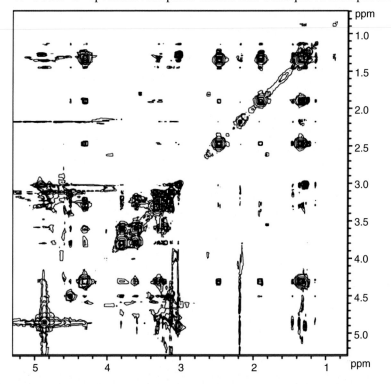

5.19 The second time variable, t_2, is the so called "real-time" variable, representing the time spent in data acquisition, while t_1 is the evolution interval between the two pulses during the experiment. The effect of t_1 can only be observed indirectly by noting its influence on t_2. It is, therefore, necessary to carry out F_2 transformation first in order to generate a series of spectra (rows of the matrix), which is then used as a pseudo-FID for F_1 transformation in the second step.

5.20 The frequency-domain spectrum is computed by Fourier transformation of the FIDs. Real and imaginary components $v(\omega)$ and $i\mu(\omega)$ of the NMR spectrum are obtained as a result. Magnitude-mode or power-mode spectra $P(\omega)$ can be computed from the real and imaginary parts of the spectrum through application of the following equation:

$$P(\omega) = \sqrt{[v(\omega)]^2 + [i\mu(\omega)]^2}$$

5.21 The noise in 2D experiments may arise from a variety of sources, such as unequal increments in evolution time, phase noise in the frequency generation system, incorrect computer handling of successive FIDs, and quantization error or digitizer noise. Symmetrization is a cleaning routine that can remove noise from the 2D plot. This process is generally used in 2D spectra that have inherent symmetry, such as COSY, NOESY, and 2D J-resolved experiments. In these spectra, the genuine signals are either symmetrically arranged about the diagonal (COSY) or symmetrically arranged about the central peak of a multiplet (2D J-resolved). The noise is random, and can be easily eliminated by this simple mathematical operation. Only peaks with a symmetry-related partner are retained.

5.22 1D spectra obtained by projecting 2D spectra along a suitable direction often contain information that cannot be obtained directly from a conventional 1D spectrum. They, therefore, provide chemical shift information of individual multiplets that may overlap with other multiplets in the corresponding 1D spectra. The main difference between the projection spectrum and the 1D spectrum in shift-correlated spectra is that the projection spectrum contains only the signals that are coupled with each other, whereas the 1D ^{1}H-NMR spectrum will display signals for all protons present in the molecule.

5.23 There are basically four types of plots, shown below:
  a. Projection plots
  b. Contour plots
  c. Stacked plots for pseudo-3D presentation
  d. Slice plots for 1D presentation

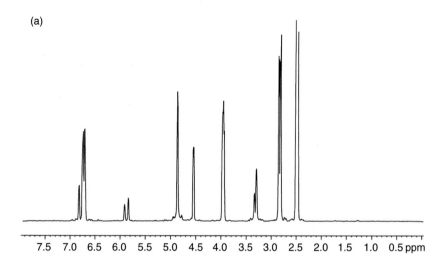

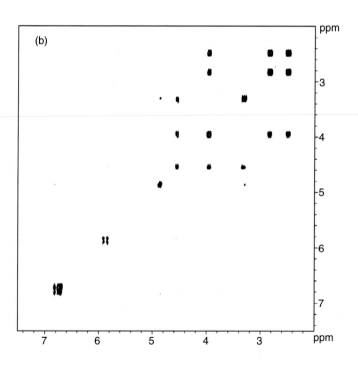

(c)

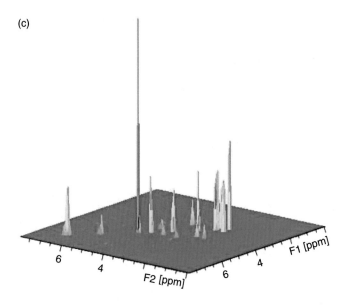

(d)

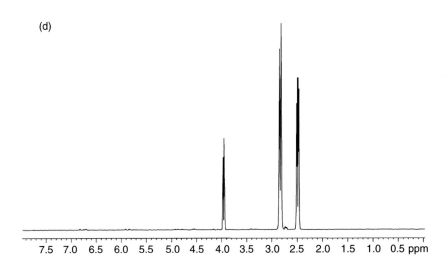

REFERENCES

Aue, W.P., Barthoidi, E., Ernst, R.R., 1976. Two-dimensional spectroscopy. Application to nuclear magnetic resonance. J. Chem. Phys. 64 (5), 2229–2246.

Bartholdi, E., Ernst, R.R., 1973. Fourier spectroscopy and the causality principle. J. Magn. Reson. 11 (1), 9–19.

Baumann, R., Kumar, A., Ernst, R.R., Wüthrich, K., 1981a. Improvement of 2D NOE and 2D correlated spectra by triangular multiplication. J. Magn. Reson. 44 (1), 76–83.

Baumann, R., Wider, G., Ernst, R.R., Wüthrich, K., 1981b. Improvement of 2D NOE and 2D correlated spectra by symmetrization. J. Magn. Reson. 44 (22), 402–406.

Bax, A., Mehlkopf, A.F., Smidt, J., 1979. Absorption spectra from phase-modulated spin echoes. J. Magn. Reson. 35 (3), 373–377.

Bax, A., Freeman, R., Morris, G.A., 1981. A simple method for suppressing dispersion-mode contributions in NMR spectra: the "pseudo echo". J. Magn. Reson. 43, 333–338.

Campbell, I.D., Dobson, C.M., Williams, R.J.P., Xavier, A.V., 1973. Resolution enhancement of protein PMR spectra using the difference between a broadened and a normal spectrum. J. Magn. Reson. 11 (2), 172–181.

Jeener, J. (1971), Paper presented at Ampere International Summer School, Basko Polje, Yugoslavia.

Keeler, J., Neuhaus, D., 1985. Comparison and evaluation of methods for two-dimensional NMR spectra with absorption-mode lineshapes. J. Magn. Reson. 63 (3), 454–472.

Levitt, M.H., Bodenhausen, G., Ernst, R.R., 1984. Sensitivity of two-dimensional spectra. J. Magn. Reson. 58 (3), 462–472.

Luo, S., Krunic, A., Kang, H.-S., Chen, W.L., Woodard, J.L., Fuchs, J.R., Swanson, S.M., Orjala, J., 2014a. Trichormamidea A and B with antiproliferative activity from the cultured freshwater *Cyanobacterium trichormus* sp. UIC 10339. J. Nat. Prod. 77 (8), 1871–1880.

Luo, Y., Pu, X., Luo, G., Zhou, M., Ye, Q., Liu, Yan., Gu, J., Qi, H., Li, G., Zhang, G., 2014b. Nitrogen-containing dihydro-β-agarofuran derivatives from *Tripterygium wilfordii*. J. Nat. Prod. 77 (7), 1650–1657.

Mevers, E., Haeckl, F.P.J., Boudreau, P.D., Byrum, T., Dorrestein, P.C., Valeriote, F.A., Gerwick, W.H., 2014. Lipopeptides from the tropical marine Cyanobacterium *symploca* sp. J. Nat. Prod. 77 (4), 969–975.

Mohamed, G.A., Ibrahim, S.R.M., Badr, J.M., Youssef, D.T.A., 2014. Didemnaketals D and E, bioactive terpenoids from a red sea ascidian. Tetrahedron 70 (1), 35–40.

States, D.J., Haberkorn, R.A., Ruben, D.J., 1982. A two-dimensional nuclear Overhauser experiment with pure absorption phase in four quadrants. J. Magn. Reson. 48 (2), 286–292.

Turner, C.J., Hill, H.D.W., 1986. Artifacts in quadrature detection. J. Magn. Reson. 66 (3), 410–421.

Chapter 6

Nuclear Overhauser Effect

Chapter Outline

6.1 WHAT IS NUCLEAR OVERHAUSER EFFECT?

The strength of an NMR signal depends on the difference in population between the ground (α) state and the excited (β) state of a given nucleus. There is a slight excess (Boltzmann excess) of nuclei in the ground state, and it is this excess that is responsible for the NMR signal. Even at 400 MHz, the difference in population between the two states is very small (about 1 in 660,000), and it gives rise to a macroscopic magnetic moment M_z^0 parallel to the static magnetic field B_0 having a magnitude corresponding to the population difference between the α and β states. This equilibrium difference magnetization M_z^0 represents the *longitudinal* magnetization, which is directed along the z-axis, i.e., parallel to the applied magnetic field, and it precesses about B_0 at its characteristic Larmor frequency. Application of an *Rf* pulse bends the direction of the equilibrium magnetization M_z^0 away from the z-axis and thus creates a transverse component of magnetization M_{xy}, which induces an alternating current within the

Solving Problems with NMR Spectroscopy. http://dx.doi.org/10.1016/B978-0-12-411589-7.00006-1

receiver coil of the NMR spectrometer that is then detected as an NMR signal. Clearly, the intensity of this signal will depend on (1) the magnetogyric ratio of the nucleus under observation, (2) the population difference between the α and β states of the nucleus that determines the intrinsic strength of the magnetizations, and (3) the angle by which the vector M_z^0 is bent by the applied Rf pulse, (since it is the magnitude of the component along the y'-axis that is detected as the NMR signal). Thus, the population difference between the upper and lower states, represented by the longitudinal magnetization, is "sampled" by the Rf pulse, which converts it into a corresponding, proportionately sized transverse magnetization before detection.

The Rf observe pulse generally has very high power, often 100 W or above. It therefore precesses with a very high frequency, typically about 25 kHz, so that the time taken to bend the equilibrium magnetization is short, often of the order of 10 ms. Such a strong pulse would excite the entire proton spectral width uniformly. Broad-band decoupling, however, does not use such high power, since it has to be applied continuously (for instance, proton noise decoupling during ^{13}C acquisition typically uses 1–10 W with fields of a few kilohertz). In contrast to such high-power Rf irradiation, selective irradiations use much lower power—a milliwatt or less, corresponding to field frequencies of a few tens of hertz. Thus, only resonances in a very narrow frequency range (*"on-resonance spins"*) are affected; resonances lying outside this narrow window (*"off-resonance spins"*) remain unaffected. Moreover, the on-resonance spins now behave differently when subjected to selective low power irradiations than when they were subjected to high Rf fields − under low Rf fields they precess at much lower frequencies (~ 10 Hz) − so the relaxation processes become very important (and indeed result in nuclear Overhauser effects (nOe), as will be seen later).

Moreover, precession under selective irradiation occurs in the longitudinal plane of the rotating frame, instead of rotation in the transverse plane which occurs during the evolution of the FID. The magnitude of the vector undergoing precession about the axis of irradiation decreases due to relaxation and field inhomogeneity effects.

6.2   nOe AND SELECTIVE POPULATION TRANSFER

In Chapter 4 (Section 4.2), we considered the phenomenon of selective population transfer (SPT). Under conditions of low-power irradiation, the lines of a multiplet may be unevenly saturated. This results in changes in the relative intensities of the multiplet components of the *coupling partner(s)* of the nucleus that has been irradiated. Thus, in an AX system, if nucleus A is irradiated and nucleus X is observed, then the intensity gained by one X line is balanced exactly by the intensity lost by another X line, so the *overall* integral of the X doublet remains unchanged (provided, of course, that there is no nOe enhancement of X). Such SPT distortions must be borne in mind when interpreting nOe difference spectra (Granger, 1990). Many methods can be used for reducing such distortions involving the use of composite pulses (Shaka et al., 1984), irradiating each line of a multiplet and coadding the results (Neuhaus, 1983), or cycling the decoupler repeatedly around the lines of a target multiplet before each scan during the preirradiation period (Williamson and Williams, 1985; Köver, 1984).

We have previously considered 1D and 2D spectra involving the coupling of nuclei through bonds (*scalar coupling*, or *J-coupling*). Nuclei can also interact with each other directly through space (*dipolar, or magnetic coupling*). This latter form of interaction is responsible for the *nuclear Overhauser effect* (nOe), which is the change of intensity of the resonance of one nucleus when the transitions of another nucleus (that lies close to the first nucleus) are perturbed by irradiation. This change in intensity may correspond to an increase, as in small, rapidly tumbling molecules, or to a decrease, as in large, slower-tumbling molecules. It can provide valuable information regarding the relative stereochemistry of the nuclei in a given structure. The perturbation is usually carried out with a weak *Rf* field, and it eliminates the population difference across the transitions of the saturated nucleus. This triggers a compensatory response from the system through a readjustment of the population difference across transitions of the close-lying nuclei; this response is measured on the NMR spectrometer in the form of the nuclear Overhauser effect (Neuhaus et al., 2012). Thus, if nucleus S is being saturated and nucleus I observed, and if the original and final (i.e., after irradiation of nucleus S) intensities of nucleus I are represented by I_0 and I', respectively, then nOe may be defined as:

$$\eta_I \{S\} = \frac{(I' - I_0) \times 100}{I_0}$$

(6.1)

where η_I is the nOe percentage at nucleus I when nucleus S is irradiated.

6.3 RELAXATION

When placed in a strong magnetic field, nuclei such as ^1H or ^{13}C, which have a spin quantum number of ½, adopt one of two quantized orientations, a low-energy orientation aligned with the applied field B_0 and a high-energy orientation aligned against the applied field. In addition, they exhibit a characteristic precessional motion at their respective Larmor frequencies. When exposed to electromagnetic irradiation from an *Rf* oscillator, the nuclei can undergo excitation when the oscillator frequency matches exactly the Larmor frequency of each nucleus. This excitation involves transfer of excess spin population from the lower energy state to the higher energy state. Certain relaxation processes then come into play as the system responds to restore thermal equilibrium. The most important of these is *spin-lattice relaxation,* T_1, in which energy is transferred from the spins to the surrounding lattice as heat energy. For such relaxation to occur, the nuclei must be exposed to local oscillating magnetic fields some of whose frequencies can match exactly their respective precessional frequencies. There can be many sources of such fluctuating magnetic fields; the major source is the magnetic moments of other protons present in the same tumbling molecule. Since the molecule is undergoing various translations, rotations, and internal motions, there is virtually a continuum of energy levels available, and energy exchange can occur readily between the nucleus and the lattice through such *dipole–dipole interactions.*

In addition to dipole–dipole interactions, nuclei may relax by other mechanisms. For example, the presence of paramagnetic materials in solution can cause very efficient relaxation, since magnetic moments of *electrons* are a thousand-fold greater than those of protons. Thus, although such interactions involve much larger internuclear distances, being intermolecular rather than intramolecular, they can still make T_1 so short that no nOe would be observed. It is for this reason that we need to remove dissolved paramagnetic oxygen carefully from solutions of the substance before undertaking nOe measurements.

In nOe we are concerned mainly with *longitudinal relaxation,* i.e., the return of longitudinal magnetization to its original value M_z^0, which involves reestablishing the original equilibrium state between the populations of the ground and excited states. This will involve changes in population in the excited and ground states *via* transitions between the two spin states, so that the original population difference is restored. To return to the original equilibrium state, the *transverse magnetization* generated must also decay to zero, by loss and dephasing of the individual contributions to the transverse magnetization, M_{xy}. But this need not concern us here, since nOe is concerned primarily with longitudinal relaxation, which occurs independent of transverse magnetization.

The number of transitions between upper and lower states depends not only on the populations N_α and N_β of the two states, but also on the efficiency of relaxation, more often called the *transition probability W,* as well as on the temperature of the surrounding lattice. Each NMR transition will be accompanied by a corresponding change in the lattice energy.

Problem 6.1

What are the two major uses of the nOe effect in NMR spectroscopy?

6.4   MECHANISM OF nOe

The *rate* at which dipole–dipole relaxation occurs depends on several factors: (1) the nature of the nucleus, (2) the internuclear distance, r, and (3) the effective correlation time, τ_c, of the vector joining the nuclei (which is inversely proportional to the rate at which the relevant segment of the molecule tumbles in solution). The *magnitude* of the dipole–dipole interaction, however, will depend only on the internuclear distance.

Let us consider two nuclei I and S that are close in space (hence, their dipole–dipole interaction is significant), but that are not scalar-coupled to each other ($J_{I,S} = 0$), with I being the one observed and S being the one irradiated.

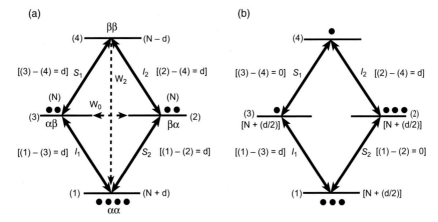

FIGURE 6.1 (a) Populations at Boltzmann equilibrium before application of a radio-frequency pulse on nucleus S. Nuclei I and S are not J-coupled. (b) Populations immediately after the application of a pulse on nucleus S. The populations connected by S transitions are readily equalized, while the *difference* of populations connected by the I transitions remains unaffected (i.e., (1)–(3) or (2)–(4) remain unchanged).

Let us also assume that both spins form part of a rigid molecule that is tumbling isotropically (i.e., without any preferred axis of rotation). Thus the only major relaxation pathway available is that *via* dipole–dipole relaxation. Figure 6.1 presents the energy-level diagram of such a system. Since the two spins I and S are not J-coupled, each appears as a single resonance line, with the line for nucleus I arising due to transitions I_1 and I_2 and that for nucleus S arising due to transitions S_1 and S_2. Since each nucleus can exist in two different energy states, α and β, there are four possible combinations of these states: $\alpha\alpha$, $\alpha\beta$, $\beta\alpha$, and $\beta\beta$. The $\alpha\alpha$ state corresponds to both nuclei in the lowest energy state, the $\beta\beta$ state represents a combination in which both are in the upper energy state, while in the $\alpha\beta$ and $\beta\alpha$ combinations one of the two nuclei is in the lower energy state while the other is in the upper energy state. A Boltzmann distribution prevails at thermal equilibrium, with the $\alpha\alpha$ state having the highest population, $\alpha\beta$ and $\beta\alpha$ states having some intermediate populations, and the $\beta\beta$ state corresponding to the lowest population. Since the nuclei I and S are not J-coupled to one another, the two single-quantum transitions I_1 and I_2 have the same energy, as do the transitions S_1 and S_2.

An interesting feature of the energy diagram is that there are two transitions that involve the *simultaneous* flip of *both* spins. These are the transitions $\alpha\alpha \leftrightarrow \beta\beta$ (a *double-quantum* process, W_2) and $\alpha\beta \leftrightarrow \beta\alpha$ (a *zero-quantum* process, W_0). Indeed, it is these two transitions that are responsible for the nOe effect observed, since it is through them that the saturation of nucleus S affects the intensity of nucleus I. Both W_2 and W_0 are "forbidden" transitions, since they cannot be produced directly by an *Rf* pulse or lead directly to detectable NMR signals; they are, however, allowed in the context of *relaxation* processes. This highlights an important feature: the selection rules governing excitation processes (i.e., an interaction of spin with an external oscillating field) are different from the selection rules governing relaxation processes (in this case involving energy exchange of the spins with the lattice).

To understand how nOe occurs, we have to consider the following situations: (1) the populations of the nucleus I prevailing at thermal equilibrium before the application of the Rf pulse on nucleus S, (2) populations immediately after the pulse is applied to nucleus S, and (3) populations after the system has had some time to respond, with either W_0 or W_2 being the predominant relaxation pathway.

For simplicity, let us assume that the $\alpha\beta$ and $\beta\alpha$ states have the same population, designated N. The upper energy state ($\beta\beta$) will then have a slightly lower population ($N - d$), and the lower energy state ($\alpha\alpha$) will have a slightly higher population ($N + d$), so that the population difference between the lowest and the highest energy states will be $2d$.

When the pulse is applied to nucleus S, the population levels connected by S transitions are rapidly equalized, but there is no *immediate* change in the *difference* in population levels connected by the I transitions. Figure 6.2 shows that on application of the pulse, levels 2 and 4 acquire the same populations as levels 1 and 3, respectively, but the population *difference* between the levels connected by the I transitions [(i.e., (1)–(3) or (2)–(4)] are unaffected (d in both cases).

Since the equilibrium state has been disturbed, the system tries to restore equilibrium. For this, it can use as the predominant relaxation pathways the double-quantum process W_2 (in fast-tumbling, smaller molecules), leading to a positive nOe, or the zero-quantum process W_0 (in slower-tumbling macromolecules), leading to a negative nOe.

Let us first consider the situation in which W_2 is the predominant relaxation pathway. The population states immediately after the pulse are represented in Fig. 6.2b. The system tries to reestablish the equilibrium state by transferring some population x from the $\beta\beta$ state, which had a population of the $N - (d/2)$ to the $\alpha\alpha$ state, which had a population of $N + (d/2)$. This transfer of population (x), shown in Fig. 6.2d (represented by a half circle), gives level 4 a somewhat reduced population $[N - (d/2) - x]$ and level 1 a correspondingly increased population $[N - (d/2) + x]$.

If we now examine the result of this population adjustment on the I transitions, we find that the upper population of each I transition has decreased by x while the lower level of each I transition has increased by x; the population difference between the upper and lower states (levels 1 and 3) connected by the I_1 transition is

$$[N + (d/2) + x] - [N - (d/2)] = d + x$$

Similarly, the population difference between the states connected by the I_2 transition (levels 2 and 4) is

$$[N + (d/2)] - [N - (d/2) - x] = d + x$$

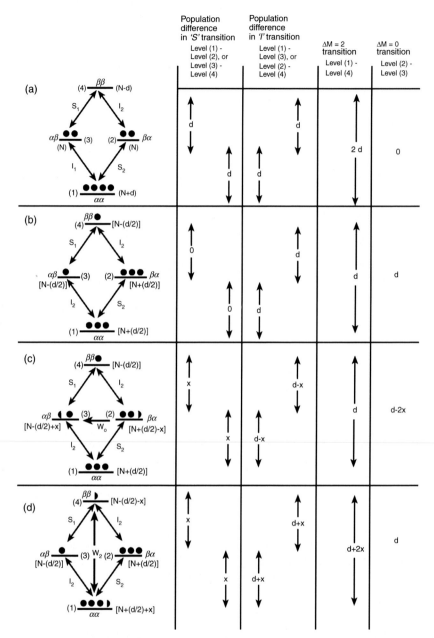

FIGURE 6.2 Energy levels and populations for an *IS* system in which nuclei *I* and *S* are not directly coupled with each other. This forms the basis of the nuclear Overhauser enhancement effect. Nucleus *S* is subjected to irradiation, and nucleus *I* is observed. (a) Population at thermal equilibrium (Boltzmann population). (b) Population immediately after the radiofrequency pulse is applied. (c) Population after the system has had time to respond, with W_0 (zero-quantum transition) being the predominant pathway, as in macromolecules. The population difference between the two energy levels connected by the *I* transitions is now less $(d - x)$ than this difference (d) at thermal equilibrium. This will lead to a *negative* nOe effect. (d) Same as (c), but with W_2 (double-quantum transition) being the predominant relaxation pathway, as in smaller molecules. The population of the lower energy level (of energy states connected by *I* transitions) is increased and that of the upper level is decreased. This gives rise to a positive nOe effect.

Hence, the population difference between the lower and upper energy states of the two I transitions becomes $d + x$, as compared to the original difference of d at equilibrium. Thus an intensification of the lines for nucleus I will be observed by an amount corresponding to this increased difference x. *This is the positive nuclear Overhauser effect that is encountered in small, rapidly tumbling molecules, in which W_2 is the predominant relaxation pathway.*

In large molecules that tumble slowly, the predominant relaxation pathway is *via* W_0. This is shown schematically in Fig. 6.2c. A part of the population x is now transferred from the $\beta\alpha$ state to the $\alpha\beta$ state. This causes an *increase* in the population of the upper level of one I transition (I_1, level 3) and a *decrease* in the lower population level of the other I transition (I_2, level 2). As a result, the population *difference* between the lower and upper levels of each I transition is reduced to $d - x$ (i.e., level 1–level 3, or level 2–level 4, becomes $d - x$). The reduction in population difference by x as compared to the equilibrium situation (Fig. 6.2a) produces a corresponding *decrease* in the signal intensity of the I nucleus. *Hence, if W_0 is the main relaxation pathway, then saturation of nucleus S will reduce the intensity of the I nucleus; i.e., a negative nuclear Overhauser effect will be observed.* In practice, the relaxation processes involve the single-quantum relaxation pathway W_1 (which causes no change in signal intensity), as well as the double-quantum W_2 and zero-quantum pathways W_0, which respectively enhance or reduce the signal intensities, with W_2 dominating in small, rapidly tumbling molecules and W_0 dominating in large, slowly tumbling molecules.

Problem 6.2

What is the significance of *sign* (positive or negative) of the nOe?

6.5   FACTORS AFFECTING nOe

In addition to the dipole–dipole relaxation processes, which depend on the strength and frequency of the fluctuating magnetic fields around the nuclei, there are other factors that affect nOe: (1) the intrinsic nature of the nuclei I and S, (2) the internuclear distance (r_{IS}) between them, and (3) the rate of tumbling of the relevant segment of the molecule in which the nuclei I and S are present (i.e., the effective *molecular correlation time, τ_c*).

Problem 6.3

Is it possible to predict the predominant mode of relaxation (zero-quantum W_0 or double-quantum W_2) by observing the *sign* of nOe (negative or positive)?

Problem 6.4

Why do we need to involve zero-quantum (W_0) or double-quantum (W_2) processes to explain the origin of the nuclear Overhauser enhancement?

The molecular correlation time τ_c represents the time taken for the relevant portion of the molecule containing the nuclei I and S to change from one orientation to another. It is chosen so that it is nearly equal to the *minimum* waiting time between various orientations rather than to the *average* waiting time between the orientations (thus waiting times shorter than τ_c do not occur frequently). The lower limit of the waiting time then corresponds to the upper limit of the frequency range of the fluctuating magnetic fields. Small, rapidly tumbling molecules may have τ_c of the order of 10^{-12} s; large molecules (or small molecules in viscous solutions) may have τ_c that is a thousand-fold longer (10^{-8}–10^{-9} s).

In order for relaxations to occur through W_1, the magnetic field fluctuations need to correspond to the Larmor precession frequency of the nuclei, while relaxation *via* W_2 requires field fluctuations at double the Larmor frequency. To produce such field fluctuations, the tumbling rate should be the reciprocal of the molecular correlation time, i.e., τ_c^{-1}, so most efficient relaxation occurs only when $v_0\tau_c$ approaches 1. In very small, rapidly tumbling molecules, such as methanol, the concentration of the fluctuating magnetic fields (*spectral density*) at the Larmor frequency is very low, so that the relaxation processes W_1 and W_2 do not occur efficiently and the nuclei of such molecules can accordingly relax very slowly. Such molecules have $\tau_c \ll 1$, and they are said to be in the *extreme narrowing limit*. The line widths are then determined by instrumental factors rather than by other fundamental considerations.

In larger, slowly tumbling molecules (or small molecules in viscous solutions), tumbling occurs very slowly, so that fields corresponding to the Larmor precession frequency v_0 (for relaxation *via* W_1) or $2v_0$ (for relaxation via W_2) cannot be generated sufficiently. Relaxation through the zero-quantum transition W_0 then becomes important, involving a *mutual spin-flip*, i.e., the shifting of energy from one spin to another ($\alpha\beta \leftrightarrow \beta\alpha$). In compounds with molecular weights of over 10,000, the nOe appears with a maximum intensity of -1. W_0 transitions occur between close-lying energy levels, and require only fields of low frequency, which are readily available from slowly tumbling molecules.

Problem 6.5

What is the molecular correlation time, τ_c?

While the *rate of change* of dipolar interaction depends on τ_c, its *magnitude* depends only on the internuclear distance and is independent of τ_c. Thus, the dipole–dipole relaxation depends on the molecular correlation time τ_c, the internuclear distance r, and the magnetogyric ratios of the two nuclei, γ_I and γ_S:

$$R_I = K\gamma_I^2\gamma_S^2(r)^{-6}\tau_c \tag{6.2}$$

The *efficiency* of the relaxation process is, therefore, governed by the r^{-6} relationship (Bell and Saunders, 1970), so that doubling of the internuclear distance decreases the relaxation by a factor of 64 (since $2^{-6} = 1/64$). Moreover, the strength of the dipole–dipole interaction depends on the square of the product of the magnetogyric ratios of the two nuclei, so that proton–proton interactions are the main relaxation processes because of the greater magnetogyric ratio of ^1H in comparison to ^{13}C. Deuterium, however, has a magnetogyric ratio 6.5 times smaller than that of a proton, so that the proton–deuterium interaction should be $(6.5)^2$, i.e., about 42 times less than for proton–proton interactions, other conditions remain the same. Carbon atoms relax mainly through attached protons, so quaternary carbons relax much more slowly than protonated carbons, and methyl carbons bearing three hydrogen atoms will normally relax faster than methine carbons that bear only one hydrogen.

6.5.1 Internuclear Distance and nOe

If the intensity of nucleus I before irradiation of the neighboring nucleus S was I_0, and its intensity after irradiation of nucleus S was I_1, then the fractional increase in intensity $\eta_I(S)$ is given by $(I_0 - I_1)/I_0$. If we assume that the relaxation of a nucleus I occurs only through dipole–dipole relaxation by interaction with nucleus S and that a steady equilibrium state has been reached, then the fractional increase in intensity η_I of nucleus I is given by the equation:

$$\eta_I(S) = \frac{(S_0 / I_0)(W_2 - W_0)}{2W_1^I + W_0 + W_2} \tag{6.3}$$

where W_0, W_1, and W_2 are zero quantum transition, single quantum transition, and double quantum transition probabilities. Thus when $W_2 > W_0$, nOe will be positive, when $W_0 > W_2$, nOe will be negative, and if $W_2 = W_0$, then there will be no nOe.

If we assume that the extreme narrowing condition exists, then the following simpler expression applies:

$$\eta_I(S) = \frac{\gamma_S}{2\gamma_I} \tag{6.4}$$

In the case of proton–proton interactions, both nuclei S and I will have the same magnetogyric ratios, and an implication of Eq. 6.4 then is that there is an upper limit of 50% on the nOe obtainable, *whatever the distance between nuclei S and I*. This means that the observation of an nOe between two nuclei does not *necessarily* mean that they are spatially close to one another, and nOe results must therefore be interpreted with caution. Similarly, as will be seen later, the absence of nOe between two nuclei does not necessarily mean that they are far apart. In the case of heteronuclear nOe, since the magnetogyric ratio of proton

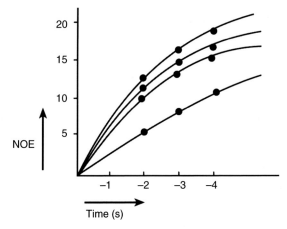

FIGURE 6.3 NOe buildup with respect to time, after irradiation of bound alanine methyl protons in ristocetin A-tripeptide complex for four different protons. *(Reprinted from D. H. Williams et al., J. Am. Chem. Soc. 105, 1332, copyright (1983), with permission from The American Chemical Society, 1155 16th Street, N.W. Washington, D.C. 20036, U.S.A.)*

(γ_S) is four times the magnetogyric ratio of carbon (γ_I), $\gamma_S./\gamma_I$ can be four times greater than that obtainable in homonuclear nOe.

While the final magnitude of nOe depends, as indicated earlier, on the relaxation pathways W_1, W_2, and W_0, *the initial rate of buildup of nOe (transient nOe) depends only on the rate of cross-relaxation between the nuclei*, and this can provide valuable information about the distance between the nuclei (r). This rate of buildup can be proportional to r^{-6}, where r is the distance between the nuclei. Thus, if the proportionality constant is determined, we can calculate an approximate distance between the two nuclei. The best results are obtained in rigid molecules when the nuclei are less than 3 Å apart. If only direct nOe's are involved in a two-spin system and no third proton takes part in the relaxation, then a plot of the magnitudes of nOe against time gives an exponential curve (Fig. 6.3) (Vögeli, 2014). If the size of the nOe at time t is h_t, and if the final steady-state value of nOe is h_∞, then a plot of $h_\infty - h_t$ against t yields a straight line of slope $-k$ (Fig. 6.4). Since $k \propto r^{-6}$, the internuclear distance can be calculated.

6.5.2 Three-Spin System

If there are more than two nuclei exerting relaxation effects on one another, then it is convenient to consider them in *pairs* and to arrive at the overall effect by adding together the effects of various possible pairs. In the case of a three-spin system, we can consider two different situations: (1) the nuclei H_A, H_B, and H_C are arranged in a straight line, and (2) they are in a nonlinear arrangement.

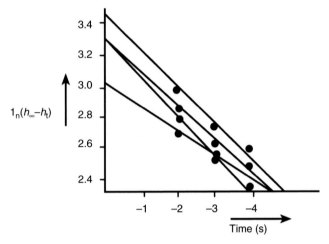

FIGURE 6.4   Data in Fig. 6.3 when plotted as $(h_\infty - h_t)$ *versus* time. The slope of the lines represents the internuclear distance r that corresponds to the rate of nOe buildup, which is directly proportional to r^{-6}. *(Reprinted from D. H. Williams et al. J. Am. Chem. Soc. 105, 1332, copyright (1983), with permission from The American Chemical Society, 1155 16th Street, N.W. Washington, D.C. 20036, U.S.A.)*

Problem 6.6

What is *transient nOe,* and why is it considered to provide a better estimate of the internuclear distance (*r*) than the normal *steady-state nOe* effect?

Problem 6.7

How can you measure the *transient nOe?*

Let us first consider the nuclei H_A, H_B, and H_C as lying in a straight line and equidistant from one another (Fig. 6.5). The central proton, H_B, can relax by interactions with two neighbors, H_A and H_C, while H_A and H_C can relax by interaction with only one neighbor, so H_B can relax twice as quickly as H_A or H_C. If we assume that relaxation can occur only through dipole–dipole relaxation, and that an equilibrium steady state has been reached, then the

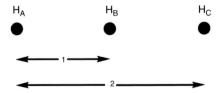

FIGURE 6.5   Three protons arranged in a straight line, equidistant from each other. Here, H_A is farther from H_C than in Fig. 6.6.

maximum nOe achieved for H_B is 50%, which can occur through relaxation *via* H_A and H_C. Assuming we are dealing with small molecules causing positive nOes, irradiation of nucleus H_A produces a sizeable positive nOe at the neighboring H_B and a small direct positive nOe at the more distant H_C. Thus, H_A and H_C can each potentially enhance H_B by 25%. Moreover, H_B would be responsible for almost all the relaxation effects on H_A or H_C since H_A and H_C are twice as far from one another as they are from H_B. Since we already know from the previous discussion that a 64-fold reduction in nOe occurs on doubling the internuclear distance ($r^{-6} = 2^{-6} = \frac{1}{64}$), then the proportion of relaxation effect of H_A on H_C (or H_C on H_A) will be $\frac{1}{64} = 0.8\%$, while the remainder (49.2%) will be due to H_B. Thus while irradiation of H_B can cause 49.2% nOe at H_A or H_C, irradiation of H_A or H_C can cause only 25% nOe at H_B, illustrating that *the nOe between two nuclei will generally differ in the two directions because of the differing relaxation pathways available to the two nuclei.* If the nuclei H_A, H_B, and H_C are arranged linearly but are not equidistant then, obviously, the nucleus nearer to H_B will contribute more to its nOe than the nucleus that is farther from it.

6.5.3 Three-Spin Effects

So far we have considered only direct nOe. However, there is an indirect effect of H_A on H_C (or of H_C on H_A) that should be considered (the so-called *three-spin effect*). Assuming that we are dealing with small molecules causing positive nOe's, irradiation of nucleus H_A causes a sizeable positive nOe at the neighboring H_B and a small positive *direct* nOe at the more distant H_C. However, since the irradiation-induced *decrease* in the population difference between the upper and lower states of H_A is causing an *increase* in the intensity (or population difference) of H_B, it is logical to assume that this increase would produce an opposite effect at H_C, i.e., a decrease of its intensity. H_C will therefore experience two opposing effects upon irradiation of H_A, a small positive direct nOe and a larger negative nOe, *so an overall negative nOe will be observed at H_C upon irradiation of H_A.* This effect is most likely occur when the three nuclei are arranged linearly. Irradiation of H_A will cause alternating +ve and −ve nOe's on other atoms (H_D, H_E, H_F, etc.) in a linear arrangement beyond H_C, though the effect decreases sharply beyond two atoms, so it is not measurable.

Problem 6.8

Following is a linear three-spin system and the observed steady-state nOes between three nuclei. The distance between nuclei A and B is double than between nuclei A and C ($\gamma_{AB} = 2\gamma_{AC}$). Explain the nOe results in terms of both relaxation and internuclear distances.

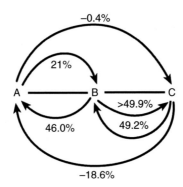

It is possible to distinguish between direct and indirect nOes from their kinetic behavior. The direct nOes grow immediately upon irradiation of the neighboring nucleus, with a first-order rate constant, and their kinetics depend initially only on the internuclear distance r^{-6}; indirect nOes are observable only after a certain time lag. We can thus suppress or enhance the indirect nOe's (e.g., at H_C by short or long irradiations, respectively, of H_A). A long irradiation time of H_A allows the buildup of indirect negative nOe at H_C, while a short irradiation time of H_A allows only the direct positive nOe effects of H_A on H_C to be recorded.

Problem 6.9

Explain what is meant by *three-spin effects,* or *indirect nOe effects.* When do such indirect effects matter?

6.5.4 Nonlinear Arrangement

If H_A, H_B, and H_C do not lie on the same line (Fig. 6.6), then as H_A comes closer to H_C, the *direct* +ve nOe between H_A and H_C will increase, and a point may come when it totally cancels the larger, *indirect* negative nOe effect exerted by H_A on H_C through H_B. Thus no nOe may be observed at H_C upon irradiation of

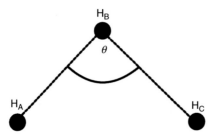

FIGURE 6.6 A three-proton angular system. H_A, H_B, and H_C are disposed at an angle θ.

H_A, *even though H_A and H_C are spatially close!* The absence of nOe between nuclei therefore does not *necessarily* mean that they are far from one another.

Molecules with a rigid central core (such as a ring system) and a freely moving side chain may exhibit significant differences in the mobility of the protons in the central ring system as compared to the side chain. These are reflected in their corresponding relaxation rates, with the protons lying on the more mobile side chain relaxing faster than those in the less mobile ring system. These differences in relaxation rates can be used to advantage. For instance, if the protons lying in the two regions have similar chemical shifts so it is difficult to distinguish them from one another, then measurement of relaxation rates may help to identify them.

So far we have been concerned with homonuclear nOe effects. nOe between nuclei of different elements can also be a useful tool for structural investigations. Such heteronuclear nOe effects—for instance, between protons and carbons—can be used with advantage to locate quaternary carbon atoms. Normally, heteronuclear nOe effects are dominated by interactions between protons and *directly bonded* carbon atoms, and they can be recorded as either 1D or 2D nOe spectra.

As stated earlier, since $\eta_I = \gamma_S/2\gamma_I$ and since the magnetogyric ratio of proton is about four-fold greater than that of carbon, then if ^{13}C is observed and ^1H is irradiated (expressed as ^{13}C {^1H}), the ^{13}C signal appears with about a three-fold enhancement of intensity due to the nOe effect, since at the extreme narrowing limit $\eta_I = 198.8\%$. This is a very useful feature. For instance, in noise-decoupled ^{13}C-NMR spectra in which C–H couplings are removed, the ^{13}C signals appear with enhanced intensities due to nOe effects.

Proton irradiation *before* acquisition of the ^{13}C-NMR spectrum results in nOe but no decoupling, whereas proton irradiation *during* ^{13}C data acquisition produces decoupling without nOe. It is, therefore, possible to separate the two effects by "gating" the decoupler on and off for appropriate time periods (Fig. 6.7). Moreover, since the power needed to induce nOe is much less than that required for decoupling, the power level of the decoupler is reduced during the preirradiation time period and then increased during the acquisition period.

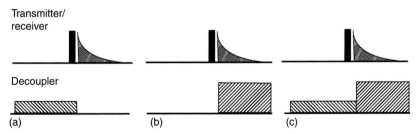

FIGURE 6.7 Pulse schemes representing separation of decoupling effects from the nOe during X nucleus acquisition. The decoupler is programmed to produce noise-modulated irradiation or composite pulse decoupling at two power levels. Suitable setting of the decoupler may produce either (a) nOe only, (b) proton decoupling only, or (c) both nOe and proton decoupling.

This cuts down the heating effects when continuous irradiation with decoupler powers of 2–3 W are employed.

The low intensities of unprotonated carbons are usually due to their long relaxation times. The addition of a paramagnetic substance such as Cr(acac)$_3$ can produce a hundred-fold increase in the relaxation rate, with the rate varying linearly with the concentrations of Cr(acac)$_3$, so that the nOe is also reduced a hundred-fold. Protonated carbons with faster relaxation rates are less affected (about 10-fold increase in relaxation rates), and the reduction of nOe is correspondingly less.

An improved procedure for recording heteronuclear nOes is to irradiate individual lines of the multiplet with a low decoupling power instead of exciting the entire multiplet with high power. This results in greater selectivity of the protons being irradiated and higher sensitivity of the carbon signals (Bigler and Kamber, 1986).

Problem 6.10

What happens in three-spin systems when the A–B–C angles θ are 180° and 78°, respectively, $r_{AB} = r_{BC} = 1$, and spin A is irradiated?

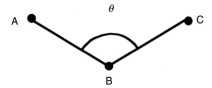

6.5.5 nOe Difference Spectra

The most widely used nOe experiment is nOe difference spectroscopy. Two different sets of experiments are recorded, one in which certain protons are subjected to irradiation and enhancements are obtained of other nearby protons, and the other without such irradiation so that normal unperturbed spectra are obtained. Subtraction of the perturbed FID from the unperturbed one, followed by Fourier transformation of the difference FID, gives the nOe difference spectrum, which ideally contains only the nOe effects (Fig. 6.8) (Stott et al., 1995).

The nOe difference spectrum has the advantage that it allows measurements of small nOe effects, even 1% or below. The experiment involves switching on the decoupler to allow the buildup of nOe (Andersen et al., 1987). It is then switched off, and a $\pi/2$ pulse is applied before acquisition. The nOe is not affected much by the decoupler being off during acquisition, since the nOes do not disappear instantaneously (the system takes several T_1 seconds to return to its equilibrium state) (Kinns and Sanders, 1984).

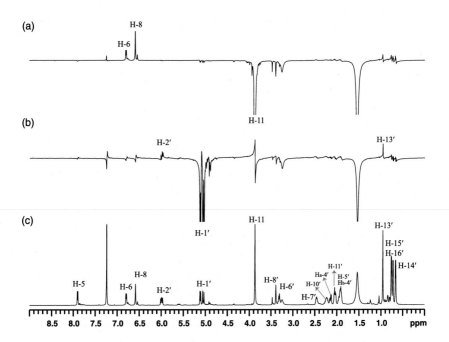

FIGURE 6.8 (a) NOe difference spectrum, after saturating proton at δ 3.87, (b) nOe difference spectrum when proton at δ 5.04 is saturated, (c) 1D ¹H-NMR spectrum of compound doremone B, isolated from *Dorema aucheri* Boiss. Steady state nOe on spatially close protons are visible in "difference" spectra a and b, generated from subtraction of nOe spectra from control spectrum c, respectively.

Problem 6.11

Explain the main advantage of nOe difference spectroscopy. Why does it involve a mathematical subtraction of the normal ¹H-NMR spectrum from the nOe-enhanced ¹H- NMR spectrum?

6.6 SOME PRACTICAL HINTS

6.6.1 Solvent

The use of a suitable solvent is an important consideration when recording nOe difference spectra. Solvents with sharp intense lock signals, such as d_6-DMSO or d_6-acetone, are preferable to a solvent such as $CDCl_3$, which has a weaker lock signal, or D_2O which has a broad lock signal. This is because the lock on FT NMR spectrometers operates by sampling continuously the dispersion-mode deuterium signal of the deuterated solvent. Any field drift would create an error signal, which, because of the dispersion-mode (half positive, half negative) shape of the deuterium signal, would be positive if the drift is on one side and negative if the drift is on the other side. The spectrometer is built to correct this error automatically, but the correction is more precise in solvents with a sharp lock signal (such as d_6-acetone) than in those with weak lock signals (e.g., $CDCl_3$) or broad lock signals (e.g., D_2O). D_2O is also not very suitable, since its chemical shift is largely temperature dependent. However, if it must be used, as in water-soluble compounds such as sugars, then addition of 2–3% of d_6-acetone for use as the lock resonance is recommended.

We normally avoid protonated solvents because the very intense solvent peak will obscure nearby protons, and the dynamic range problem will also reduce the quality of the spectrum. If the solvent has exchangeable protons (such as H_2O or HOD), then saturation transfer processes from solute to solvent or solvent to solute can complicate the spectra.

Problem 6.12

Why is degassing of sample solutions in NMR tubes essential before nOe experiments, and why are aqueous solutions (solutions in D_2O) not generally degassed in the NMR tubes?

Problem 6.13

Explain the dependence of the nOe on molecular motion (tumbling).

6.6.2 Temperature Gradient

Temperature gradients within the probe in solution state NMR can affect the quality of the spectrum. Temperature gradients can have affect on various physical parameters including chemical shifts as different temperatures across the sample can lead to different chemical shift values (chemical shift gradient). This can, in turn, produce line broadening (Loening and Keeler, 2002).

There may be a systematic deviation between the actual temperature of the sample and the temperature reported by the spectrometer. In addition, temperature gradients are often present along the axis of the sample tube. The size of the temperature gradients depends on several factors including the room temperature, heating due to radiofrequency pulses, and most profoundly due to the flow rate of cooling and heating gases. Most of the modern NMR spectrometers are fitted with a sample temperature controller which uses a stream of temperature-regulated air fed from the bottom of the probe. However when the temperature of the air stream is above the room temperature, the air cools as it ascends, creating temperature gradients all along the sample volume.

If the temperature gradient is large, convective flow (concerted and collective movement of molecules within the solution) can occur in the NMR tube. This sample convection can have a strong effect on the B_0-field gradient, and lead to distortions in diffusion coefficient-based NMR experiments as well as in nOe enhancement studies. The key impact is on the line broadening, which adversely affects the quality of the nOe difference spectrum.

To record good quality nOe difference spectra, it is important to lower the temperature gradients to a level of 0.05°C or less. This can be achieved by decreasing the sample-tube diameter, decreasing the sample height, or increasing sample viscosity. Additionally, temperature gradients and the resulting thermal convective flow can be prevented by using a probe temperature control system where the sample is heated by air flow from above (Jerschow and Müller, 1998).

6.6.3   Sample Purity

Impurities that can lead to sharp decreases in spin-lattice relaxation times, such as paramagnetic metal ions or dissolved oxygen, need to be removed. Paramagnetic ions can be removed by complexation with ethylenediamine tetraacetate (EDTA) by filtering solutions through a chelating resin. Oxygen should be removed through repeatedly freezing the contents of a specially constructed NMR tube by dipping it in liquid nitrogen, evacuating the tube, and then thawing. Such tubes are commercially available. Simply bubbling an inert gas through the tube is not enough for proper degassing of the sample.

The nOe difference spectrum is highly demanding, since even the slightest variation in the spectra recorded with and without preirradiations will show up as artifacts in the difference spectrum (Fig. 6.9). The errors can be random, due to phase instability caused by temperature effects on the *Rf* circuits, variations in spinner speed, etc. The problem of phase instability is reduced in the latest generation of instruments with digital frequency synthesizers. If the spinner speed variation is a serious problem, then the difference spectrum can be recorded with the spinner off.

The long-term changes due to variations in temperature, field drift, etc. can be minimized by acquiring alternately the pre-irradiation data and the control data, at the shortest time intervals possible, and co-adding the data later. Typically, a cycle would consist of two to four dummy scans, followed by 8 or 16 data acquisitions at each pre-irradiation frequency. If there are only two frequencies in the cycle, then each cycle would take about a minute. In automatic multiple-scan

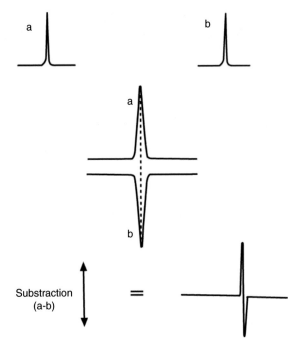

FIGURE 6.9   A slight variation in the positions of peaks (a) and (b) will not lead to mutual or complete cancellation on subtraction, and a peak of characteristic line shape may result.

experiments with several irradiation frequencies, each cycle would take a correspondingly longer time, depending on the number of frequencies and the number of data acquisitions at each frequency. The dummy scans are necessary to ensure that the effects of irradiation at the previous frequency in the cycle have disappeared before data acquisition. The need for multiples of four or eight scans arises as a requirement of the CYCLOPS phase cycle.

One needs to avoid having moving metal objects near the magnet when carrying out nOe difference experiments, to prevent random variations in frequency. A small line-broadening ($\sim$2 Hz) can also be applied to the spectra before or after subtraction to reduce subtraction artifacts.

Problem 6.14

What precautions are normally taken during sample preparation for the nOe experiment?

6.6.4   Importance of NMR Tube Quality

Along with the above-cited parameters, the quality and type of the NMR tube can play an important role in the success of nOe experiments. Only good quality NMR tubes should be used for nOe measurements. The use of a poor quality NMR tube will generate large spinning side bands and bad line shapes that cannot be fixed by shimming, no matter how well the shimming is performed, or how many scans are acquired. The NMR tube quality also can affect resolution. Thus for nOe and NOESY spectra (Sections 6.5.5 and 6.8), a high-quality NMR tube is required. Special NMR tubes are not required for nOe measurement, since good quality sample tubes, used in general NMR measurements, work well. The borosilicate glass used in good quality NMR tubes should have the least possible paramagnetic contamination, and endurance to be safely operated at temperatures up to 230°C. The inner surface of such tubes must be resistant to strong acid and base at ambient temperature. Similarly, the thickness of the borosilicate NMR tube should allow maximum volume to be placed.

Special tubes and inserts, invented by Shigemi, can be used to restrict the active sample volume, and hence reduce the amount of solvent without causing lineshape problems. In macromolecular NMR spectroscopy, where the quantity of sample may be a limiting factor, Shigemi tubes work very well for nOe measurements.

6.6.5 Potential Interference of Chemical Exchange Process on nOe

The process of chemical exchange can profoundly affect the outcome of an nOe experiment in a multitude of ways (Fig. 6.10). For example if a nucleus is involved in both the Overhauser effect and chemical exchange, the intensities of the nOe signals may be influenced by the chemical exchange rate (Macura et al., 1981). Moreover, if the chemical exchange rate is six to seven times faster than the cross relaxation rate, the nOe signals will be too weak to be detected, regardless of the internuclear distances.

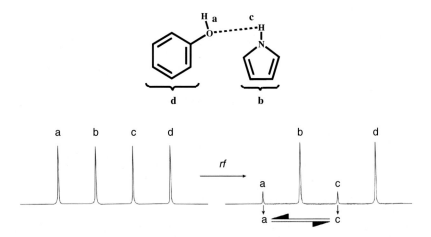

FIGURE 6.10 Schematic representation of the effects of chemical exchange on the nOe experiment. Saturation of nuclei a will lead to the simultaneous saturation of nuclei c, if nuclei a and c experience mutual chemical exchange during the saturation period.

Similarly, the exchangeable spins always display the same or dependent sign behavior. This means that if spin (I) is in chemical exchange with a partner that is saturated by a selective *Rf* pulse (I), it will also experience simultaneous saturation, and thus their signs will show mutual dependency. In the case of positive nOe (low molecular weight compounds), peaks with opposite sign appear due to the chemical exchange partner, while in the case of negative nOe (high molecular weight compounds), the peaks with the same sign appear. As a result, the identification of responses arising from the chemical exchange-based saturation, rather than genuine nOe, becomes a major challenge.

6.7   LIMITATIONS OF 1D nOe METHODS

A number of parameters affecting the successful outcome of nOe experiments are discussed in Section 6.6. The nOe difference measurements are normally made in 1D or 2D modes. 1D nOe measurements can have several limitations such as:

1. The 1D nOe difference spectrum measures the *steady state nOes* which arise by saturating S spin transitions by selective application of weak *Rf* irradiation to the S nucleus. However, steady state nOe between the two nuclei cannot be translated into internuclear distance, because its magnitude depends on a balance between the influences of all neighboring nuclei. At best, they can only provide a rough approximation or relative internuclear distances. The % nOe value, therefore, only has a relative meaning in cases where multiple nOes, originating from the same ^{1}H, are compared.

2. The selective nature of 1D nOe methods can be a limitation, as only nOe between *pairs* of nuclei can be observed in one experiment. Nucleus (S) needs to be selectively saturated/irradiated in order to observe its dipolar couplings with a spatially close nucleus (I). Sometimes, it is difficult to selectively irradiate a nucleus due to its close chemical shift with other nuclei. In such cases, the non-selective irradiation leads to the appearance of several peaks in the nOe difference spectra which do not belong to the target nucleus. This makes interpretation rather complex and often misleading.

3. In the 1D nOe technique, several experiments need to be performed, by selectively irradiating individual nuclei (one ^{1}H at a time) one after the other, in order to decipher the dipolar coupling network in a molecule. Thus, several carefully planned experiments need to be performed, which is an inefficient and time-consuming process.

4. 1D nOe methods are only suitable for small molecules. They fail in the case of macromolecules, where the resulting nOes are negative, and there is too much overlap in chemical shifts as there are too many nuclei with chemical shifts very close to one another.

These limitations in classical 1D nOe were largely overcome with the advent of Nuclear Overhauser Effect SpectroscopY (NOESY), a two-dimensional (2D) version of nOe spectroscopy, but with many fundamental and conceptual differences.

6.8 TWO-DIMENSIONAL NUCLEAR OVERHAUSER EFFECT SPECTROSCOPY (NOESY)

The NOESY technique is discussed in Chapter 7, along with its applications in solving stereochemical (conformational and configurational) problems. In the NOESY spectrum, the cross-peaks connect resonances from nuclei that are spatially close (dipolarly coupled), rather than those that are coupled through bonds (scalarly or J-coupled) to each other. These cross-peaks result from the process of *cross-relaxation* between two close lying nuclei, i.e., from processes in which both the dipolar coupled spins flip simultaneously.

NOESY experiments are now widely used for the determination of homonuclear (^{1}H–^{1}H) nOe effects, in both small and large molecules. The use of NOESY is a key factor for the success of NMR spectroscopy in macromolecular structural chemistry (structural biology).

The NOESY experiment is useful for observing transient nOes (from the rate at which the nOe grows), which can be directly related to internuclear distances. The NOESY experiment has a number of advantages over its 1D counterpart NOESY experiments are non-selective in nature and thus all nOes can be mapped in a single experiment. Similarly as compared to 1D nOe techniques, NOESY experiments are simple to set up and less time demanding. The NOESY technique is capable of handling both small and large molecules. Most importantly, the NOESY spectrum is easy to interpret in a qualitative manner, as the *"COSY-like cross-peaks"* in the square plot represent dipolar couplings between two or more protons (implying short distances between them). These factors have made NOESY a favorite technique for structural chemists and biologists.

Like other 2D NMR techniques, the NOESY pulse sequence has preparation, evolution, mixing, and acquisition periods. It is during the mixing period that the nuclear Overhauser cross-relaxation between the nuclear spins takes place which eventually establishes the correlations. Cross-relaxation only occurs between the nuclei that are close together. The spectrum obtained is similar to COSY, with diagonal and off-diagonal peaks (cross-peaks). A cross-peak is readily interpretable as it occurs at the cross section of the chemical shifts of those nuclei that are close in space to one another. The relationship between the strength of the cross-peaks and the internuclear distance is not straightforward, and in most cases, it is just a rough approximation of the internuclear distance which should be less than 5 Å of the nuclei to be observable.

The basic pulse sequence of the NOESY experiment (Fig. 6.11) comprises three 90° pulses. After excitation by a 90° ^{1}H pulse (preparation pulse), the transverse magnetization evolves in the $x'y'$-plane during a free variable evolution time t_1 period. A second 90° ^{1}H pulse then creates the longitudinal magnetization. During the NOE mixing time, magnetization transfer *via* cross-relaxation or chemical exchange can take place. A final 90° ^{1}H pulse then creates transverse magnetization which is detected as FIDs (Macura and Ernst, 1980).

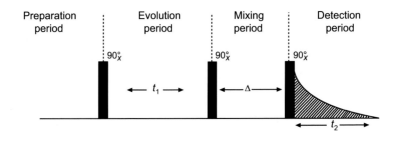

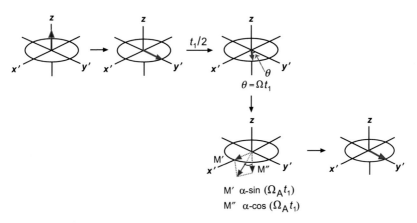

M′ α-sin $(\Omega_A t_1)$
M″ α-cos $(\Omega_A t_1)$

FIGURE 6.11   Pulse sequence and vector representation of a NOESY experiment.

Let us assume that there are two close-lying protons A and X that are not J-coupled together. The first $90°_x$ pulse bends the longitudinal z-magnetization of nucleus A to the y-axis. In the subsequent t_1 time period, this magnetization precesses in the $x′y′$-plane so that after time t_1 it will have traveled by an angle $\Omega_A t_1$, where Ω_A is the precession frequency of nucleus A. The second $90°_x$ pulse will rotate this magnetization from the $x′y′$-plane to the $x′z$-plane. At this time, the $x′$-component of magnetization M′ will be proportional to sin $(\Omega_A t_1)$. The $-z$-component M″ will be proportional to $-\cos (\Omega_A t_1)$. The $-z$-component is of primary interest since it is this component that will be converted (after modulation during the mixing period) to detectable transverse magnetization by the third $90°_x$ pulse. During the mixing period, magnetization transfer occurs between spins A and X so that the z-magnetizations of nuclei A and X are modulated by each other. Thus, spin A magnetization will be modulated by cos $(\Omega_X t_1)$ and spin X magnetization will be modulated by cos $(\Omega_A t_1)$. These modulations will appear as cross-peaks on either side of the diagonal, i.e., at the coordinates $(\Omega_1, \Omega_2) = (\Omega_A, \Omega_x)$ and (Ω_x, Ω_A), thereby establishing the existence of nOe between them. The unmodulated signals appear on the diagonal, as in COSY spectra. A typical NOESY spectrum is shown in Fig. 6.12.

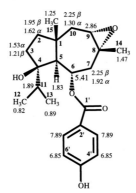

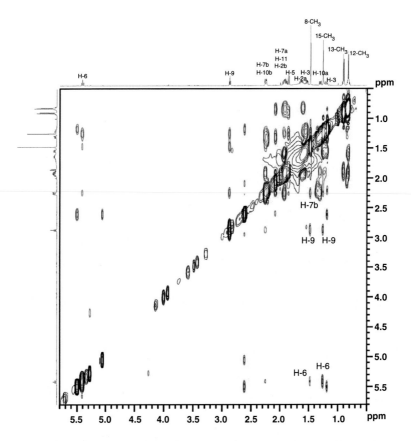

FIGURE 6.12 The 2D NOESY spectrum of compound ferutinin-8,9-α-epoxide *Ferula ovina* (Bioss.) Bioss. The cross-peaks shown in red (off-diagonal peaks) represent through space couplings between the dipolar coupled protons.

The diagonal and cross-peaks (off-diagonal peaks) in the NOESY spectrum have either the same phase or exactly opposite phases, depending on the magnitude of the molecular correlation time τ_c, so NOESY spectra are normally recorded in the pure-absorption mode. Some "COSY peaks" (i.e., peaks due to coherent magnetization transfer from scalar coupling) can appear as artifacts in NOESY spectra; therefore, either we need to cycle the phases of the first two pulses, keeping the receiver phase constant, or we should cycle the last pulse together with the receiver phase to remove COSY-type signals.

In contrast to NOESY where an exchange of *longitudinal magnetization* between close lying protons occurs, in the <u>r</u>otating-frame <u>O</u>verhauser <u>e</u>nhancement <u>s</u>pectroscop<u>y</u> (ROESY) experiment an exchange of *transverse magnetization* is involved. As discussed earlier, large, slow-tumbling molecules exhibit *negative sign nOe*, while small rapidly tumbling molecules yield *positive sign nOe*. This means that various nOe experiments (nOe, nOe difference, NOESY, 1D NOESY) are mostly suited for small organic (~1000 amu or smaller) and large biomolecules (~3000 amu). However, for intermediate weight molecules (between 1000 and 3000 amu), normal nOe is generally very small, i.e., close to zero. In such a situation, ROESY is effectively used. In the ROESY spectrum, cross-peaks always have a sign opposite to the diagonal peaks, irrespective of the molecular weight. The ROESY technique uses a spin lock to align the spins in the rotating frame. Cross-relaxation takes place during the spin lock time, and develops at a rate comparable to T_1. Cross-peak intensity in the ROESY spectrum is generally not prominent, and therefore a large number of scans are required. The key difference between the NOESY and ROESY techniques, and applications of ROESY are presented in Section 7.3.3.

6.8.1 One-Dimensional Selective Transient nOe (1D NOESY) Experiments

NOe difference spectroscopy (Section 6.5.5) has been a *work horse* in structural chemistry for decades. Though this technique is still valid and used, it has been largely superseded by 1D selective transient nOe experiments in the last decade, which are now extensively used in structural chemistry, particularly involving small molecules. 1D Transient nOe experiments, also known as 1D NOESY, are finding applications in cases where quantitative or qualitative measurements of distances between nuclei are required.

Before discussing the key advantages of 1D NOESY over nOe difference spectroscopy, it is useful to briefly discuss the differences between the *steady-state* and *transient* nOes. *Steady-state nOe*, measured by classical 1D nOe difference experiments, involves the use of extended selective ^{1}H saturation of the S-spin. In these experiments, a long on- and off-resonance saturation period (>1–3 longest T_1 in the molecule) is used for accurate measurement of % nOe in the difference spectrum. During the long saturation period, a *"steady-state"* (equilibrium) condition is reached for the I-spin population (dipolarly coupled nucleus), on the time scale of T_1. If the saturation and recovery periods are not set to a few T_1, the selective saturation is not enough to create the full nOe and spins are not fully relaxed back to equilibrium before saturation. In such cases, significant errors may contribute in respect of the measured % nOe. A long saturation period, however, also creates other issues, such as hardware stability, selectivity, development of through-bond coupled dispersive signals, etc. which contribute to artifacts in the spectrum. For these reasons, steady-state 1D nOe experiments are often the trickiest to handle. The *steady-state nOe* between the two nuclei is quoted as a percentage, the maximum possible increase (% nOe) being 50%. At best, the *steady-state nOe* experiments can only provide information about the *relative* internuclear distances, and they are not an absolute measurement of internuclear separation. In contrast, in the *transient nOe* or 1D NOESY experiments, the nOe is built up during a short mixing period ($\sim$0.5 second) with the spins relaxing along the z-axis. In these experiments with careful calibration, fairly accurate distance information can be obtained between the dipolar coupled nuclei, particularly through a nOe built-up curve measurement. *Transient nOe* experiments are therefore most effective in determining the relative proximity of the nuclei (typically <5– 6 Å) based on the rate of increase of the nOe intensity. In all such cases, the gradient enhanced 1D NOESY experiments are ideal (Stott et al, 1997). The delay between the scans can be short ($\sim$1 second) in these experiments, and the artifacts are much smaller than in nOe difference spectra.

A useful transient nOe experiment is presented here. The 1D selective transient nOe pulse sequence is called the *D*ouble *P*ulsed *F*ield *G*radient *S*pin *E*cho experiment (DPFGSE) (Fig. 6.13) (Stott et al., 1997). The pulse sequence involves the selective excitation of the selected resonances by using the selective pulse field gradient or DPFGE block of pulses. This is followed

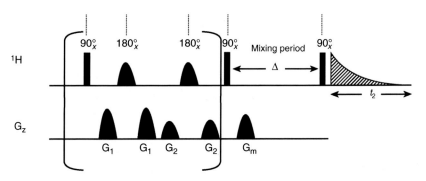

Selective excitation (DPFGSE)

FIGURE 6.13 The gradient-selected DPFGS-NOE pulse sequence for 1D selective transient or 1D ge-NOESY experiments.

by a mixing period, consisting of the basic 90° (¹H) pulses with a delay in between. During this mixing period, in-phase polarization transfer to other dipolar coupled spins *via* nOe occurs. Purging gradients are usually applied during
the mixing period in order to remove any residual transverse magnetization.
At the end, proton detection is carried out as usual. The resulting spectra are
ultra-clean and artifact-free, and therefore easy to interpret. Many variants of
1D NOESY are now available.

   In selective transient nOe or 1D NOESY spectra, through space connectivities from selected spin (nucleus) can be traced out in a very simple way
(Fig. 6.14).

Problem 6.15

What factors determine the choice between 1D selective NOESY *versus* 2D
NOESY?

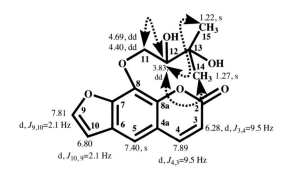

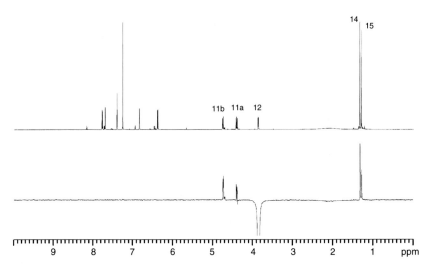

FIGURE 6.14   Selected 1D ge-NOESY spectra of heraclenol isolated from *Ferula ovina* (Bioss.)
Bioss. recorded with a mixing time of 120 ms.

6.9   SATURATION TRANSFER DIFFERENCE (STD) NMR SPECTROSCOPY

While nOe techniques are extensively used in the structure determination of both small organic and large bio-molecules, another important use of nOe is in the study of ligand-receptor interactions. When ligand molecules bind to a receptor protein, the nOes become clearly visible. The change in the intensity of selected protons of ligand molecule provides a quantitative measure of ligand-receptor binding, and also characterizes the part of the ligand molecule which is in contact with the receptor protein (so-called epitope). This technique is called Saturation Transfer Difference (STD) NMR spectroscopy (Mayer and Meyer, 1999). STD NMR is therefore based on a simple principle of observation of the transferred nOe effect from the receptor protein to the binding ligand.

As discussed in section 6.5, low molecular weight compounds (molecular weight <1000) (ligands) have short molecular correlation times (τ_c) and exhibit positive nOe. In contrast, high molecular weight molecules (such as receptor proteins) have long molecular correlation times (τ_c) and thus exhibit strong negative nOe. When low molecular weight ligand tightly and irreversibly binds to a large molecular weight receptor protein, it behaves as an integral component of the receptor molecule and starts exhibiting a negative nOe. This change in sign and size of nOe (nOe behavior) is called as *transferred* nOe. Transferred nOes can be quantitatively measured by recording difference ^{1}H-NMR spectra of the ligand molecules in the presence and absence of the receptor protein. The main advantages of the transferred nOe techniques are that they do not need large amounts of the receptor proteins (μM protein concentration), no labelling of the receptor protein is required, and there is no limitation of the size of the protein. In contrast, if the ligand is loosely and reversibly interacting with the protein, it exhibits a positive nOe which can be measured by STD NMR spectroscopy.

STD NMR spectroscopy has emerged as a powerful application of transferred nOe that allows the robust determination of ligand orientation of the bound ligand in the binding pocket of the receptor proteins. STD NMR experiments are now used for characterization of ligand binding to immobilized proteins, to viruses, to membrane-integrated proteins, and for epitope mapping, *etc* (Meyer and Peters, 2003). Interestingly STD effects can also be measured in intact human and bacterial cells (between a protein present in intact cell and an external ligand interacting with it). Another exciting application of STD NMR spectroscopy is in lead identification and optimization during the drug discovery process (target based drug discovery) (Viegas et al., 2011).

An STD NMR experiment starts with the selective irradiation of the protons of the large biomolecule, such as a protein, by using a train of Gaussian *Rf* pulses. The resulting *Rf* saturation is then rapidly propagated across the receptor protein through a spin diffusion effect. When a low molecular weight ligand binds to the receptor protein, *Rf* saturation is also transferred to the bound ligand by a cross-relaxation process at the ligand-protein interface, leading to

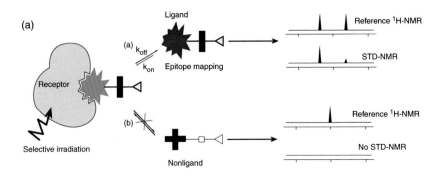

(a)

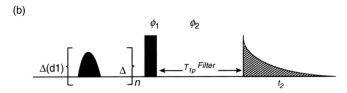

(b)

FIGURE 6.15 (a) Schematic representation of STD NMR spectroscopy. Initially, the protons of the receptor proteins are irradiated with a train of selective *Rf* Gaussian shaped pulses for 1 to 2 seconds. This *Rf* saturation is then transferred from the receptor protein to the ligands that is in exchange between a bound and free forms. The ¹H-NMR spectra of the ligand molecule, without and with receptor protein, are then recorded. The difference of these spectra shows signals of only those protons of the ligand that have received saturation from the receptor protein through binding. (b) Pulse sequence of STD NMR spectroscopy.

an increase in the intensity *of only those ligand resonances that are in close proximity of the receptor protein.* The STD difference spectra are obtained by subtracting the resulting *on-resonance* spectrum from the reference spectrum (*off-resonance* without saturation of protein). This simple concept as well as the pulse sequence of STD NMR experiment is shown in Fig. 6.15.

Problem 6.16

Why has STD NMR spectroscopy emerged as an important tool for early phase drug discovery?

STD NMR methods measure the difference of two experiments, as shown in Fig. 6.16. In the *on-resonance* experiment, the receptor proton magnetization is selectively saturated by a train of selective *Rf* pulses (saturation time 1 to 2 second with a train of Gaussian pulses, each 50 ms long with a spacing of 1 ms). The *Rf* pulses are applied to the frequency region where majority of the methyl resonances of various amino acids in receptor proteins appear, and as a result the entire protein is saturated uniformly (e.g. $-1.0 - -2.0$ ppm in a folded protein). The selection of *on-resonance* irradiation frequency values is critically important for successful STD experiments. *This range should not overlap*

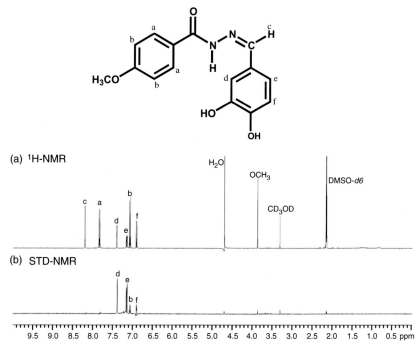

(a) ¹H-NMR

H₂O

OCH₃

DMSO-d6

c a   b

d f

e

CD₃OD

(b) STD-NMR

d

e

b f

9.5 9.0 8.5 8.0 7.5 7.0 6.5 6.0 5.5 5.0 4.5 4.0 3.5 3.0 2.5 2.0 1.5 1.0 0.5 ppm

FIGURE 6.16 The STD NMR spectra of N'-(3,4-dihydroxybenzylidene)-4-methoxybenzohydrazide (ligand) interacting with urease enzyme (receptor) as an inhibitor. (a) Normal ¹H-NMR spectrum displaying all protons of the ligand molecule. (b) STD spectrum showing only the signals of those ligand protons which experience close interaction with the receptor protein.

with any of the ligand resonances. Irradiation frequency values between −1 to −2 ppm are most suitable as it is unlikely to have any ligand protons resonating in this region. For a non-aromatic ligand, the *on-resonance* frequency values can be placed between 7-8 ppm or even further downfield.

The *Rf* saturation propagates from the selected receptor protons to other receptor protons through the vast network of intramolecular ¹H-¹H cross relaxation pathways (intramolecular nOe). This process, also called spin diffusion, is quite efficient due to the typically large molecular weight of the receptor protein. The *Rf* saturation of the receptor protein is then transferred to the bound form of the low molecular weight ligand through intermolecular ¹H-¹H cross-relaxation (intermolecular nOe), and chemical exchange mechanisms at the ligand-receptor interface. The small molecular ligand then dissociates back into solution where a saturated state persists due to the small longitudinal relaxation rate constants (T_1). Since the T_1 relaxation rate is much slower than the rate of dissociation of ligands from the saturated receptor protein, the protein has the capacity to saturate many ligand molecules. This means that "fresh" unsaturated ligands continue to *exchange on and off the receptor*, while saturation energy continues to be pumped into the system through the sustained application of *Rf* pulses. This process increases the population of saturated free ligands in solution.

A reference *off-resonance* spectrum is then recorded in which identical *Rf* pulses are applied far from the *on-resonance* region (typically 25-50 ppm away from the *on-resonance* experiment where no protein proton signal is expected to resonate). The *on-* and *off-resonance* experiments are recorded in an interleaved fashion to avoid any experimental inconsistencies and to minimize the effect of any *Rf* induced sample temperature changes. The resulting difference spectrum shows only those resonances of the ligand that have experienced saturation, namely, the receptor and the binding ligand resonances. Because of the minimal concentration of the receptor protein (usually 50 − 100-fold less than the ligand concentration), receptor protein resonances are normally not visible. The outcome of a successful STD NMR experiment is a simple 1D ^{1}H-NMR spectrum that reveals only the binding protons of the ligand and thus provides valuable information about interactions between small molecules and large biomolecules.

SOLUTIONS TO PROBLEMS

6.1   The nOe experiment is one of the most powerful and widely exploited methods for structure determination of small organic molecules, as well as large biomolecules such as proteins. The nOe difference (NOED) or the 2D experiment, NOESY, is used extensively for stereo-chemical assignments. These experiments provide an indirect way to extract information about internuclear distances. Another use of nOe is in signal intensification in certain NMR experiments, such as in broad-band decoupled ^{13}C-NMR experiments.

6.2   The magnitude and *sign* of nOe is the direct result of competition between various cross-relaxation pathways. When the high energy W_2 pathway ($\alpha\alpha \leftrightarrow \beta\beta$) is dominant, *positive sign nOe* is observed. In contrast when low energy W_0 pathway ($\alpha\beta \leftrightarrow \beta\alpha$) dominates, *a negative sign nOe* is observed. Only these cross-relaxation pathways contribute to the generation of nOe effect. W_2–W_0 transitions determine the *sign* of the observed nOe for a molecule. The magnitude and sign of nOe is, therefore, directly relevant to the molecular correlation time or molecular motion. Small molecules in low viscosity solvents have rapid molecular tumbling and short correlation time τ_c. These molecules have predominately the W_2 cross-relaxation pathway operational, and thus exhibit *positive sign* nOes. Large molecules (or small molecules in highly viscous solvents) have slow molecular tumbling and long correlation times τ_c, thus favouring the W_0 pathway and display *negative sign nOes*. The *sign* of nOe, therefore, has a special significance in determining the molecular motion and size, and it is extensively used in dynamic and molecular recognition studies.

6.3   The positive nOe observed in small molecules in non-viscous solution is mainly due to double-quantum W_2 relaxation, whereas the negative nOe observed for macromolecules in viscous solution is due to the predominance of the zero-quantum W_0 cross-relaxation pathway.

6.4 If only single-quantum transitions (I_1, I_2, S_1, and S_2) are active as relaxation pathways, saturating nucleus S would not affect the intensity of nucleus I; in other words, there will be no nOe at I due to S. This is fairly easy to understand with reference to Fig. 6.2. After saturation of S, the population difference between levels 1 and 3 and that between levels 2 and 4 will be the same as at thermal equilibrium. At this point W_0 or W_2 relaxation processes act as the predominant relaxation pathways to restore somewhat the equilibrium population difference between levels 2 and 3 and between levels 1 and 4 leading to negative or positive nOe, respectively.

6.5 The molecular correlation time τ_c is the average time required for the molecule to rotate through an angle of 1 radian about any axis. τ_c roughly corresponds to the interval between the two successive reorientations or positional changes of the molecule due to complex molecular motions in solution, including translations and rotations. These motions may be intermolecular and intramolecular in nature. Intramolecular motions mean that various parts of the same molecule move in different ways and at different speeds. All of these motions are represented by a single parameter, the molecular correlation time τ_c. The parameter τ_c depends largely on the molecular weight, the viscosity of the solution, and the temperature. Hydrogen bonding and pH also influence τ_c. This means that molecules with higher molecular weights tend to move more slowly in a solution (longer τ_c) than the low molecular weight molecules (shorter τ_c). The variations in τ_c therefore determine the dominant cross-relaxation pathway(s), and influence the size and sign of the resulting nOe.

6.6 *Transient nOe* represents the *rate* of nOe buildup. The nOe effect (so-called equilibrium value) itself depends only on the competing balance between various complex relaxation pathways. However, the initial rate at which the nOe grows (so-called *transient nOe)* depends only on the rate of cross-relaxation τ_c between the relevant dipolar coupled nuclei, which in turn depends on their internuclear distance (r).

6.7 The rate of growth of nOe (transient nOe) can be measured readily by the following pulse sequence:

$$(180°)\text{Sel} - \tau_m - 90° - \text{acquisition}$$

The procedure involves applying a selective inversion pulse on a selected resonance as the initial perturbation. This is followed by a variable waiting period τ_m by application of a 90° pulse before acquisition. During these variable waiting periods, the nOe will build up on the dipolar coupled nuclei. A plot of the intensities of the observed nucleus *versus* the waiting period gives a graph in which the slope represents the transient nOe. The following are transient nOe spectra obtained using the pulse sequence shown above with a variable delay, along with the normal ^{1}H-NMR spectrum.

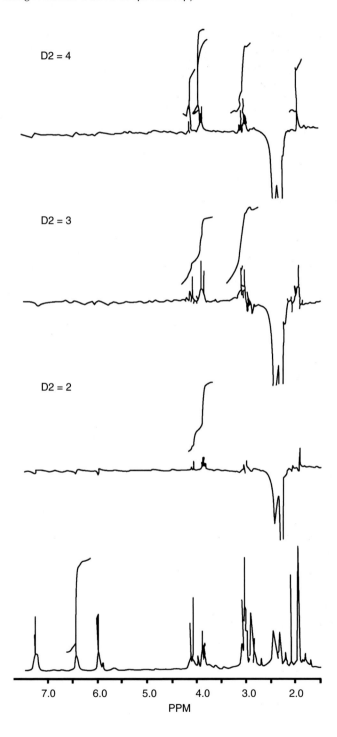

D2 = 4

D2 = 3

D2 = 2

7.0   6.0   5.0   4.0   3.0   2.0

PPM

6.8   From the figure it is obvious that the central nucleus, B, relaxes almost twice as fast as nuclei A and C because it has two nuclei providing the relaxation pathways. However, nucleus B contributes to virtually all the relaxation of both A and C. The saturation of B results in about 49.0% nOe of both A and C. Since the internuclear distance between A and B is greater than that between B and C, the nOe effects experienced at B by saturation of A and C are not equal. When C is saturated, the nOe enhancement at B is almost 50% (actually, 49.2%). In contrast, saturation of A gives only a very small nOe at B. This is because of the fact that A contributes very little to the relaxation of B relative to C. The negative enhancements shown in the figure are quite different from the general negative nOe's that arise *via* a different mechanism. For example, saturation of spin A directly increases the net magnetization of spin B and also, to some extent, of spin C. In other words, decreasing the population difference in A increases the population difference in B and, to a lesser extent, in C by the direct effect. The increased population difference in B, however, will lead to a more pronounced decrease in the population difference in C, thereby canceling the direct positive nOe effect of nucleus A on C leading to an overall negative nOe at C.

6.9   Three-spin effects arise when the non-equilibrium population of an enhanced spin itself acts to disturb the equilibrium of other spins nearby. For example, in a three-spin system, saturation of spin A alters the population of spin B from its equilibrium value by cross-relaxation with A. This change, in turn, disturbs the whole balance of relaxation at B, including its cross-relaxation with C, so that its population disturbance is ultimately transmitted also to C. This is the basic mechanism of *indirect nOe*, or the three-spin effect.

The magnitude of the indirect nOe effect depends on the geometry of the three-spin system. It is maximum with a linear geometry of the spin system ($\theta = 180°$). When θ decreases, the distance r_{AC} also decreases, and the direct enhancement at C becomes more and more significant. At $\theta = 78°$, the direct and indirect nOe effects become equal. With smaller values of θ, the direct contribution rapidly starts to dominate the nOe at C, and a strong positive enhancement results. This means that indirect nOe effects are to be expected mainly when the spins are close to having a linear geometry.

6.10   In a system, when $\theta = 180°$, the distance between spins A and C (r_{AC}) will be a maximum, and the nOe between A and C a minimum. When θ decreases from 180°, r_{AC} decreases and the direct nOe between A and C increases. As a result of the consequently more effective A–C relaxation, the B–C relaxation process becomes relatively less important, and the indirect negative A–C contribution is correspondingly decreased. When $\theta = 78°$, $r_{AC} = 1.26$, so there will be no net nOe between

A and C, even though the A and C spins are very close to each other. This is because the direct (positive) and indirect (negative) nOe effects are equal and opposite.

6.11 In the nOe difference spectrum, only the nOe effects of interest remain, while the unaffected signals are removed by subtraction. It does not, therefore, matter if the nOe responses are small or buried under the unaffected signals, since they show up in the difference spectrum. The main benefit of nOe difference spectroscopy is that it converts the changes in intensity into a form that is more readily recognizable. The difference spectrum is obtained by a process in which a control (normal) spectrum is subtracted from a spectrum acquired with the irradiation of a particular signal.

6.12 Paramagnetic materials, such as dissolved oxygen and other gases, contribute to the relaxation of nuclei after their initial perturbation. It is, therefore, necessary to exclude paramagnetic materials, principally oxygen, from the solutions in NMR tubes in order to derive a true picture of internuclear cross-relaxations during the nOe experiment. This is generally achieved by degassing the solutions through freeze-drying. Aqueous solutions are difficult to freeze without the risk of cracking the NMR tubes. They are therefore degassed in a separate flask and then transferred to the NMR tubes.

6.13 When the zero-quantum W_0 transition is greater than the double-quantum W_2 transition, the nOe enhancements will be negative. Similarly, when W_2 is greater than W_0, the resultant nOe will have a positive sign. The predominance of W_2 and W_0 over one another depends on the molecular motion. It is known that the W_0 transition is maximal when the molecule tumbles at a rate of about 1 kHz, while the W_2 transition is maximal at a tumbling rate of about 800 MHz. On this basis, a rough idea of the sign of nOe can be obtained. For example, small molecules in nonviscous solvents tumble at a faster rate than larger molecules. It is, therefore, reasonable to expect that for small molecules, W_2 will dominate over W_0 and the sign of the enhancement will therefore be positive.

6.14 Following are some important precautions for sample preparation.
   1. Removal of solid particles by filtration.
   2. Removal of paramagnetic impurities, such as paramagnetic metal ions and molecular oxygen. These metal ions can be removed easily be adding a small amount of ethylenediamine tetraacetate (EDTA). Molecular oxygen can be removed by degassing the sample. Paramagnetic impurities lead to rapid spin-lattice relaxation and therefore reduce the intensities of enhancement.
   3. The concentration of solute in the sample should not be very large to avoid aggregation.

6.15 The choice between 2D-NOESY *versus* 1D selective NOESY depends on the amount of material available, and the information needed (i.e. protons that need to be studied for their spatial orientation). A single 2D NOESY experiment provides all NOE information simultaneously, whereas the 1D experiments provide nOes one at a time. The minimum amount of time required for 2D and 1D experiments also differs. The 2D NOESY spectrum requires relatively longer times (normally 30 minutes to 1.5 hours), whereas the minimum time required for a single 1D selective NOESY spectrum is much less (often about 2 minutes). Thus, if more quantity of sample is available (0.04 M or more), and you are interested in less than five nOe points, then the 1D NOESY versions can be preferred. If only little material (< 0.005 M) is available or many nOe points need to be studied, then the 2D version should be chosen. This also means that for small molecules 1D NOESY is the usual choice, while for large molecules, NOESY is preferred. The 1D version has the advantage of easy and fast processing.

6.16 In early phase lead drug discovery, low molecular weight compounds (e.g. organic compounds, peptides, oligosaccharides, etc.) are screened against a biological target (receptors) by using biochemical methods. The binding affinities between the ligands and receptors are measured through ELISA-based colorimetric or flourimetric methods. The same screening can now be performed by using extremely efficient and robust STD NMR techniques. In STD NMR-based studies, the same biochemical screening assays are performed, but the binding affinities between the small molecular ligands and the target protein are measured through nOe-based intensity enhancements of the epitope signals of the ligands.

STD NMR spectroscopy has emerged as a powerful NMR tool in pharmaceutical research. It identifies binding epitope(s) of small molecular ligands at the atomic resolution, while interacting with their biological receptors, such as proteins and/or nucleic acids. The method is now widely used to screen bioactive substances (drug molecules), while at the same time ranking them in a qualitative manner based on their binding affinities. STD NMR is applicable to a variety of high molecular weight drug targets, such as enzymes, viruses, platelets, intact human and bacterial cells, lipopolysaccharide micelles, membrane proteins, recombinant proteins, and dispersion pigments. The most successful use of the STD NMR-based drug discovery has been in the discovery of clinical inhibitors of enzymes, associated with various diseases, such as acetylcholinesterase (Alzheimer's disease), urease (peptic ulcer and urolithiasis), α-glucosidase (post-prandial hyperglycemia), etc.

REFERENCES

Andersen, N.H., Nguyen, K.T., Hartzell, C.J., Eaton, H.L., 1987. Rapid acquisition of NOE difference spectra. The relative merits of transient *versus* driven experimental protocols. J. Magn. Reson. 74 (2), 195–211, (1969).

Bell, R.A., Saunders, J.K., 1970. Correlation of the intramolecular nuclear Overhauser effect with internuclear distance. Can. J. Chem. 48 (7), 1114–1122.

Bigler, P., Kamber, M., 1986. Heteronuclear Overhauser effect measured with selective proton irradiation. Application of a new method. Magn. Reson. Chem. 24 (11), 972–976.

Granger, P., 1990. Polarization transfer INEPT, DEPT, NOE. Multinucl. Magn. Reson. Liq Solids: Chem. Appl. 322, 63–79.

Jerschow, A., Müller, N., 1998. Convection compensation in gradient enhanced nuclear magnetic resonance spectroscopy. J. Magn. Reson. 132 (1), 13–18.

Kinns, M., Sanders, J.K.M., 1984. Improved frequency selectivity in nuclear Overhauser effect difference spectroscopy. J. Magn. Reson. 56 (3), 518–520, (1969).

Köver, K.E., 1984. A simple method for selective saturation of multiplets in 1H-{1H} NOE measurements. J. Magn. Reson. 59 (3), 485–488, (1969).

Loening, N.M., Keeler, J., 2002. Temperature accuracy and temperature gradients in solution-state NMR spectrometers. J. Magn. Reson. 159 (1), 55–61.

Macura, S., Ernst, R.R., 1980. Elucidation of cross relaxation in liquids by two-dimensional NMR spectroscopy. Mol. Phys. 41 (1), 95–117.

Macura, S., Huang, Y., Suter, D., Ernst, R.R., 1981. Two-dimensional chemical exchange and cross-relaxation spectroscopy of coupled nuclear spins. J. Magn. Reson. 43 (2), 259–281, (1969).

Mayer, M., Meyer, B., 1999. Characterization of ligand binding by saturation transfer difference NMR spectroscopy. Angew. Chem. Int. 38 (12), 1784–1788.

Meyer, B., Peters, T., 2003. NMR spectroscopy techniques for screening and identifying ligand binding to protein receptors. Angew. Chem. Int. 42 (8), 864–890.

Neuhaus, D., 1983. A method for the suppression of selective population transfer effects in NOE difference spectra. J. Magn. Reson. (1969) 53 (1), 109-114.f.

Neuhaus, D., 2012. Nuclear Overhauser effect. In: Harris, R. K., Wasylishen, R.E. (Eds.), Encyclopedia of NMR, 6, Wiley, New York, pp. 3020–3035.

Shaka, A.J., Bauer, C., Freeman, R., 1984. Selective population transfer effects in nuclear Overhauser experiments. J. Magn. Reson. 60 (3), 479–485, (1969).

Stott, K., Keeler, J., Van, Q.N., Shaka, A.J., 1997. One-dimensional NOE experiments using pulsed field gradients. J. Magn. Reson. 125 (2), 302–324.

Stott, K., Stonehouse, J., Keeler, J., Hwang, T.L., Shaka, A.J., 1995. Excitation sculpting in high-resolution nuclear magnetic resonance spectroscopy: Application to selective NOE experiments. J. Am. Chem. Soc. 117 (14), 4199–4200.

Viegas, A., Manso, J., Corvo, M.C., Marques, M.M., Cabrita, E.J., 2011. Binding of ibuprofen, ketorolac, and diclofenac to COX-1 and COX-2 studied by saturation transfer difference NMR. J. Med. Chem. 54 (24), 8555–8562.

Vögeli, B., 2014. The nuclear Overhauser effect from a quantitative perspective. Prog. Nucl. Magn. Reson. Spectrosc. 78, 1–46.

Williams, D.H., Williamson, M.P., Butcher, D.W., Hammond, S.J., 1983. Detailed binding sites of the antibiotics vancomycin and ristocetin A: Determination of intermolecular distances in antibiotic/substrate complexes by use of the time-dependent NOE. J. Am. Chem. Soc. 105 (5), 1332–1339.

Williamson, M.P., Williams, D.H., 1985. 1H-NMR studies of the structure of ristocetin A and of its complexes with bacterial cell wall analogues in aqueous solution. J. Chem. Soc., Perkin Trans. 1, 949–956.

Chapter 7

Important 2D NMR Experiments

Chapter Outline

Solving Problems with NMR Spectroscopy. http://dx.doi.org/10.1016/B978-0-12-411589-7.00007-3

265

7.1 HOMONUCLEAR AND HETERONUCLEAR SHIFT-CORRELATION SPECTROSCOPY

7.1.1 Homonuclear Shift-Correlation Spectroscopy (COSY)

Before the advent of two-dimensional (2D) NMR spectroscopy, the classical procedure for determining proton–proton connectivities was by homonuclear proton spin decoupling experiments. Such experiments can still serve to determine some $^1H/^1H$ connectivities in simple molecules.

A useful type of 2D NMR spectroscopy is shift-correlated spectroscopy (COSY), in which both axes describe the chemical shifts of the coupled nuclei, and the cross-peaks obtained tell us which nuclei are coupled to which other nuclei. The coupled nuclei may be of the same type – for example, protons coupled to protons, as in *homonuclear* 2D shift-correlated experiments – or of different types – for example, protons coupled to ^{13}C, or ^{15}N nuclei, as in *heteronuclear* 2D shift-correlated spectroscopy. Thus, in contrast to *J*-resolved spectroscopy, in which the nuclei were being modulated (i.e., undergoing a repetitive change in signal amplitude) by the *coupling frequencies* of the nuclei to which they were coupled, in shift-correlated spectroscopy the nuclei are modulated by the *chemical shift frequencies* of the nuclei to which they are coupled. So, if a proton H_A (having a chemical shift of 2 ppm, i.e., 200 Hz from TMS on a 100-MHz instrument) is coupled to a H_B (having a chemical shift of 4 ppm, i.e., 400 Hz from TMS on the same instrument) with a coupling constant of 8 Hz, then in *J*-resolved spectroscopy the protons H_A and H_B will both be modulated as a function of 8 Hz, while in shift –correlated spectroscopy H_A will be modulated as a function of δ_{H_B} – i.e., 400 Hz – while H_B will be modulated as a function of δ_{H_A} – i.e., 200 Hz. The variation of signal amplitude with changing evolution time t_1 is modulated as shown in Fig. 7.1. If the signal amplitude is modulated by more than one frequency, then a different pattern would result. Fig. 7.2 shows the amplitude being modulated by two different chemical shift frequencies.

When a Jeener pulse sequence ($90°_x - t_1 / 2 - 90°_x - t_1 / 2 - $ Acq.) is applied to coupled protons, the second $90°_x$ pulse causes the magnetization arising from one proton transition during t_1 to be distributed among all other transitions associated with it. For instance, in an AX spin system, the 1D spectrum will show two lines for proton A, corresponding to the transitions A_1 and A_2, and two lines for proton X, corresponding to the transitions X_1 and X_2. The energy-level diagram for such a spin system in which only single-quantum transitions are shown in Fig. 7.3. The second $90°_x$ pulse causes the magnetizations arising from each transition to be distributed among all the four transitions. Thus, the magnetization arising from the A_1 transition will be partly retained in the A_1 transition and partly distributed among the A_2, X_1, and X_2 transitions. A similar redistribution of magnetization will occur for the A_2, X_1, and X_2 transitions. A line arising from

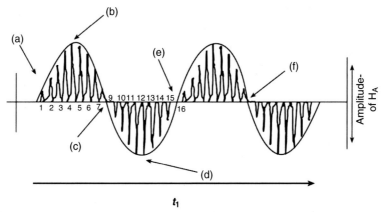

FIGURE 7.1 Variation in the amplitude of a proton H_A with changing evolution time t_1.

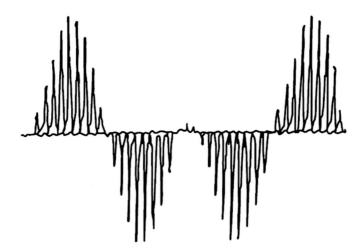

FIGURE 7.2   The amplitude of the lines is seen to be modulated by at least two different frequencies.

the A_1 transition will therefore have its amplitude modulated by the frequency components of all the other transition lines of the AX spin system, giving rise to corresponding cross-peaks in the COSY spectrum.

As is evident from Fig. 7.4, we can have either a sinusoidal modulation or a cosinusoidal modulation of the signals, depending on whether the two $90°_x$ pulses have the same phase or different phases. In Fig. 7.4a, both pulses have the same phase. This means that in the first t_1 period, the magnetization vector

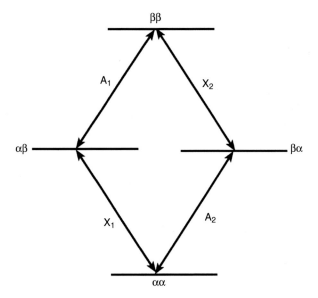

β β

A₁   X₂

αβ   βα

X₁   A₂

αα

FIGURE 7.3   Energy levels for an AX spin system representing only single-quantum transitions.

FIGURE 7.4 (a) The two 90°_x pulses are applied with an intervening variable t_1 period, which ▶ is systematically incremented. The first 90°_x pulse bends the magnetization from its equilibrium position along the z-axis in (i) to the y' -axis, as in (ii). During the t_1 interval, the magnetization vector precesses in the $x'y'$ -plane, and it may be considered to be composed of two component vectors, **M'** and **M''**, along the x'- and y'-axes, respectively. The magnitudes of the component vectors **M'** and **M''** are **M'** = M sinθ and **M''** = M cosθ, where θ is the angle of deviation of **M** from the y'-axis. The second 90°_x pulse bends the component **M''** lying along the y'-axis to the $-z$-axis so that it is no longer detectable, leaving the sinusoidal modulations of the **M'** vector along the x'-axis to be recorded. Therefore, if both the 90° pulses have the same phase, a *sinusoidal* modulation of the signals is observed. (b) Effect of two 90° pulses that are 90° out-of-phase with each other on the magnetization vector, **M₀**. (i) The first 90°_x pulse bends **M₀** to the $-x'$-axis. (ii) In the subsequent delay, t_1, the magnetization **M** precesses in the $x'y'$ -plane. After time t_1 it will have precessesd away from the $-x'$-axis by an angle θ, and its two components along the x'- and y'-axes will have magnitudes **M'** =M cos θ and **M''** =M sin θ, respectively. (iii) The second 90°_x pulse removes the vector component along the y'-axis and bends it to the $-z$-axis. This leaves only one vector component along the $-x'$-axis (**M'** = M cos θ) to be measured. Consinusoidal modulation of the magnetization is, therefore, observed during different t_1 increments when the two 90° pulses applied are 90° out-of-phase with one another. (c) If the first pulse is a 90°_x pulse and the second pulse a 90°_y pulse, then a *cosinusoidal* modulation of the signals is observed. Now the second 90°_y pulse bends the component **M''** (=M sin θ) to the z-axis, leaving only the cosinusoidally modulated component **M'** (= M cos θ) in the transverse ($x'y'$) plane to be detected.

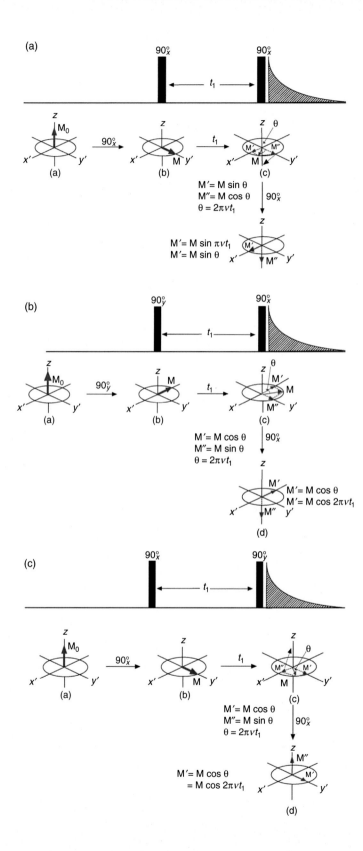

precesses in the $x'y'$-plane, so if it has precessed by an angle θ away from the y'-axis by the end of the t_1 period, it will have two components, $\mathbf{M}'$ and $\mathbf{M}''$ along the x'- and y'-axes, respectively, with magnitudes $\mathbf{M}' = \mathbf{M} \sin \theta$ and $\mathbf{M}'' = \mathbf{M} \cos \theta$. The second 90°_x pulse bends the component $\mathbf{M}''$ lying along the y'-axis to the $-z$-axis so that it will no longer be detected. This will leave the component vector along the x'-axis (with magnitude $M \sin \theta$) to be detected. In other words, *sinusoidal* modulations of the magnetization vector $\mathbf{M}'$ will be observed if *both* the first and the second 90° pulses have the same phase. If, however, they have opposite phases (Fig. 7.4b and c), then the second 90° pulse will remove the sinusoidally modulated component to the z-axis, leaving the cosinusoidally modulated component to be detected. Hence, with appropriate phase cycling, it is possible to obtain *phase-sensitive COSY spectra* in which the absorption- and dispersion-mode signals are nicely separated; i.e., the peaks appearing on or near the diagonal will differ in phase by 90° from the cross peaks. Normally, we adjust the phasing so as to show up the cross-peaks in the absorptive mode and the diagonal peaks in the dispersive mode, since it is the cross-peaks that are of real interest (Fig. 7.5).

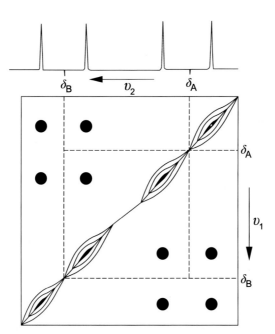

FIGURE 7.5 Schematic diagram of a COSY spectrum illustrating how intrinsically different cross-peaks are symmetrically located in either side of the diagonal peaks. The circles represent the cross-peaks in an AB coupled system. The diagonal peaks have dispersion shape, and the cross-peaks have an absorption shape though they alternate in sense. The diagonal peaks are centered at (δ_A, δ_A) and (δ_B, δ_B); the cross-peaks appear at (δ_A, δ_B) and (δ_B, δ_A).

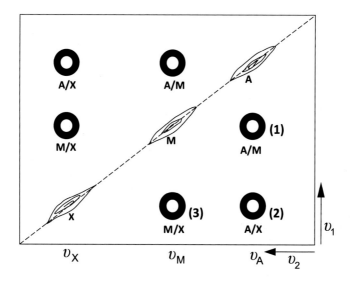

FIGURE 7.6 A schematic drawing of an AMX spin system representing coupling interactions recorded at low digital resolution so that no fine splittings are visible. Note the symmetrical appearance of the cross-peaks on either side of the diagonal.

An advantage of phase-sensitive COSY spectra, as compared to absolute-mode spectra, is that when recorded at enough digital resolution they allow us to determine coupling constants. Let us consider an AMX spin system. A schematic drawing of a phase-sensitive COSY spectrum of an AMX spin system is shown in Fig. 7.6. The coupling between H_A and H_M gives rise to the cross-peak (1) marked A/M. This is designated as the *active coupling* and contains information about J_{AM}. However, H_A is also coupled to H_x and the fine structure in the A/M cross-peak (1) will also contain information about J_{AX} (*passive coupling*). Similarly, cross-peak (2), arising from the coupling of A with X, will contain the active coupling information J_{AX} as well as the fine structure for J_{AM}. The fine splitting can be seen if the phase-sensitive COSY spectrum of fenchone is recorded at a high digital resolution (Fig. 7.7). The active coupling shows an antiphase disposition of cross-peaks, while passive couplings appear in-phase. This is illustrated in the schematic drawing of a cross-peak obtained at v_A, v_X in an AMX spin system (Fig. 7.8). If we read the multiplet horizontally, the distance between the center points of peaks 1 and 2 (or 3 and 4) corresponds to the active coupling J_{AX} (note antiphase disposition).

Similarly, the distance between peaks 1 and 3 (or 2 and 4) represents J_{AM}. If we read the multiplet vertically, the distance between peaks 1 and 5 again corresponds to J_{AX}, while the distance between peaks 1 and 9 provides J_{MX}. If we read horizontally, J_{AX} is active and J_{AM} passive, if we read vertically, J_{AX} is active and J_{MX} is passive. If the antiphase character of the cross-peaks is not readily

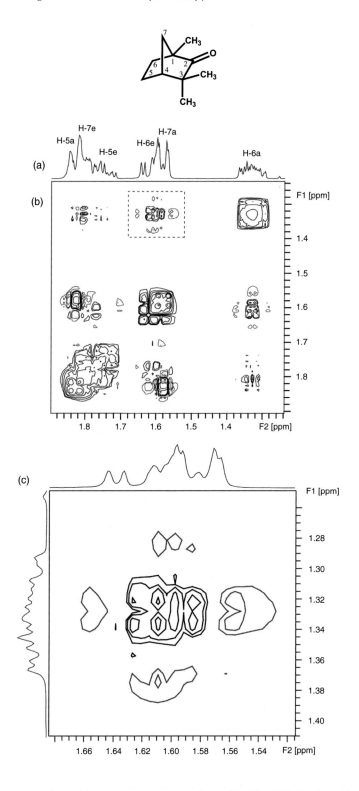

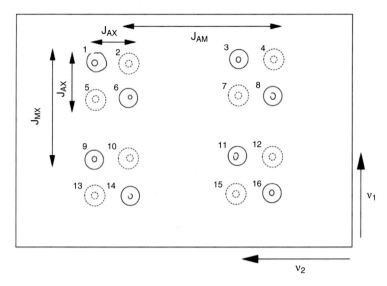

FIGURE 7.8 Drawing of a spectrum with cross-peak at v_A, v_X in an AMX system. The active couplings display an antiphase disposition of peaks. If we read horizontally, J_{AX} is active and J_{AM} passive; if we read vertically, J_{AX} is active and J_{MX} passive.

discernible due to peak overlap, then this may be read from the slices taken at different values of v_1 and v_2, which allows the phase patterns to be seen with greater clarity. This is illustrated in Fig. 7.9. In the phase-sensitive COSY spectrum of ethyl glucoside (Fig. 7.9b), the C-2' methyl protons (δ 1.22) show vicinal coupling with the C-1' oxygenated methylene protons, resonating at δ 3.96 and 3.60, represented by cross-peaks A and B, respectively, in the spectrum. The cross-peak C shows the geminal coupling between the C-1' methylene protons. The cross-peak D corresponds to the coupling of C-1 anomeric proton (δ 4.24) with the C-2 proton (δ 3.15). The vicinal interactions between methine protons of glucose moiety are presented as cross-peaks E–H in the spectrum.

7.1.1.1 COSY-45° Spectra

If the width of the second pulse in the COSY pulse sequence is reduced from 90° to 45°, the mixing of populations between transitions that do not share the same energy level (*unconnected transitions*) is reduced. Only directly connected transitions then show cross-peaks, resulting in a simplification of the multiplet structure and, to some extent, a decrease in the area occupied by the multiplet.

FIGURE 7.7 (a) The ^{1}H-NMR spectrum of fenchone, isolated from *Illicium religiosum* Sieb. (b) A phase-sensitive COSY spectrum recorded at higher digital resolution in which H-6a shows geminal coupling with H-6e and vicinal coupling with H-5a and H-5e. (c) Expansion of a cross-peak identified by a dashed area in (b).

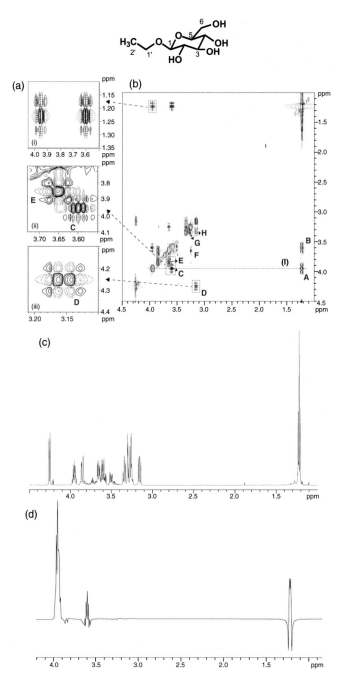

FIGURE 7.9 (a) The cross-peaks in (b) are expanded in (i), (ii), and (iii) to show up the fine structures. (b) Phase-sensitive COSY spectrum of ethyl glucoside. (c) 1D ^{1}H-NMR spectrum of the same compound. (d) A cross-section (slice) along the row marked (I) in the COSY spectrum.

The intensity of the diagonal peaks is also reduced. In such spectra, the direction in which the cross-peak responses are tilted gives the relative signs of the coupling constant. This can be useful in distinguishing positive vicinal coupling constants from negative geminal couplings. The example given in Fig. 7.10 (COSY spectrum of (+)-5-hyroxyisopinocampheol) shows the connectivities between the various protons. The most downfield proton H-3 (δ 4.05) shows coupling interactions with H-2 (δ 1.77) and H$_2$-4 (δ 2.41 and 1.79). H-4a, H-7a show geminal couplings with their protons, i.e., H-4b (δ 1.79) and H-7b (δ 1.47). The C-10 methyl protons (δ 1.13) show a cross-peak with H-2 (δ 1.77).

In general, the ratio of intensities of cross-peaks arising from connected transitions to those arising from unconnected transitions is given by $\cot^2(\alpha/2)$, where α is the flip angle of the last mixing pulse. With a mixing pulse of 45°, the connected transitions give five to eight times stronger cross-peaks than unconnected transitions. A better signal-to-noise ratio will result if the mixing pulse has an angle of 60° (a COSY-60° experiment), with the cross-peaks from connected transitions appearing three-fold stronger than those from unconnected transitions.

Since many of the signals in COSY spectra are in antiphase, they may not show up as cross-peaks due to the intrinsic nature of the polarization transfer experiment. The intensities of cross-peaks in COSY spectra may be represented by an antiphase triangle (Fig. 7.11b), in contrast to multiplet intensities in 1D spectra, in which the multiplicities are represented by a Pascal triangle (Fig. 7.11a). This is illustrated by a schematic drawing of the coupling of H$_4$ and H$_6$ with H$_5$ in 3-bromobenzaldehyde (Fig. 7.12). The COSY cross-peaks are missing at the center of the triplet since its intensity ratio in the polarization transfer experiment is represented by +1, 0, and −1.

Problem 7.1

How can we correlate protons that are coupled to each other (proton coupling network)?

Problem 7.2

How do the peaks appearing on the diagonal differ from the "off-diagonal" cross-peaks in COSY spectra? Why do they arise?

As in 1D spectra, the folding of signals may occur in 2D spectra if the spectral width chosen is too small, giving rise to artifacts. This can be readily corrected by recording the spectra with a wider spectral width. Such artifacts as fold-over peaks or t_1-noise lines running across the spectrum can be corrected by the process of *symmetrization* (Fig. 7.13). Since genuine cross-peaks appear as symmetrically disposed signals on either side of the diagonal, the process of symmetrization will remove all random noise and other non-symmetrically disposed signals on the two sides of the diagonal. Of course, the cosmetic improvement due to symmetrization must be undertaken with caution, for any

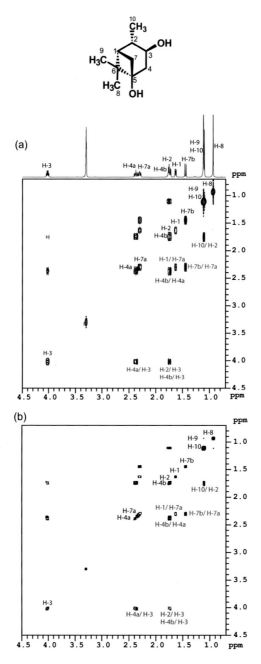

FIGURE 7.10 (a) COSY-90° spectrum of (+)-5-hydroxyisopinocampheol obtained by biotransformation of (+)-isopinocampheol with *Cunninghamella blakesleeana*. (b) COSY-45° spectrum of the same compound.

(a)

```
              1
           1     1
        1     2     1
     1     3     3     1
  1     4     6     4     1
2     5    10    10     5     1
1
```

(b)

```
              1
           1    -1
        1     0    -1
     1     1    -1    -1
  1     2     0    -2    -1
1     3     2    -2    -3    -1
```

FIGURE 7.11 (a) Pascal triangle. (b) Antiphase triangle.

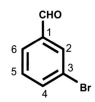

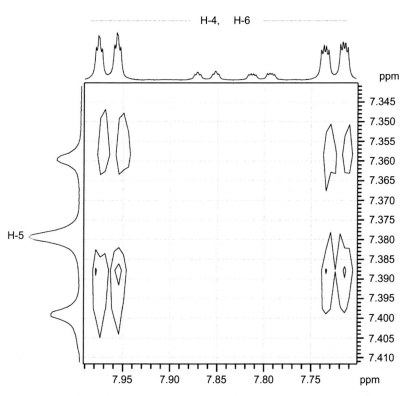

FIGURE 7.12 Representation of the coupling interactions of the H_4, H_5, and H_6 protons of 3-bromo-benzaldehyde. The H_4 and H_6 are split into double doublets due to their couplings with H_5 and H_2; H_5 is split into a "triplet" (double doublet) by the two *ortho* protons. Note that the COSY cross-peaks are missing at the central peak of the triplet due to the intensity ratio of the triplet in the polarization transfer experiment being +1, 0, −1.

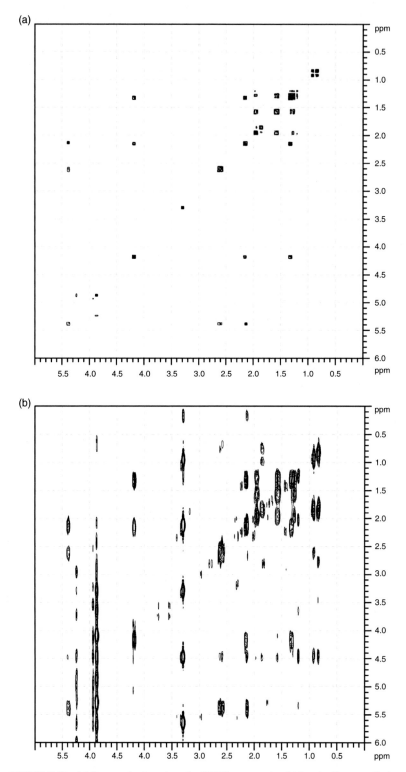

FIGURE 7.13 (a) Symmetrized version of a COSY spectrum in which noise lines are eliminated. (b) Unsymmetrized version of the same spectrum.

noise signals that appear symmetrically on either side of the diagonal will not be removed and they will show up as a false "cross-peak." The application of shaping functions for the improvement of 2D spectra has already been discussed in Section 5.3.2 and will therefore not be repeated here.

Problem 7.3

Following are the 1D ^{1}H-NMR and COSY-45° spectra of citroside B, isolated from *Osyris wightiana* Wall. ex Wight. Interpret the spectra and assign chemical shifts to various protons.

Problem 7.4

Is it possible to obtain proton–proton coupling information from a 1D ^{1}H-NMR experiment?

Problem 7.5

Give the advantages of the COSY-45° experiment over the COSY-90° experiment.

7.1.1.2 Decoupled COSY Spectra

When there is serious overlapping of cross-peaks because the coupled protons lie close to one another, then we can narrow the area of the cross-peaks by decoupling in the v_1 dimension. This is illustrated in Fig. 7.14. All the protons appear as fully decoupled signals at their respective chemical shifts along the v_1 axis, while the projection on the v_2 axis corresponds to the normal coupled spectrum. This provides a useful procedure for obtaining the fully decoupled NMR spectrum and for obtaining the chemical shifts as well as the number of different protons in a molecule – information that may not be readily obtained from the 1D ^{1}H-NMR spectrum because of the overlapping of multiplets. Since the cross-peaks are now fairly thin, the information about all the coupling partners of a given proton (cross-section along v_2) is also obtained much more readily. Many other modifications of the COSY experiment have been developed; some of the more important ones will be presented here.

Problem 7.6

How is decoupling in the v_1 dimension achieved in decoupled COSY spectra, and what are the advantages of this technique?

7.1.1.3 Double Quantum Filtered COSY

One problem associated with COSY spectra is the dispersive character of the diagonal peaks, which can obliterate the cross-peaks lying near the diagonal. Moreover, if the multiplets are resolved incompletely in the cross-peaks, then because of their alternating phases an overlap can weaken their intensity or even cause them to disappear. In *double-quantum* filtered COSY spectra, both the diagonal and the cross-peaks possess antiphase character, so that they can be phased simultaneously to produce pure 2D absorption line shapes in both. In practice, this is done by applying a third 90°_x pulse immediately after the second (mixing) pulse. This third pulse converts the double-quantum coherence generated by the second pulse into detectable single-quantum coherence. *Only* signals due to double-quantum coherence are inverted by the third 90° pulse, and if the receiver phase is correspondingly inverted, then only signals due to double-quantum coherence will be detected (in the case of triple-quantum coherence spectra, the signal inversion will be achieved by 60° phase shift). Since all other coherences are filtered out, only the desired double-quantum coherences will

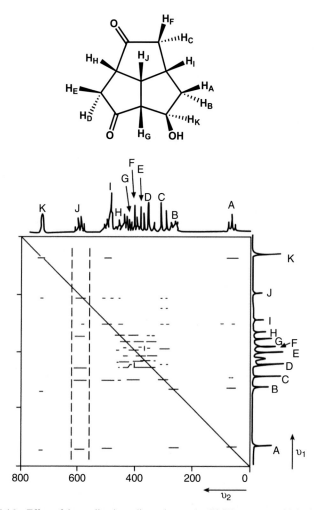

FIGURE 7.14   Effect of decoupling in v_1 dimension on the COSY spectrum of 9-hydroxytricyclo-decan-2,5-dione. The 1D ^{1}H-NMR spectrum on v_2 is fully coupled, while along v_1 it is fully decoupled. The region between the dashed lines represents a band of signals associated with proton H_J and its connectivities with other protons by projection onto the v_1-axis. *(Reprinted from Bax and Freeman (1981) J. Magn. Reson., **44**, 542–561, copyright with permission from Academic Press, Inc.).*

be modulated as a function of t_1 and yield cross-peaks in the 2D spectrum. This will significantly simplify the COSY spectrum. Thus, in a double-quantum filtered spectrum, couplings due to AB and AX spin systems will show up, but solvent signals – which cannot generate double-quantum coherence – or signals due to triple-quantum coherence (as in a three-spin system) will not appear. Direct connectivities are represented in double-quantum spectra by pairs of signals situated symmetrically on the two sides of the diagonal; remotely connected or magnetically equivalent protons will appear as lone multiplets.

The sensitivity of multiple-quantum spectra is less than that of unfiltered COSY spectra. In the case of double-quantum filtered (DQF) COSY spectra, the sensitivity is reduced by a factor of 2, though this is usually not a serious drawback, since in homonuclear COSY spectra we are normally concerned with the more-sensitive ^{1}H/^{1}H coupling interactions. Fig. 7.15 presents the

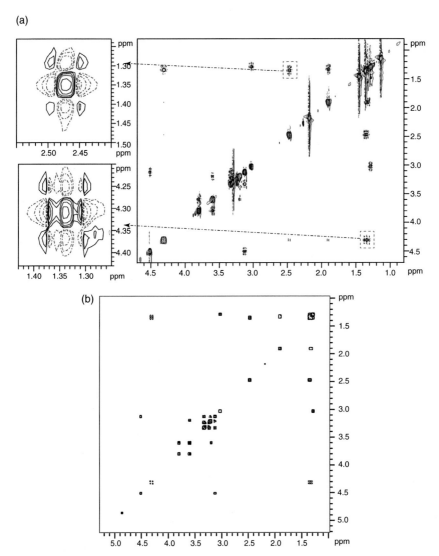

FIGURE 7.15   (a) Conventional phase-sensitive COSY spectrum of citroside B, isolated from *Osyris wightiana* Wall. ex Wight. (b) Double-quantum filtered (DQF) phase-sensitive COSY spectrum of the same compound. Note how clean the spectrum is, especially in the region near the diagonal line.

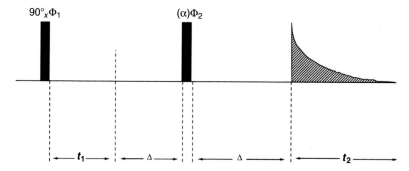

FIGURE 7.16 Pulse sequence for delayed COSY – a modification of the COSY experiment. The fixed delays at the end of the evolution period t_1 and before the acquisition period t_2 allow the detection of long-range couplings between protons.

phase-sensitive COSY and DQF-COSY spectra. The peaks lying near the diagonal are no longer obliterated in the phase-sensitive DQF spectrum. The interpretation of COSY-45° spectrum of citroside B has been discussed in Problem 7.3.

7.1.1.4 Delayed COSY

In normal COSY spectra, the stronger coupling interactions appear as prominent cross-peaks and the weaker long-range couplings may not appear. Introducing a fixed-delay Δ before acquisition results in the weakening (and often even the disappearance) of the stronger coupling interactions and an intensification of the smaller long-range couplings. The pulse sequence used is shown in Fig. 7.16. The value of the fixed delay is optimized between 0.3 and 0.5 second. To improve the sensitivity, the last mixing pulse is set at 50–60°. The pulse field gradient (PFG) COSY pulse sequence is presented in Fig. 7.17 (also see Section 2.4.2).

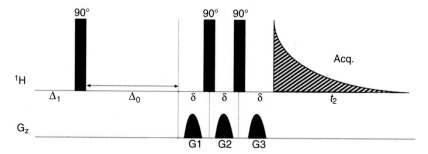

FIGURE 7.17 The PFG-COSY pulse sequence.

When is the delayed-COSY experiment advantageous over the standard COSY experiment?

7.1.1.5   E. COSY

The measurement of coupling constants is possible in multiple-quantum COSY spectra with phase-sensitive display. The recording of spectra in the absolute-value mode is necessary to remove the phase-twist caused by the mixing of absorption and dispersion lines. Pseudoecho weighting functions can be employed, but with some loss of sensitivity.

Many procedures have been developed (States et al., 1982) for measuring coupling constants in 2D spectra; only one will be mentioned here. This is a modification of the COSY experiment that incorporates a mixing pulse β of variable angle in the pulse sequence $90^{\circ}_x - t_1 - (\beta)_y - t_2$. In practice, it is more convenient to use the variable parameter β as a phase shift in the equivalent sequence $(90^{\circ})_\beta - t_1 - (90^{\circ})_\beta (90^{\circ})_{-x} - t_2$. In this $\underline{E}$xchange $\underline{C}$orrelation $\underline{S}$pectroscopy (E. COSY), many experiments with different values of β are combined, so that multiplet components between unconnected transitions are eliminated and only multiplet components arising from connected transitions are detected. Thus, in contrast to conventional COSY spectra, in which spin systems comprising N mutually coupled spins give 2^{2N-2} multiplet components, in E. COSY spectra only 2^N correlations will appear from connected transitions, resulting in a simplification of the multiplet components.

How are coupling constants measured from the E. COSY spectrum?

7.1.1.6   SECSY

Another 2D homonuclear shift correlation experiment that provides the coupling information in a different format is known as $\underline{S}$pin-$\underline{E}$cho $\underline{C}$orrelation $\underline{S}$pectroscop$\underline{Y}$ (SECSY). It is of particular use when the coupled nuclei lie in a narrow chemical shift range and nuclei with large chemical shift differences are not coupled to one another. The experiment differs from the conventional COSY experiment in that the acquisition is not done immediately after the second mixing pulse but is delayed by time $t_1/2$ after the mixing pulse (i.e., the mixing pulse is placed in the middle of the t_1 period). Because the observation now begins at the top of the coherence transfer echo, the experiment is called "spin-echo correlation spectroscopy."

The untransferred magnetization now appears along a horizontal line (instead of along a diagonal, as in COSY). The cross-peaks appear above and below horizontal line at distances equal to half the chemical shift differences between the two coupled nuclei. The identification of cross-peaks is facilitated by the fact that they occur on slanted parallel lines that make the same angle with the horizontal line and by the fact that each pair of cross-peaks is equidistant above and below the horizontal line. To determine the connectivities, we drop lines vertically from the cross-peaks onto the horizontal line; the points at which they meet the horizontal line give the chemical shifts of the coupled protons. This is illustrated in the SECSY spectrum of (+)-5-hydroxyisopinocampheol (Fig. 7.18). The coupling of the most downfield oxygenated methine proton (H-3) resonating at δ 4.05 is observed with the C-2 proton (δ 1.77) and C-4 protons (δ 2.41, and 1.79). The C-4 and C-7 methylenic protons show interactions between geminal protons resonating at δ 2.41 and 1.79, and, 2.33 and 1.47, respectively. The C-1 proton resonating at δ 1.67 affords a cross-peak with the C-7 proton (δ 2.33). The upfield C-10 methyl protons (δ 1.13) also show through-bond interactions with the C-2 protons resonating at δ 1.77. The various interaction based on the SECSY experiment are presented on the spectrum (Fig. 7.18).

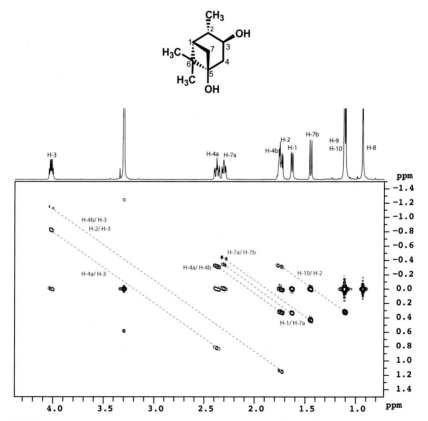

FIGURE 7.18 The structure and SECSY spectrum of (+)-5-hydroxyisopinocampheol, a transformed product of (+)-isopinocampheol with *Cunninghamella blakesleeana*. Cross-peaks represent coupling interactions between various protons. These interactions are also presented around the structure.

What are the major differences between the COSY and SECSY experiments?

7.1.2 Heteronuclear Shift-Correlation Spectroscopy

In *homonuclear* shift-correlation experiments like COSY we were concerned with the correlation of chemical shifts between nuclei of the same nuclear species, for example, ^1H with ^1H. In *heteronuclear* shift-correlation experiments, however, the chemical shifts of nuclei belonging to different nuclear species are determined (e.g., ^1H with ^{13}C). These may be one-bond chemical shift correlations, for example, between directly bound ^1H and ^{13}C nuclei, or they may be long-range chemical shift correlations, in which the interactions between protons with carbon atoms that are 2, 3, or 4 bonds away are detected.

The basic pulse sequence employed in the heteronuclear 2D shift-correlation (or HETCOR) experiment is shown in Fig. 7.19. The first 90_x° ^1H pulse bends the ^1H magnetization to the y'-axis. During the subsequent evolution period, this magnetization precesses in the $x'y'$-plane. It may be considered to be made up of two vectors corresponding to the lower (α) and higher (β) spin states of carbon to which ^1H is coupled. These two vectors diverge in the first half of the evolution period due to their differing angular velocities. The 180° ^{13}C pulse applied at the center of the evolution period reverses their direction of rotation so that during the second half of the evolution period they converge and are refocused at the end of the t_1 period. The angle α by which the vector has diverged from the y'-axis corresponds to its chemical shift, with the downfield protons having a larger angle α then the upfield protons. A constant mixing delay period Δ_1 occurs after the variable evolution period t_1, and the length of the mixing delay is set to the average value of J_{CH}. After the first $\frac{1}{2}J$ period, the vectors diverge and become aligned in opposite directions. The second 90_x° mixing ^1H pulse is then applied, which rotates the two vectors toward the $+z$- and $-z$-axes. One vector now points toward the $+z$-axis (equilibrium position), while the other points toward the $-z$-axis (non-equilibrium position).

What we have succeeded in doing is to invert *one* of the proton lines of the CH doublet (corresponding to the vector pointing toward the $-z$-axis), amounting to a selective population inversion, as in the INEPT experiment, described earlier (Section 4.2.1). This correspondingly intensifies the ^{13}C signals due to the polarization transfer. A 90° ^{13}C pulse is then applied that samples the evolved spin states. Clearly, the extent of the polarization transfer will depend on the disposition of the proton multiplet components when the second 90° mixing pulse is applied (which in turn depends on the ^1H chemical shifts) as well as on the duration of the evolution period t_1. The amplitude of the ^{13}C signals detected during t_2 is, therefore, modulated as a function of t_1 by the frequencies of ^1H spins, giving rise to $^1H/^{13}C$ shift correlation spectra.

If broad-band decoupling is not applied during acquisition, then a coupled spectrum is obtained in both dimensions. This complicates the spectrum, and

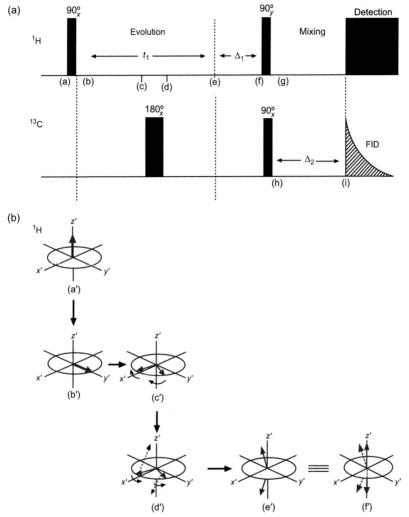

FIGURE 7.19 (a) Pulse sequence for the 2D heteronuclear shift-correlation experiment. (b) Effect of the pulse sequence in (a) on ^1H magnetization vectors of CH.

the splitting of the signals into multiplets also results in a general lowering of intensity. Since the multiplet components are in anti-phase, an overlap of multiplets can also cause their mutual cancellation. It is therefore usual to eliminate the heteronuclear couplings in both dimensions. In the case of non-equivalent geminal protons of CH_2 groups, two different cross-peaks will appear at the same carbon chemical shift but at two different proton chemical shifts, thereby facilitating their recognition.

The HETCOR spectrum of a naturally occurring isoprenyl coumarin is shown in Fig. 7.20. The spectrum displays one-bond heteronuclear correlations

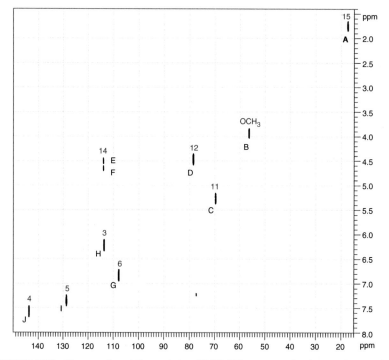

FIGURE 7.20 Heteronuclear shift correlation (HETCOR) spectrum of an isoprenyl coumarin.

of all protonated carbons. These correlations can easily be determined by drawing vertical and horizontal lines starting from each peak. For example, peak A represents the correlation between a proton resonating at δ 1.76 and the carbon at δ 17.3. Similarly, cross-peaks E and F show that the protons at δ 4.66 and 4.51 are coupled to the same carbon, which resonates at δ 114.1; i.e., these are the non-equivalent protons of an exo-methylenic group. Cross-peak J represents the one-bond heteronuclear correlation between the proton at δ 7.60 and the carbon at δ 143.6. The chemical shift assignments to various protonated carbons, based on the HETCOR experiment, are presented around the structure in Fig. 7.20.

If the delays Δ_1 and Δ_2 are adjusted so they correspond to J_{CH} of 130–150 Hz, then only the direct one-bond couplings between sp^3- and sp^2-hybridized ^{13}C–^1H bonds will be observed. However, if we wish to observe long-range ^1H couplings with ^{13}C nuclei, then we adjust the value of the delay to correspond to J_{CH} of 5–10 Hz. The cross-peaks will then correspond to coupling interactions between ^1H and ^{13}C nuclei separated by two, three, or even four bonds. In such a *long-range hetero COSY* experiment, the one-bond magnetization transfer is suppressed, and long-range heteronuclear couplings observed.

Many improved modifications of the basic heteronuclear shift-correlation experiment have been developed (Quast et al., 1987; Zektzer et al., 1987a, b, 1986; Kessler et al., 1986, 1985, 1984a; Schenker and Philipsborn, 1985; Wernly and Lauterwin, 1985; Wimperis and Freeman, 1985, 1984; Welti, 1985; Bauer et al., 1984; Bax, 1983; Garbow et al., 1982). A discussion of these is beyond the scope of this text. For long-range shift correlations, COLOC (*CO*rrelation spectroscopy *via LO*ng-range *C*oupling) has proved particularly useful (Kessler et al., 1986, 1985, 1984a, b). However, as inverse polarization transfer experiments are now widely available, the *H*eteronu clear *M*ultiple-*Q*uantum *C*oherence (HMQC) and *H*eteronuclear *S*ingle-*Q*uantum *C*oherence (HSQC) experiments are the experiments of choice for determining one-bond C–H shift correlations. The *H*eteronuclear *M*ultiple-*B*ond *C*onnectivity (HMBC) experiment is now widely employed for determining long-range C–H connectivities, particularly when the sample quantity is limited so that maximum sensitivity is required. These inverse experiments are discussed in a later section (Section 7.6).

Problem 7.10

How can we determine the carbon–carbon connectivities in a molecule through a combination of homo- and heteronuclear shift correlation experiments?

7.2 HOMO- AND HETERONUCLEAR *J*-RESOLVED SPECTROSCOPY

7.2.1 Heteronuclear 2D *J*-Resolved Spectroscopy

Heteronuclear 2D *J*-resolved spectra contain the chemical shift information of one nuclear species (e.g., ^{13}C) along one axis, and its coupling information with another type of nucleus (say, ^1H) along the other axis. 2D *J*-resolved spectra are therefore often referred to as *J,δ*-spectra. The heteronulear 2D *J*-resolved spectrum of di-*O*-methylcrenatin, isolated from *Osyris wightiana* Wall. Ex. Wight., is shown in Fig. 7.21. On the extreme left is the broad-band ^1H-decoupled ^{13}C-NMR spectrum, in the center is the contour plot, and on the right is the stacked plot of 2D *J*-resolved spectrum. The multiplicity of each carbon can be seen clearly in the contour and stacked plots.

The mechanics of obtaining a 2D spectrum have already been discussed in the previous chapter (Sections 5.2, and 5.3). A 1D ^1H-coupled ^{13}C-NMR

Di-*O*-methylcrenatin

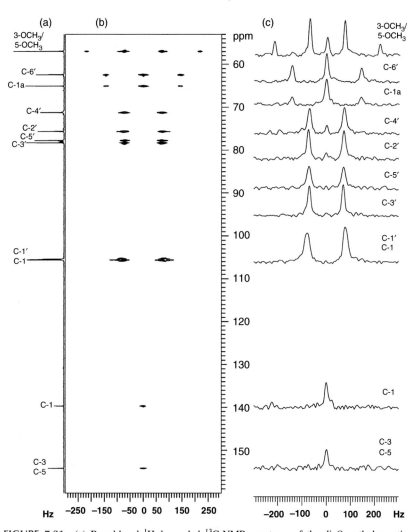

FIGURE 7.21 (a) Broad-band ¹H-decoupled ¹³C-NMR spectrum of the di-*O*-methylcrenatin, isolated from *Osyris wightiana* Wall. ex. Wight.; (b) 2D heteronuclear *J*-resolved spectrum of the same compound in the form of a contour plot; (c) stacked plot presentation of 2D heteronuclear *J*-resolved spectrum of di-*O*-methylcrenatin.

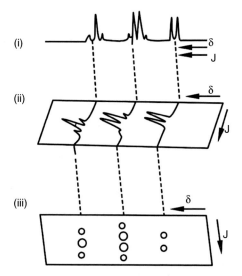

FIGURE 7.22 Schematic representation of 2D J-resolved spectra. In the 1D plot (i), both δ and J appear along the same axis, but in the 2D J-resolved spectrum (ii), the multiplets are rotated by 90° at their respective chemical shifts to generate a 2D plot with the chemical shifts (δ) and coupling constants (J) lying along two different axes. (iii) the 2D J-resolved spectrum as a contour plot.

spectrum contains both the chemical shift and coupling information along the same axis (Paudel et al., 2013; Lupulescu et al., 2012; Wang and Wong, 1985; Jin-shan et al., 1982; Becker et al., 1969). Let us consider what would happen if we could somehow swing each multiplet by 90° *about* its chemical shift so that the multiplet came to lie at right angles to the plane containing the chemical shift information. Thus, if the 1D spectrum was drawn in one plane, that defined by the NMR chart paper, then the multiplets would rotate about their respective chemical shifts so that each multiplet came to lie above and below the plane of the paper, perpendicular to it, crossing the paper at the chemical shift of the nucleus. The chemical shifts would then be defined by an axis lying in one plane (the plane of the paper), and the coupling information would be defined by an axis lying perpendicular to that plane (Fig. 7.22). This is precisely what happens as a result of the 2D J-resolved experiments. Since the various multiplets now no longer overlap with another (because the individual nuclei usually have differing chemical shifts), the multiplicity, and J-values can easily be read in two separate planes (Giraudeau and Akoka, 2007).

To understand how the chemical shift and multiplicity information is separated, we need to reconsider the spin-echo sequence described earlier (Section 4.1).

Let us consider the magnetization vector of a ^{13}C nucleus of a CH group, for purpose of illustration. A 90°_{x} ^{13}C pulse will bend the magnetization by 90° to the y'-axis. During the subsequent evolution period, the magnetization vector will be split into two component vectors under the influence of the coupling (J_{CH}) with the attached protons, and these vector will oscillate in the $x'y'$-plane.

If, for simplicity, we observe the behavior of these vectors in the rotating frame with respect to the chemical shift frequency of the CH carbon (i.e., if the frame itself is rotating at the frequency of the CH carbon), then they will appear to separate from one another and rotate away in opposite directions in the $x'y'$-plane with an angular velocity of $\pm J/2$ in the rotating frame. As the vectors precess away from the y'-axis, the magnitude of their net resultant vector along the y'-axis will decrease till it becomes zero (when they point along the $+x'$- and $-x'$-axis). As they precess away further toward the $-y'$-axis, the resultant will have a growing negative intensity that will reach a maximum when they coalesce along the $-y'$-axis. The two vectors then move back $+y'$-axis, with a corresponding increase in positive signal intensity. This corresponds to the perturbation of the intensity of the ^{13}C *signal as a function of the scalar coupling constant, J* (Fig. 7.23e. i). A series of spectra are acquired by Fourier transformation of the FIDs at various values of t_1 (Fig. 7.23b). The data are then transposed by the computer so that the spectra are arranged in rows, one behind the other. Fourier transformation is then again carried out along columns of modulated peaks to produce the 2D J-resolved spectrum in which the chemical shift information lies along one axis, and the coupling information lies along the other axis. This is illustrated in Fig. 7.23c and d.

In the case of CH_2 groups, since there are now two protons coupled to the ^{13}C nucleus, it will split into triplet (δ 62.4 and 65.1, CH_2-6' and CH_2-1a, respectively). If we again position the reference frequency at the center of the triplet, then the middle vector will appear stationary, while the outer vectors will precess away in opposite directions with angular velocities of $+J$ and $-J$. The resultant along the y'-axis will again oscillate but with a different periodicity (depending on δ and J) along the y'-axis (Fig. 7.23e, ii). Similarly, in the case of a CH_3 group, if the reference frequency is positioned at the center of the quartet, the two vectors corresponding to the outer peaks will possess angular velocities of $+3/2\ J$ and $-3/2\ J$, while the middle vectors will have angular velocities of $+J/2$ and $-J/2$ (Fig. 7.23e, iii). Clearly, the periodic perturbation of CH_3, CH_2, and CH carbons with varying t_1 values will be characteristically different, depending on the number of vectors and their respective angular velocities.

A simplifying assumption was made in the earlier discussion that the frame was rotating at the chemical shift frequency of the ^{13}C nucleus so that as the vectors diverged from one another, the position of the *net* magnetization remained unchanged, aligned with the y'-axis, and only its magnitude changed. If, however, we consider that the rotating frame is rotating at the chemical shift frequency of the TMS reference frequency, then both of the ^{13}C vectors of the CH group will rotate away in the same direction but with angular velocities determined by the chemical shift frequency of the CH group and that differ from each other by J in Hz. This will be due to the difference in chemical shift frequency from the reference frequency of the rotating frame. The

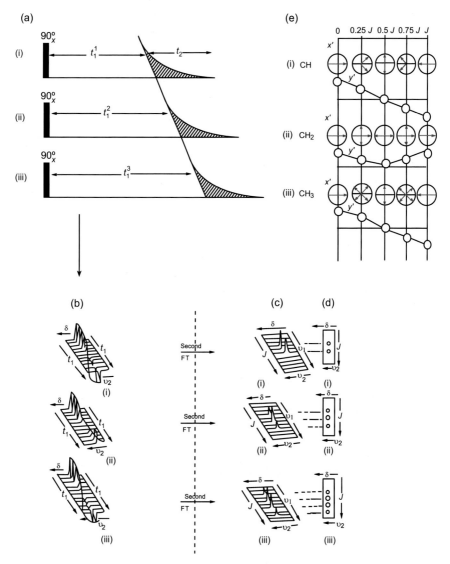

FIGURE 7.23 (a) Many FIDs are recorded at incremented evolution periods, t_1, as the first step involved when recording 2D spectra. (b) The first series of Fourier transformations (FT) affords a corresponding series of 1D spectra in which the individual peaks exhibit periodic changes in signal amplitudes at various incremented t_1 values. The second series of Fourier transformations then yield the 2D plot as stacked plots (c) or as contour plots (d). In the 2D J-resolved spectra thus obtained, one axis defines the chemical shift (δ), while the other axis defines the coupling constant (J). (e) Different positions of magnetization vectors of (i) CH, (ii) CH_2, and (iii) CH_3 carbons with changing evolution time t_1. The periodic changes in the positions of the magnetization vector result in corresponding periodic changes in signal amplitudes, as shown in (e) (i), (ii), and (iii).

net magnetization will therefore actually not remain static along the y'-axis, but will precess in the $x'y'$-plane with a phase shift of v_0/J. This is illustrated in Fig. 7.24. The net magnetization is represented by the dotted line at angle α to the y'-axis. As the vectors rotate in the $x'y'$-plane, this angle α increases; when it reaches a value of 180°, the net magnetization points towards the $-y'$-axis. Thus, the *resultant* along the y'-axis oscillates periodically, having a maximum positive value along the $+y$-axis immediately after the 90_x° pulse (when both vectors are aligned along the $+y'$-axis), and maximum negative amplitude when both vectors are aligned with the $-y'$-axis. This oscillation of the resultant is detected as corresponding oscillations of the NMR signal at different values of the evolution period, t_1.

Figure 7.25 shows the heteronuclear 2D J-resolved spectrum of fenchone, isolated from *Illicium religiosum* Sieb. The broad-band ^{1}H-decoupled ^{13}C-NMR spectrum is plotted alongside it. This allows the multiplicity of each carbon to be read without difficulty, the F_1 dimension containing only the coupling information, and the F_2 dimension only the chemical shift information. If, however, proton broad-band decoupling is applied in the evolution period t_1, then the 2D spectrum obtained again contains only the coupling information in the F_1 domain, but the F_2 domain now contains *both* the chemical shift and the coupling information (Fig. 7.26). Projection of the peaks onto the F_1 axis, therefore, gives the ^{1}H-decoupled ^{13}C spectrum; projection onto the F_2 axis produces the fully proton-coupled ^{13}C spectrum.

7.2.2 Homonuclear 2D *J*-Resolved Spectroscopy

In the heteronuclear 2D J-resolved spectra described earlier, the chemical shifts of the ^{13}C nuclei were located along one axis and the ^{1}H/^{13}C couplings along the other. If the coupled nuclei of the *same* type are required to be observed, then homonuclear 2D J-resolved spectra are obtained. For instance, if the nuclear species being observed is ^{1}H, then the ^{1}H chemical shifts will be spread along the F_2 axis, and the coupling constants of the ^{1}H nuclei with other neighboring ^{1}H nuclei will be spread along the F_1 axis. This offers a powerful procedure for unraveling complex overlapping multiplets.

Let us consider, for instance, an overlapping triplet and a quartet, so that seven peaks integrating for 2H appear in a certain region of the NMR spectrum (Fig. 7.27). If the coupling constants of the two multiplets are not significantly different, it may be difficult to recognize the overlapping signals readily.

It is very unusual to find that such overlapping multiplets which have *exactly* the same chemical shifts. This difference in chemical shifts may be exploited if we could somehow rotate the triplet and the quartet *about* their respective chemical shifts by 90° so that, instead of lying flat in a horizontal plane, they come to lie above and below the plane, perpendicular to it. Each multiplet is then positioned vertically so that it passes through the plane of the paper at its chemical shift. Clearly, if the two multiplets have different chemical shifts,

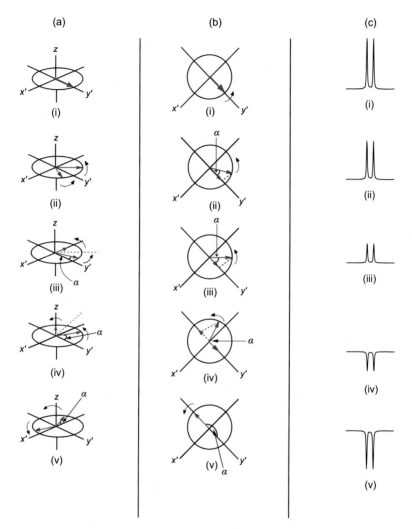

FIGURE 7.24 (a) The two ^{13}C vectors (representing the doublet of a CH group) move away from each other as well as away from the y'-axis (y'-axis is assumed to be rotating with angular velocity equal to TMS). The distance between the two vectors therefore grows with the delay till their angle of separation reaches 180° before it starts decreasing. The vectors therefore progressively go out-of phase and come-in phase. (b) The corresponding positions of the *net* magnetization vectors along the y'-axis. (c) Signal amplitudes with respect to the position of magnetization in the $x'y'$-plane.

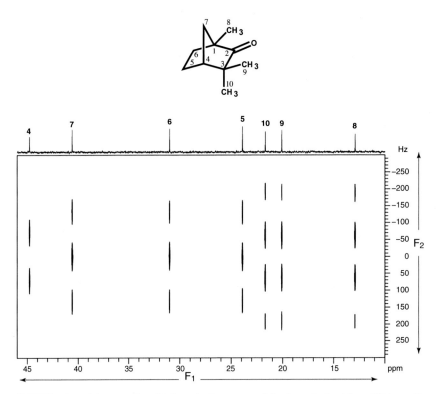

FIGURE 7.25   A heteronuclear 2D *J*-resolved spectrum of fenchone, isolated from *Illicium religiosum* Sieb, along with a broad-band decoupled spectrum. Only δ 10–45 region is shown, containing signals of protonated carbons. Quaternary carbons, C-2 and C-3, resonate further downfield as singlets (not shown here).

the overlap present before the 90° rotation will disappear, with the chemical shifts lying along one axis, and the coupling information appearing along a perpendicular axis. This is illustrated in Fig. 7.28.

The pulse sequence used in homonuclear 2D *J*-resolved spectroscopy is shown in Fig. 7.29a. Let us consider a proton, A, coupled to another proton, X. The 90°_x pulse bends the magnetization of proton A to the y'-axis. During the first half of the evolution period, the two vectors $\mathbf{H_F}$ (faster vector) and $\mathbf{H_S}$ (slower vector) of proton A precess in the $x'y'$-plane with angular velocities of $\Omega/2 + (J/2)$ and $\Omega/2 - (J/2)$, respectively. The 180°_x ^{1}H pulse results in (a) the vectors' adopting mirror image positions across the x'-axis so that $\mathbf{H_S}$ comes to lie ahead of $\mathbf{H_F}$ when viewed in a clockwise fashion. However, since the 180° ^{1}H pulse was not selective, it also affects neighboring proton X to which proton A was coupled. Thus, if $\mathbf{H_F}$ was coupled to the α state of proton X before the 180° pulse, it exchanges its identity with $\mathbf{H_S}$ and becomes coupled to the β state of proton X. This exchange of identity (or "relabeling")

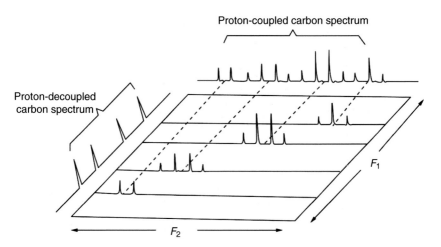

FIGURE 7.26 If proton broad-band decoupling is applied in the evolution period, t_1, then the resulting 2D spectrum contains only chemical shift information in the F_1 domain, while both chemical shift and coupling information is present in the F_2 domain. Projection onto the F_1-axis therefore gives the ^{1}H-decoupled ^{13}C spectrum, whereas projection along F_2 gives the fully coupled ^{13}C spectrum.

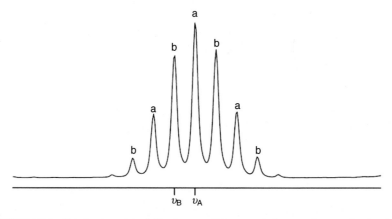

FIGURE 7.27 Overlapping triplet "a" and quartet "b", respectively. v_A and v_B are the chemical shifts of the two protons.

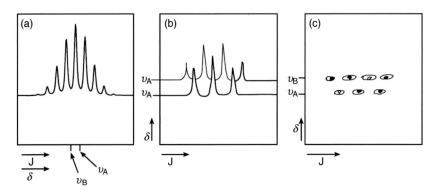

FIGURE 7.28   (a) Overlapping triplet and quartet of protons A and B, with the chemical shifts and coupling constants along the same axis. (b) Triplet and quartet of the same protons, shown with the chemical shifts lying along the vertical axis, and coupling constants along the horizontal axis separately. (c) Overhead view of the same peaks as in (b) but in the form of contours.

of the vectors $\mathbf{H_F}$ and $\mathbf{H_S}$ has $\mathbf{H_F}$ again becoming positioned ahead of $\mathbf{H_S}$ (Fig. 7.29b, c), so that in the second half of the evolution period t_1, the two vectors continue to diverge from each other. The extent of this divergence depends both on the time t_1 for which they are allowed to diverge and on the coupling constant J. The first series of Fourier transformations give the 1D spectra in v_1 and F_2. Transposition of the data so that they are arranged in rows, one behind the other, and Fourier transformation of the individual columns of oscillating signals along the F_2 (vertical) axis leads to the 2D plots in the F_1 and F_2 frequency dimensions, with the chemical shifts being defined by the F_2 axis and the coupling frequencies by the F_1 axis. The J-modulation of doublets and triplets with changing evolution time is shown in Fig. 7.30.

In homonuclear 2D J-resolved spectra, couplings are present during t_2; in heteronuclear 2D J-resolved spectra, they are removed by broad-band decoupling (Sakhaii et al., 2013). The multiplets in homonuclear 2D J-resolved spectra appear on the diagonal, and not parallel with F_1. If the spectra are plotted with the same Hz/cm scale in both dimensions, then the multiplets will be tilted by 45° (Fig. 7.31). So if the data are presented in the absolute-value mode and projected on the chemical shift (F_2) axis, the normal, fully coupled 1D spectrum will be obtained. To make the spectra more readable, a tilt correction is carried out with the computer (Fig. 7.32), so that F_1 contains only J information and F_2 contains only δ information. Projection onto the F_2 (chemical shift) axis would then produce a fully decoupled spectrum.

Peaks in homonuclear 2D J-resolved spectra have a "phase-twisted" line shape with equal 2D absorptive and dispersive contributions. If a 45° projection is performed on them, the overlap of positive and negative contributions will mutually cancel and the peaks will disappear. The spectra are therefore presented in the absolute-value mode.

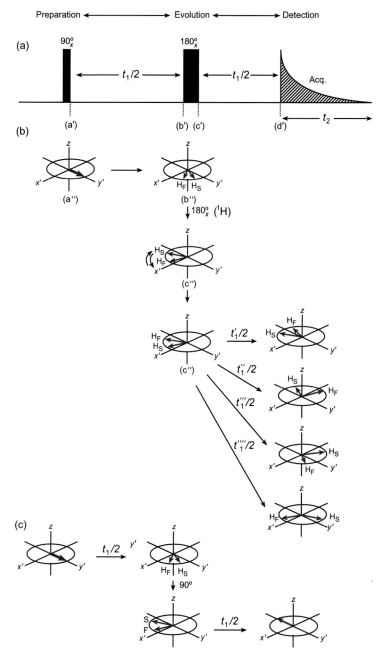

FIGURE 7.29   (a) Pulse sequence for homonuclear 2D *J*-resolved spectroscopy. (b) Effect of 90°_x ^{1}H and 180°_x ^{1}H pulses on an ^{1}H doublet. (c) In the absence of coupling, the vectors are refocused by the 180°_x ^{1}H pulse after t_1. This serves to remove any field inhomogeneities or chemical shift differences.

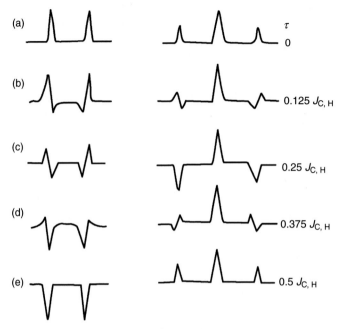

FIGURE 7.30   Modulation effects of a doublet and triplet by a spin-echo pulse sequence, $90^{\circ}_x - \tau - 180^{\circ}_y - \tau - $ echo.

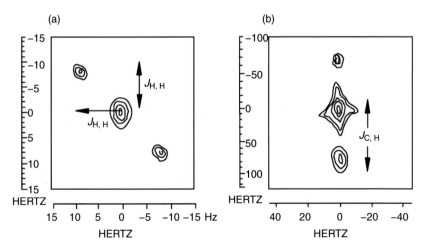

FIGURE 7.31   (a) Homonuclear J-spectrum. Since the coupling is both in v_1 and v_2 dimensions, a 45° tilt is observed. (b) Heteronuclear J-spectrum in which coupling appears only in one dimension.

(a)

(b)

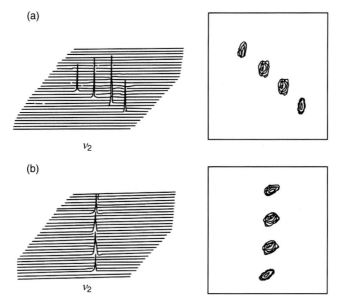

FIGURE 7.32 (a) The methylene protons of ethyl alcohol appear as a quartet with a 45° tilt in the homonuclear 2D J-resolved spectrum. (b) The same, but after tilt correction. The stacked plot is presented on the left; the corresponding contour plot appears on the right.

Homonuclear 2D J-resolved spectra often show artifact signals due to second-order effects appearing in strongly coupled systems (Fig. 7.33). Such artifact signals generally do not appear in weakly coupled spin system ($\Delta\delta/J > 4$), which are largely first-order. They are usually readily recognized, as they appear at chemical shifts midway between strongly coupled protons, and they occur as wavy contours spread across the spectrum (Thrippleton et al., 2005). Such artifact signals may disappear if the spectra are recorded at higher fields, due to the greater first-order character of the spectra recorded on high-field instruments. A common artifact in J-resolved spectra, as with other 2D spectra, is the presence of t_1-noise ridges, particularly when sharp, intense signals are present, like those due to methyl groups. These noise ridges appear parallel to F_1 before the tilt correction is performed, but after the tilt correction they appear at 45°, provided both dimensions are recorded with the same scale. Symmetrization can be carried out to remove the t_1-noise.

Problem 7.11

What are the essential ingredients of 2D J-resolved NMR spectroscopy?

Problem 7.12

What is the difference between homo- and heteronuclear 2D J-resolved spectroscopy?

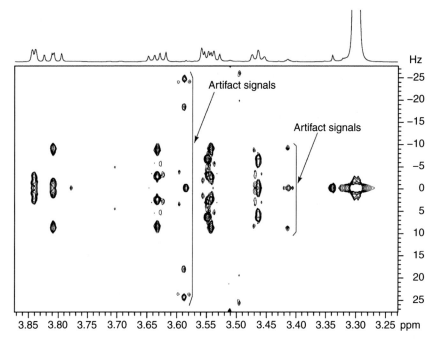

FIGURE 7.33 Artifact signals appear due to strongly coupled nuclei, as shown by the vertical lines of contours at about δ 3.59 in the spectrum.

7.3 TWO-DIMENSIONAL NUCLEAR OVERHAUSER SPECTROSCOPY

7.3.1 NOESY

A large number of 1D nOe experiments may have to be performed if the spatial relationships among many protons in a molecule are to be determined. In such cases, instead of employing multi-pulse 1D nOe experiments, we can opt for the 2D nOe spectroscopy (NOESY) experiment. If many protons have close chemical shifts, then NOESY may be particularly advantageous, since it is difficult to carry out irradiation in 1D experiments selectively without affecting other nearby protons. In the NOESY experiment, all inter-proton nOe effects appear simultaneously, and the spectral overlap is minimized due to the spread of the spectrum in two dimensions. A typical NOESY spectrum of (+) catechin is shown in Fig. 7.34. In this spectrum, the C-2′ and C-6′ protons (δ 6.82 and 6.71) in ring B show interaction with H-2 (δ 4.55) and H-3 (δ 3.96). The H-2 and H-3 show mutual cross-peaks as well. Another strong correlation is observed between H-4a (δ 2.84) and H-3 (δ 3.96). These interactions indicate that above-mentioned protons are in close proximity in space. Different NOESY interactions are shown on structure and spectrum (Fig. 7.34). This topic is extensively discussed in Chapter 6 (Section 6.8).

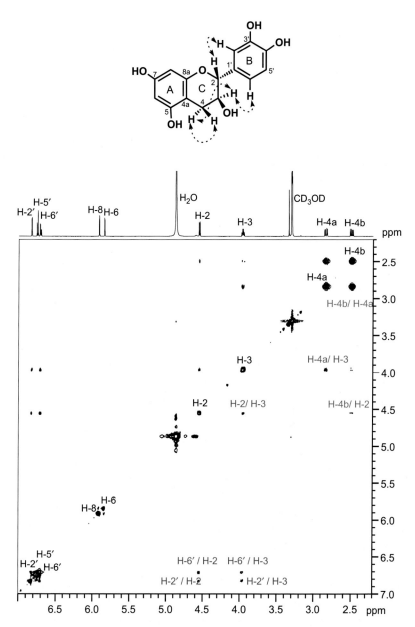

FIGURE 7.34   A typical NOESY spectrum of (+)-catechin, isolated from *Osyris wightiana* Wall. Ex. Wight., the off-diagonal cross-peaks represent the nOe interactions between various nuclei.

7.3.2 Phase-Sensitive NOESY

Phase-sensitive NOESY spectra are generally significantly superior in resolution to the absolute-mode spectra. As in phase-sensitive COSY spectra, discussed in Section 7.1.1, there are "four phase quadrants" arising from the various combinations of real (R) and imaginary (I) parts corresponding to quadrants RR (real F_1, real F_2), RI (real F_1, imaginary F_2), IR (imaginary F_1, real F_2), and II (imaginary F_1, imaginary F_2). Phasing of the 2D NOESY spectrum is carried out to obtain the pure double-absorption phase in the RR quadrant; the other three quadrants are then discarded. Figure 7.35 shows a phase-sensitive absorption-mode spectrum as compared to an absolute-value-mode spectrum. A slice through the spectrum at any given chemical shift provides the equivalent of a 1D NOE spectrum.

7.3.3 CAMELSPIN, or Rotating-Frame Overhauser Enhancement Spectroscopy (ROESY)

When molecules have the molecular correlation time τ_c close to the reciprocal of the angular Larmor frequency ($\tau_c = 1/v_L$), then nOe signals may be completely absent or too weak to be detected. In such cases, it is advisable to record the CAMELSPIN (or ROESY) experiment instead of the NOESY experiment, since it always gives a positive value for the nOe effect, even when $\tau_c = 1/v_L$. Another advantage of the ROESY experiment is that cross-peaks due to relayed nOe (i.e., magnetization of one spin relayed *via* an intermediate spin to another spin) have the same phase as the diagonal peaks, so that they can be readily distinguished from direct nOe signals which have the opposite phase relative to the diagonal signals.

The pulse sequence used for the ROESY experiment is shown in Fig. 7.36. A 90°_x pulse bends the magnetization by 90°. At the end of the evolution period t_1, a strong *Rf* field is applied for the duration of the mixing time τ_m, so that the magnetization vectors precessing in the $x'y'$-plane become "spin-locked" with the applied *Rf* field and precess synchronously with it. As soon as the vectors of two different nuclei have acquired an identical precessional frequency, a transfer of *transverse* magnetization can occur between them, i.e., an exchange of *transverse magnetization* (in the $x'y'$-phase) takes place. This is in contrast to the nOe experiment, in which an exchange of *longitudinal magnetization* occurs (i.e., in nOe-difference or NOESY experiments, a change of z-magnetization of one nucleus changes the z-magnetization of another, close-lying nucleus, which is observed). Hence, ROESY is a 2D *transverse* nOe experiment. The ROESY spectrum of gelseminol, isolated from *Lyonia ovalifolia* (Wall.) Drude, is shown in Fig. 7.37. In this spectrum, the methyl protons (δ 0.98) show through space coupling with H-7b (δ 1.50) and H-9 (δ 1.82), which indicates their proximity in space. H-7a and H-8 also show a cross-peak indicating their close disposition. The coupling between H-4 and H-5 also indicates their close proximity in space. The key ROESY interactions are shown on the structure and in the spectrum (Fig. 7.37).

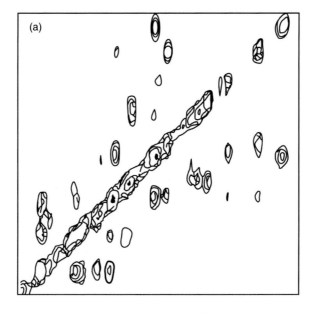

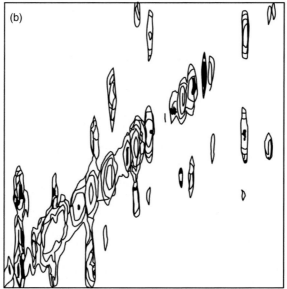

FIGURE 7.35   (a) Phase-sensitive absorption-mode NOESY spectrum of bovine phospholipase A_2 in D_2O. (b) The same spectrum in the absolute-value mode. The phase-sensitive spectra are generally superior in resolution to the absolute-value mode spectra. *(Reprinted from Neuhaus and Williamson, 1989 The Nuclear Overhauser Effect in structural and conformational analysis, copyright (1989), 278, with permission from the VCH Publishers, Inc.).*

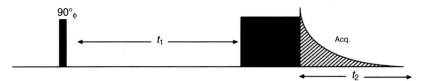

FIGURE 7.36 Pulse sequence used in the ROESY experiment. The data obtained from odd and even scans are stored separately.

Problem 7.13

What are the key differences between NOESY and ROESY methods?

7.3.4 Heteronuclear nOe Spectroscopy (HOESY)

Heteronuclear nOe experiments are similar to homonuclear experiments, discussed earlier. Normally, protons are irradiated and the enhancement of the heteronucleus is recorded, since if the reverse were done, the enhancements would usually be very small (the magnetogyric ratios of most nuclei are much smaller than those of ^1H). An exception to this is ^{19}F, so $^1H\{^{19}F\}$ experiments have been reported (Fig. 7.38).

7.3.5 Relayed nOe Experiments

Both homonuclear and heteronuclear versions of relayed nOe experiments are known. The homonuclear *relayed* NOESY experiment involves both an *incoherent* transfer of magnetization between two spins H_k and H_l that are not coupled but close in space, and a *coherent* transfer of magnetization between two spins H_l and H_m that are J-coupled together. The magnetization pathway may be depicted as:

$$H_k \xrightarrow{\text{nOe}} H_l \xrightarrow{J} H_m$$

In a heteronuclear nOe experiment, the first step may be identical to the homonuclear nOe experiment (i.e., involving an incoherent transfer of coherence from H_k to H_l), while the second step could involve a coherence transfer from ^1H to ^{13}C nucleus by an INEPT sequence. These methods suffer from poor sensitivity and have therefore not been used extensively.

7.4 TWO-DIMENSIONAL CHEMICAL EXCHANGE SPECTROSCOPY

It is possible to transfer magnetization between nuclei that are undergoing slow exchange. When the exchange process is slow on the NMR time scale, then each form contributing to the exchange process will produce a separate signal, and irradiation of one nucleus will cause the magnetization to be transferred

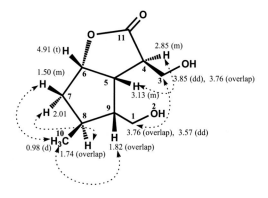

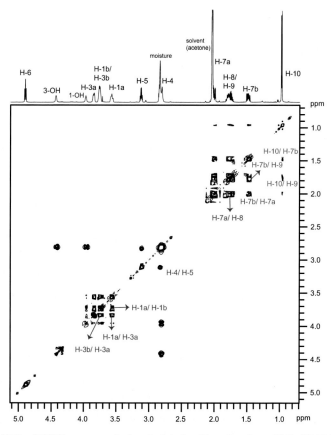

FIGURE 7.37 ROESY spectrum of gelseminol, isolated from *Lyonia ovalifolia* (Wall.) Drude.

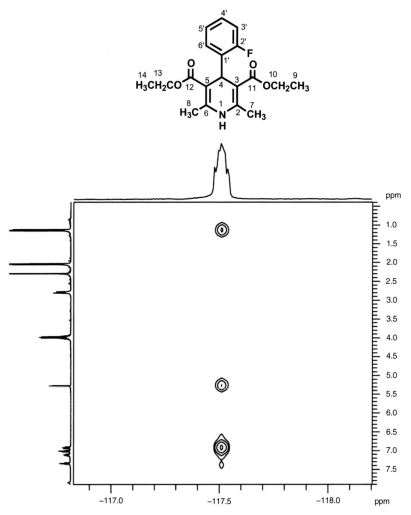

FIGURE 7.38 HOESY spectra of diethyl 4-(2-fluorophenyl)-2,6-dimethyl-1,4-dihydropyridine-3,5-dicarboxylate.

to other nuclei by an exchange process – a phenomenon known as *transfer of saturation* (or *saturation transfer*). This is exemplified by the classical case of *N,N*-dimethylacetamide (Fig. 7.39). The methyl groups attached to the nitrogen are non-equivalent and slowly undergo exchange by rotation about the amide bond. They appear at differing chemical shifts, and a 2D exchange spectrum would show up in the exchange process as cross-peaks between them. Again, as in the NOESY experiment, it is the modulations during the mixing period that give rise to the cross-peaks.

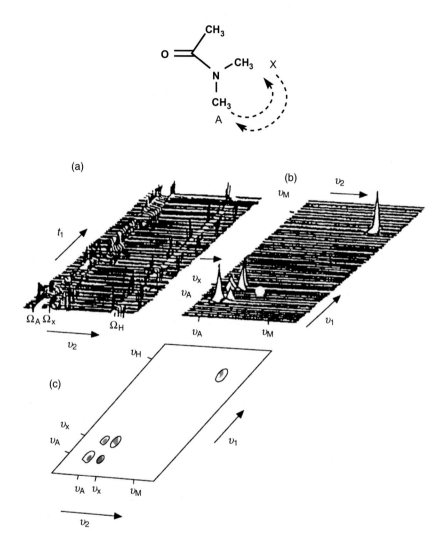

FIGURE 7.39   2D exchange spectrum of *N,N*-dimethylacetamide and its generation. (a) The first set of spectra results from the first series of Fourier transformations with respect to t_2. The modulation of signals as a function of t_1 is observed. (b) The second set of spectra is obtained by the second series of Fourier transformations. The unmodulated signals appear on the diagonal at (v_A, v_A), (v_X, v_X), and (v_M, v_M), whereas the modulations due to exchange show up as cross-peaks on either side of the diagonal at (v_A, v_X) or (v_X, v_A). (c) A contour plot representation of (b). *(Reprinted from Bax, and Lerner, (1986). Two-dimensional nuclear magnetic resonance spectroscopy, Science,* **232**, *960–967, copyright with permission from Science-AAAS, c/o Direct Partners Int., P.O. Box 599, 1200 AN Hilversum, The Netherlands).*

If we designate the two methyl groups as A and X, then that proportion of the X resonance that has been exchanged with the A resonance will experience the modulation. The cross-peaks that appear at (v_X, v_A) therefore represent that portion of the total magnetization that has undergone a change in precessional frequency during the mixing period, but the unmodulated signals (i.e., those experiencing no change in precessional frequency) appear on the diagonal.

7.5 HOMONUCLEAR HARTMANN–HAHN SPECTROSCOPY (HOHAHA), OR TOTAL CORRELATION SPECTROSCOPY (TOCSY)

When protons are subjected to *strong* irradiation, then the chemical shift differences between them are effectively scaled to zero and *isotropic mixing* occurs. In other words, when the two scalar-coupled nuclei are subjected to identical field strengths through the application of a coherent *Rf* field, the Zeeman contributions are removed and the Hartmann–Hahn matching condition is established, so that an oscillatory exchange of "spin-locked" magnetization (with a periodicity of 1/*J*) occurs. Such nuclei experience strong coupling effects, even if there are several intervening nuclei between them, as long as they are a part of the same spin network. The magnetization diffuses to the neighboring nuclei within the spin network at a rate dependent on the magnitude of the scalar coupling (or dipolar coupling in solids). At short mixing times (<0.1 *J*), normally only the vicinal and geminal protons are affected, and cross-peaks are found largely between them so that a COSY-like spectrum is obtained. With longer mixing intervals the magnetization initially transferred to the immediately neighboring protons is *relayed* further down the chain within that spin network. Net magnetization transfer occurs and a phase-sensitive 2D spectrum is obtained.

In practice, it is convenient to record HOHAHA spectra at, say, 20, 40, 60, 80, and 100 ms. Slices are then taken at specific chemical shifts at the various time intervals. This yields highly informative plots of the spread of magnetization from key points in the molecule to adjoining regions in the same spin system, thereby providing a powerful tool for structure elucidation. It is thus possible to divide an NMR spectrum into many sub-spectra, with each sub-spectrum corresponding to the NMR spectrum of a separate spin system. This simplifies the interpretation greatly, since the overlap between protons belonging to different spin systems is removed in such sub-spectra. Thus, we can rather neatly divide a molecule into various fragments corresponding to discrete sub-spectra. The pulse sequence employed is shown in Fig. 7.40. Pulse field gradient version of TOCSY experiment is presented in Fig. 7.41 (also see Section 2.4.2).

Figure 7.42 shows the five HOHAHA spectra obtained at 20, 40, 60, 80, and 100 ms of ferticin, a sesquiterpene isolated from *Ferula ovina* (Boiss.) Boiss. The various short-range and long-range homonuclear connectivities established on the basis of HOHAHA spectra of ferticin provided a powerful tool for structure elucidation. The spectrum obtained with a mixing time of 20 ms

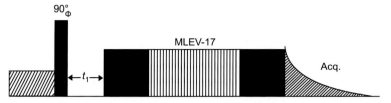

FIGURE 7.40 The pulse sequence employed in 2D homonuclear Hartmann–Hahn spectroscopy. An extended MLEV-16 sequence is used. Trim pulses are applied before and after this pulse sequence in order to refocus the magnetization not parallel to the x-axis.

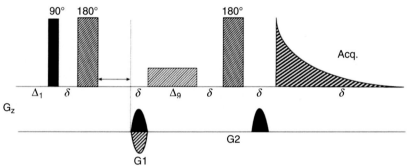

FIGURE 7.41 The PFG-TOCSY pulse sequence.

closely resembled the COSY-45° spectrum, showing mainly direct connectivities (geminal and vicinal). With longer mixing intervals, the magnetization is seen to spread to more distant protons within the individual spin systems. This spread of magnetization with increase in the mixing delay can be seen in the projections of the HOHAHA spectra, in which slices have been taken at δ 0.99, 1.62, and 6.10 (Fig. 7.43).

7.5.1 Heteronuclear Hartmann–Hahn Spectroscopy

The cross-polarization in the rotating frame described earlier is not confined to nuclei of the same nuclear species. Indeed, the concept was originally introduced to transfer polarization between different nuclear species in solids. In such cases, two strong *Rf* fields are applied simultaneously to the two nuclear species, I and S, at their precessional frequencies, and the nuclei I and S then exchange their spin energies at a rate dependent on the magnitude of the *Rf* fields. This is done in practice by applying a 90°_y pulse to create transverse magnetization of the I spins, and then "spin-locking" it in the rotating frame by applying a long *Rf* pulse that is phase-shifted by 90°. Simultaneously, a long *Rf* pulse is applied on the S spins so that the Hartmann–Hahn matching conditions are reached and an oscillatory exchange of spin energy takes place. The two spins I and S are now said to be "in contact." Once an equilibrium state has been established, they are said to possess the same "negative spin temperature." During this contact

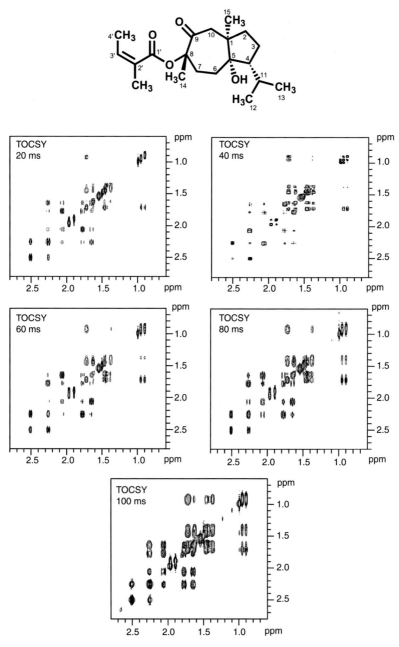

FIGURE 7.42   TOCSY spectra (selected regions) of ferticin, a sesquiterpene isolated from *Ferula ovina* (Boiss.) Boiss., recorded with mixing intervals of 20, 40, 60, 80, and 100 ms.

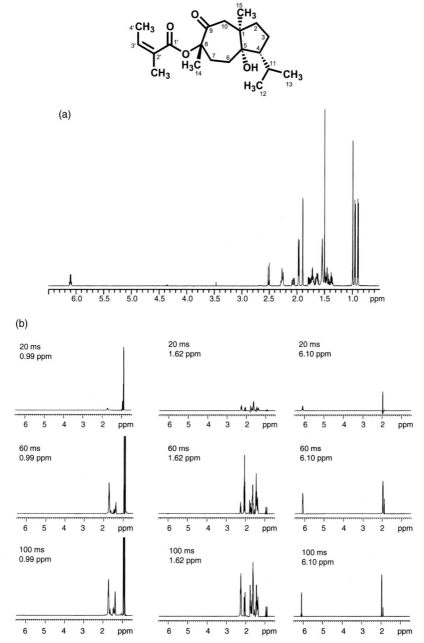

FIGURE 7.43   (a) 1D ^{1}H-NMR spectrum of ferticin, a sesquiterpene isolated from *Ferula ovina*. (b) HOHAHA spectra of the same compound recorded at 20, 60, and 100 ms. The projection spectra are taken at (1) δ 0.99, (2) δ 1.62, and (3) δ 6.10.

period, polarization transfer occurs from the I spin (the abundant spin, with a high magnetogyric ratio) to the S spin (the dilute spin, with a low magnetogyric ratio) in the IS two-spin system. During the contact period $\tau = 1/J_{IS}$, the complete transfer of polarization takes place, resulting in a sensitivity enhancement that is determined by the ratio of the magnetogyric ratios of the two spins. This produces an enhancement of about 4 if the polarization transfer is occurring from ^1H to ^{13}C nuclei and an enhancement of about 10 if polarization is being transferred from ^1H to ^{15}N nuclei.

7.6   "INVERSE" NMR SPECTROSCOPY

7.6.1 ^1H-Detected Heteronuclear Coherence Spectra

A considerable enhancement of sensitivity is obtainable when nuclei with a low magnetogyric ratio like ^{13}C and ^{15}N are detected through their effects on the more sensitive nuclei, like ^1H, and ^{19}F. This approach is widely used because it is based on proton detection, offering high sensitivity when compared with the conventional carbon-detected 2D HETCOR experiment. Protons not directly coupled to the ^{13}C (or ^{15}N) nuclei are suppressed by numerous methods, including incorporating pulse field gradients, or by using an older but still useful method of initial bilinear (BIRD) pulse, so that only protons directly bonded to the ^{13}C nuclei produce cross-peaks in the 2D spectrum.

There are two principle heteronuclear single-bond correlation techniques, HMQC and HSQC; each has its own pulse sequence, and merits and demerits.

7.6.1.1 ^1H-Detected Heteronuclear Multiple-Quantum Coherence (HMQC) Spectra

A considerable enhancement of sensitivity is obtainable when nuclei with a low *magnetogyric* ratio like ^{13}C are detected through their effects on the more sensitive nuclei, like ^1H. Protons not coupled to the ^{13}C nuclei are suppressed by an initial bilinear (BIRD) pulse so that only protons directly bonded to the ^{13}C nuclei produce cross-peaks in the 2D spectrum. The *H*eteronuclear *M*ultiple-*Q*uantum *C*oherence (HMQC) experiment (Müller, 1979) has provided a highly sensitive procedure for determination of one-bond proton–carbon shift correlations. The pulse sequence for the HMQC spectrum is shown in Fig. 7.44. The BIRD pulse is followed by a delay τ whose magnitude is adjusted so that the inverted magnetizations of protons not bound to the ^{13}C nuclei pass through zero amplitude (as they change from the original to the final positive amplitude) when the first 90°_x pulse is applied. The relaxation time T_1 of protons directly bonded to ^{13}C nucleus is short, because of the presence of adjacent ^{13}C nuclei. So such nuclei relax efficiently, and the time between successive scans can be kept quite short (about $1.3T_1$ of the time between protons). Pulse field gradient version of HMQC experiment is presented in Fig. 7.45 (also see Section 2.4.2).

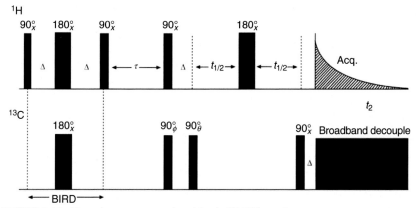

FIGURE 7.44 The pulse sequence employed for the HMQC experiment.

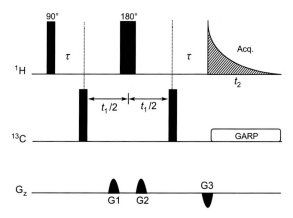

FIGURE 7.45 The PFG-HMQC pulse sequence.

The HMQC experiment is about 16 or 100 times more sensitive for ^{13}C and ^{15}N detections, respectively, as compared to the conventional heteronuclear shift-correlated experiment. The sensitivity advantage in this experiment (as well as in HMBC experiment described later) is due to the fact that it relies on the equilibrium magnetizations derived from *protons*. Since this magnetization is proportional to the larger population difference in *proton energy levels* (rather than the smaller population difference of ^{13}C energy levels), a stronger NMR signal is obtained. Moreover, since the strength of the NMR signal increases with the frequency of observation, a larger signal will be obtained at the higher proton observation frequencies than at the lower ^{13}C observation frequencies. An HMQC spectrum of quercetin is shown in Fig. 7.46. The spectrum contains one-bond heteronuclear shift correlations. The C-8 (δ 93.0), C-6 (δ 97.7), C-2' (δ 114.6), C-5' (δ 115.0), and C-6'(δ 120.7) carbons

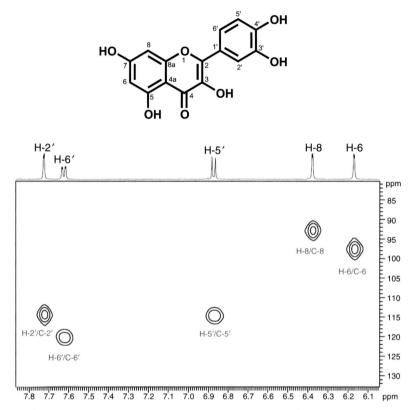

FIGURE 7.46   HMQC spectrum of quercetin isolated from *Lyonia ovalifolia* (Wall.) Drude.

show one-bond interactions with the methine protons H-8 (δ 6.38), H-6 (δ 6.16), H-2' (δ 7.72), H-3' (δ 6.87), and H-6' (δ 7.62), respectively. The one-bond heteronuclear shift interactions based in the HMQC spectrum are shown on the spectrum (Fig. 7.46).

7.6.1.2   *¹H-Detected Heteronuclear Single-Quantum Coherence (HSQC) Spectra*

Like HMQC, the HSQC experiment is a 2D heteronuclear chemical shift correlation map between *directly-bonded* (*J*-coupled) ¹H and X-heteronuclei (such as¹³C and ¹⁵N). HSQC techniques are frequently used in structure elucidation of organic molecules, as well as in protein NMR spectroscopy (Bodenhausen, and Ruben, 1980). The resulting spectrum is a 2D plot with protons (¹H) on one axis and a heteronucleus X (an atomic nucleus other than a proton), which is usually ¹³C or ¹⁵N. On the other axis, the spectrum contains a cross-peak for each unique proton attached to the heteronucleus being considered.

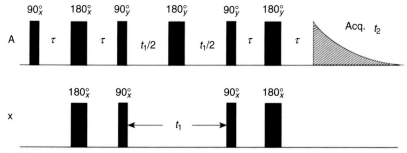

FIGURE 7.47 The pulse sequence employed for the HSQC experiment.

The HSQC experiment is a highly sensitive 2D-NMR experiment and was first described in a ^{1}H–^{15}N system, but it is also applicable to other nuclei, such as ^{1}H–^{13}C system. The experiment involves the transfer of magnetization from the proton to the second nucleus X (^{15}N or ^{13}C), *via* an _I_nsensitive _N_uclei _E_nhanced by _P_olarization _T_ransfer (INEPT) step. After a time delay (t_1), the magnetization is transferred back to the proton *via* a retro-INEPT step and the signal is then recorded. In HSQC, a series of experiments is recorded while the evolution time delay t_1 is incremented. The ^{1}H signal is detected in the directly measured dimension in each experiment, while the chemical shift of ^{15}N or ^{13}C is recorded in the indirect dimension from the series of experiments.

The basic 2D HSQC pulse sequence (Fig. 7.47) consists of four independent blocks of pulses. An initial INEPT pulse train transfers polarization from ^{1}H to the less sensitive nucleus X *via* $^1J_{(XH)}$ (INEPT block). The antiphase ^{13}C magnetization evolves during the variable evolution t_1 period under the effect of X chemical shift. The heteronuclear ^{1}H–X couplings are refocused by a 180° ^{1}H pulse at the middle of this period (^{1}H-decoupled X evolution block). A retro-INEPT pulse train converts the X magnetization to in-phase ^{1}H magnetization (reverse-INEPT block). Proton acquisition is performed with X decoupling (^{1}H-acquisition with X-decoupling). Due to a large number of pulses, setting up of an HSQC requires special care, i.e., the probe needs to be properly tuned and matched with the sample, and pulse lengths need to be carefully calibrated. The proton pulse should be determined on each sample individually, and the ^{13}C and ^{15}N pulses should be set up correctly. The pulse field gradient (PFG) pulse sequence of the HSQC experiment is shown in Fig. 7.48 (also see Section 2.4.2).

The HSQC spectrum of quercetin, along with its chemical structure, is shown in Fig. 7.49. The horizontal F_2 dimension provides ^{1}H chemical shifts, while the F_1 dimension gives the ^{13}C chemical shifts. Cross-peaks in the HSQC plot, representing the direct ^{1}H–^{13}C correlations, appear as singlets.

Both HMQC and HSQC techniques correlate coupled heteronuclear spins (^{1}H-X) across a single bond. They are both inverse techniques and are thus far more sensitive than the equivalent direct heteronuclear techniques (such as

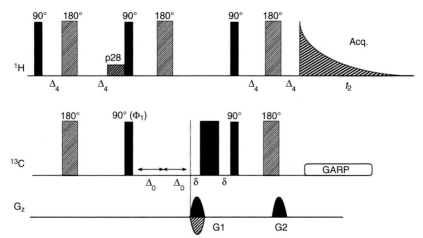

FIGURE 7.48   The PFG-HSQC pulse sequence.

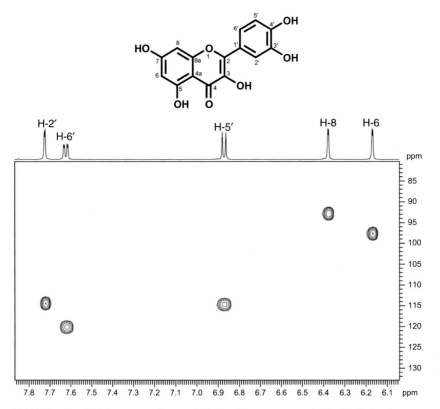

FIGURE 7.49   HSQC spectrum of quercetin, isolated from *Lyonia ovalifolia* (Wall.) Drude.

HeteroCOSY) (Reynolds et al., 1997). However, both techniques have their *pros and cons,* such as:

1. An HSQC spectrum gives a better resolution in the indirect dimension (^{13}C and ^{15}N) as compared to an HMQC spectrum, and thus it is better suited for crowded spectra.
2. The HSQC signals are singlets, while the HMQC spectra usually give multiplets due to homonuclear (^{1}H–^{1}H) couplings, complicating the interpretation of crowded regions of the HMQC spectra. Therefore the HSQC experiment is preferred over HMQC experiments for large biomolecules.
3. HMQC, due to ease of experimental set-up, is a robust technique for routine structural elucidation.
4. HSQC experiment uses significantly more pulses than the HMQC experiment, and therefore it is more sensitive to experimental imperfections.

Problem 7.14

Why is HSQC preferred over HMQC in ^{1}H–^{15}N measurements in protein NMR spectroscopy?

7.6.2 Heteronuclear Multiple-Bond Connectivity (HMBC) Spectra

In contrast to the HMQC experiment, which provides connectivity information about directly bonded ^{1}H–^{13}C interactions (i.e., one-bond coupling), the *H*etero-nuclear *M*ultiple-*B*ond *C*onnectivity (HMBC) experiment provides information about long-range ^{1}H coupling interactions with ^{13}C nuclei. Like the HMQC experiment, the HMBC experiment is also an "inverse" experiment (i.e., the ^{13}C population differences are measured indirectly through their effects on ^{1}H nuclei), and it is therefore significantly more sensitive than the conventional long-range heteronuclear shift-correlated or 2D COLOC experiments. The pulse sequence used is shown in Fig. 7.50. The first 90° ^{13}C pulse serves to remove the one-bond $^{1}J_{CH}$ correlations so that cross-peaks due to direct connectivities do not appear, allowing long-range ^{1}H–^{13}C connectivities to be recorded. The second 90° ^{13}C pulse creates zero- and double-quantum coherences, which are interchanged by the 180° ^{1}H pulse. After the last 90° ^{13}C pulse, the ^{1}H signals resulting from ^{1}H–^{13}C multiple-quantum coherence are modulated by ^{13}C chemical shifts and by homonuclear proton couplings. The PFG pulse sequence of the HMBC experiment is shown in Fig. 7.51.

It is easier to obtain a higher digital resolution in 2D experiments in the F_2 domain than in the F_1 domain, since doubling the acquisition time t_2 results in little overall increase in the experiment time. This is so because we can compensate for the increase by decreasing the relaxation delay between successive experiments. However, if we double the maximum time reached during t_1, we have to double the number of t_1 increments, thereby not only increasing the total

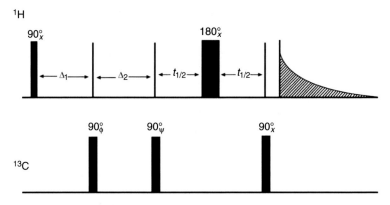

FIGURE 7.50   The pulse sequence for the HMBC experiment.

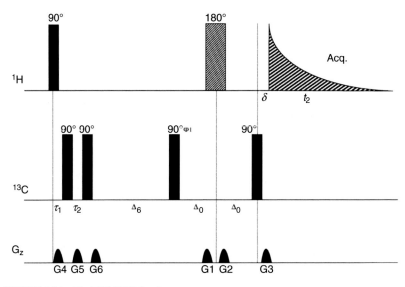

FIGURE 7.51   The PFG-HMBC pulse sequence.

time taken for the experiment but also suffering a loss of signal due to relaxation during t_1. This problem is avoided in the inverse experiment, because C–H correlation is detected *via* the proton spectrum, and the crowded proton region (requiring a higher resolution) now lies in the F_2 dimension, which has a high digital resolution, whereas the better-dispersed ^{13}C spectrum (which required a lower digital resolution) lies in the F_1 dimension.

The HMBC experiment, like the 2D COLOC experiment, is particularly useful for locating the quaternary carbons by identifying the various protons

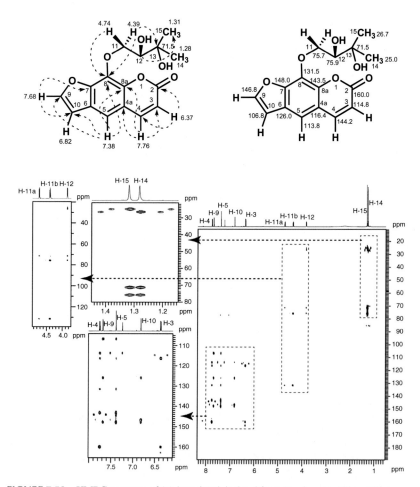

FIGURE 7.52 HMBC spectrum of (±)-heraclenol, isolated from *Ferula ovina* (Bioss.) Bioss.

interacting with them through two-bond ($^2J_{CH}$), three-bond ($^3J_{CH}$), and occasionally four-bond ($^4J_{CH}$) interactions. An HMBC spectrum of heraclenol, isolated from *Ferula ovina* (Bioss.) Bioss., is presented in Fig. 7.52. The spectrum was particularly helpful in joining individual structural fragments (established on the basis of COSY, HOHAHA, etc.) into a complete structure. In the HMBC spectrum of (±) heraclenol one pair of doublets at δ 7.76 and 6.37 is assigned to H-4 and H-3 based on their HMBC interactions with carbon signals resonating at δ 160.0 (C-2), 116.4 (C-4a), 143.5 (C-8a), and 131.5 (C-8). Similarly, the other pair of doublets at δ 7.68 and 6.82 show their interactions with C-6 (δ 126.0) and C-7 (δ 148.0); therefore, these doublets can be assigned to C-9 and C-10. The singlet at δ 7.38 can be assigned to C-5 due to its couplings with C-4a, C-8a, C-8,

C-10, and C-9. The presence of an aliphatic chain can be inferred from the coupling of pair of double doublet δ 4.74 (H-11a), 4.39 (H-11b) with C-8 (δ 3.15). The HMBC correlations are also shown on structure (Fig. 7.52). Other examples of the HMBC experiment are presented as problems at the end of this chapter.

7.7   INADEQUATE

The NMR techniques discussed so far provide information about proton–proton interactions (e.g., COSY, NOESY, SECSY, 2D *J*-resolved), or they allow the correlation of protons with carbons or other hetero atoms (e.g., hetero COSY, COLOC, hetero *J*-resolved). The resulting information is very useful for structure elucidation, but it does not reveal the carbon framework of the organic molecule directly. One interesting 2D NMR experiment, *I*ncredible *N*atural *A*bundance *D*oubl*E* *QUA*ntum *T*ransfer *E*xperiment (INADEQUATE), allows the entire carbon skeleton to be deducted directly *via* the measurement of ^{13}C–^{13}C couplings.

Carbon is a mixture of two isotopes in its natural abundance. The major isotope, ^{12}C, occurs in a natural abundance of 98.9%, but it is insensitive to the NMR experiment. The minor isotope, ^{13}C, occurs in a natural abundance of 1.1%; it is with this isotope that we are concerned in the INADEQUATE and many other NMR experiments. Thus, there will be only about one molecule in a hundred molecules with a particular carbon bearing the ^{13}C isotope. To find two adjacent carbons bearing ^{13}C isotope would be even less likely (i.e., 1.1% $\times$ 1.1% $\approx$ about 12 molecules in 100,000 molecules). The INADEQUATE experiment, which is based on the detection of ^{13}C–^{13}C interactions in such molecules, is therefore very insensitive, requiring about 500 mg of a compound of molecular weight 400–500 amu, and scanning for 24–36 h before an spectrum with an acceptable signal-to-noise ratio can be recorded.

The INADEQUATE experiment is a homonuclear NMR technique in which ^{13}C–^{13}C coupling satellites are detected after suppressing the stronger ^{13}C–^{1}H couplings. The theory behind the technique can be explained by exploring two particular properties of the coherence processes: (1) multiple-quantum coherence, or MQC, can occur only in a coupled-spin system. (2) Multiple-quantum coherences have significantly different phase properties than single-quantum coherences.

MQC can be understood by analogy with the more familiar single-quantum coherence. For instance, a spin $-\frac{1}{2}$ nucleus (such as ^{1}H or ^{13}C) has two energy levels, α and β. The energy difference ΔE between these two levels can be related to the absorption of frequency, $\Delta E = h\nu$ (Fig. 7.53). When the population is promoted from the lower to the upper level after absorption of energy, the process is called a *transition*. In the NMR process, we do not detect such transitions directly. Instead, we detect the evolution of magnetization in the *xy*-plane. This *single-quantum coherence* can be created by the application of a pulse on the equilibrium *z*-magnetization.

In contrast, in a two-spin system the two nuclei coupled with each other by the coupling constant, *J*, will have four energy levels available for

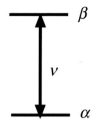

FIGURE 7.53 Single-spin energy-level diagram.

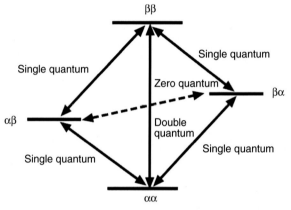

FIGURE 7.54 Energy-level diagram representing two nuclei in a two-spin system coupled with each other.

transitions (Fig. 7.54). Such a system not only has single-quantum coherences $(\alpha\alpha \rightarrow \beta\alpha; \alpha\alpha \rightarrow \alpha\beta; \alpha\beta \rightarrow \beta\beta; \beta\alpha \rightarrow \beta\beta)$ but also has a double-quantum transition $(\alpha\alpha \rightarrow \beta\beta)$ and a zero-quantum transition $(\alpha\beta \rightarrow \beta\alpha)$. These transitions cannot be detected directly but can be detected by indirect procedures: forcing the magnetization to evolve at the double-quantum (or zero-quantum) frequency during a certain time period before it is converted by a mixing pulse into a single-quantum coherence for detection.

A 2D INADEQUATE experiment therefore requires a pulse sequence capable of producing the following four changes in the spin and energy states:

1. It should create double-quantum coherence through an appropriate pulse sequence.
2. It should allow the double-quantum coherence to evolve during the incremented delays.
3. The evolution is then converted into observable magnetization by a "mixing" or "detect" pulse.
4. The FIDs are acquired and processed through double Fourier transformation in the usual fashion.

In the 2D INADEQUATE spectrum, the two adjacent ^{13}C nuclei (say A and X) appear at the following coordinates:

$$\delta_A; \quad \delta_A + \delta_X \text{ and } \delta_X; \quad \delta_A + \delta_X$$

Thus, if the vertical axis representing the double-quantum frequencies v_{DQ} is defined by v_1, and the horizontal axis representing the chemical shifts of the two carbons A and X is defined by v_2, then both pairs of ^{13}C satellites of nuclei A and X will lie on the same row (at $v_A + v_X$), equidistant from the diagonal. This greatly facilitates identification. The advantage is that no overlap can occur, even when J_{CC} values are identical, since each pair of coupled ^{13}C nuclei appear on different horizontal rows.

The basic INADEQUATE pulse sequence is:

$$90^{\circ}_{\emptyset1} - \tau - 180^{\circ}_{\emptyset2} - \tau - 90^{\circ}_{\emptyset1} - t_1 - \theta_{\emptyset2} - Acq$$

The delay is generally kept at $^{1\!/\!4}J_{CC}$. The coupling constant J_{CC} for directly attached carbons is usually between 30 and 70 Hz. The first two pulses and delays $(90^{\circ}_{\emptyset1} - \tau - 180^{\circ}_{\emptyset2} - \tau)$ create a spin echo, which is subjected to a second 90°_x pulse (i.e., the second pulse in the pulse sequence), which then creates a double-quantum coherence for all directly attached ^{13}C nuclei. Following this is an incremented evolution period t_1, during which the double quantum coherence evolves. The double-quantum coherence is then converted to detectable magnetization by a third pulse $\theta_{\emptyset2}$ and the resulting FID is collected. The most efficient conversion of double-quantum coherence can be achieved through a flip angle of 90°, but this makes the determination of the sign of the optimum double-quantum coherence frequency difficult. Selection of the optimum double-quantum coherence transfer signal can also be made by cycling phase $\emptyset_2$ of the pulse θ, but it requires a phase shift of 45°, which is difficult to implement. We can, however, detect the desired type of response selectively by designing the appropriate phase cycle.

The 2D INADEQUATE spectrum of α-picoline (Fig. 7.55) provides carbon–carbon connectivity information. The pairs of coupled carbons appear on the same horizontal line, allowing them to be identified readily. A diagonal line is drawn between the satellite peaks, which occur at equal distances on either side of the diagonal.

In order to interpret the spectrum, let us start with the most downfield carbon resonance and trace the subsequent carbon–carbon connectivities. For instance, a horizontal line is drawn from satellite peak F, which corresponds to the most downfield carbon resonating at δ 158.6 (signal "f"), to another carbon resonance C at δ 123.0 (signal "c"), establishing the connectivities between the C-2 and C-3 carbons. A vertical line is then drawn upward from C to satellite peak C', which has another satellite peak, D, at a mirror image position across the diagonal line on the horizontal axis. This establishes the connectivity of the C-3 carbon with C-4 (δ 135.9). Similarly, satellite peak D', lying on the same vertical

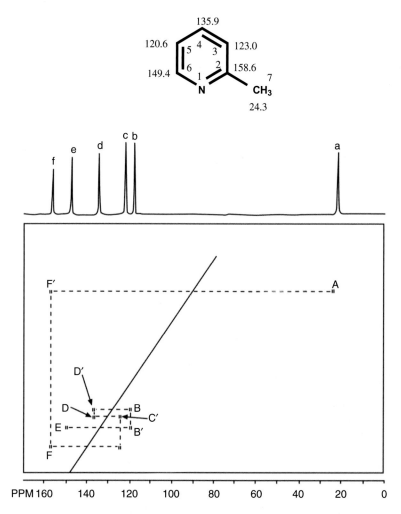

FIGURE 7.55   2D INADEQUATE spectrum of α-picoline.

axis, shows connectivity with another satellite peak, B, corresponding to the carbon resonating at δ 120.6. By dropping a vertical line from B, we reach the point B′, which has a mirror image partner, E, on the other side of the diagonal line, indicating connectivities of the C-5 and C-6 carbons (δ 120.6 and 149.4, respectively).

A line is then drawn vertically upward from F to satellite peak F′. The mirror image partner of F′, appearing on the same horizontal axis, is A, which establishes the connectivity of the C-2 quaternary carbon with the attached methyl carbon (δ 24.3). The carbon–carbon connectivity assignments based on the 2D INADEQUATE experiment are presented around the structure.

¹H/¹H Homonuclear Shift Correlation by COSY-DQF Experiment

The COSY-DQF spectrum of a lignin, podophyllotoxin and its ¹H-NMR data are shown. Assign and interpret the ¹H/¹H cross-peaks in the COSY spectrum.

δ_H	No. of Protons	J in Hz
2.76, m	1H	-
2.83, dd	1H	4.2, 13.8
3.73, s	6H	-
3.79, s	3H	-
4.07, t	1H	9.6
4.58, m	1H	-
4.59, m	1H	-
4.76, d	1H	9.0
5.95, d	1H	1.2
5.97, d	1H	1.2
6.35, s	2H	-
6.49, s	1H	-
7.09, s	1H	-

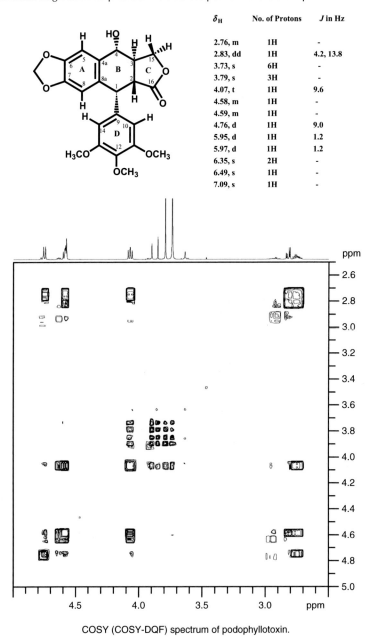

COSY (COSY-DQF) spectrum of podophyllotoxin.

Problem 7.16

¹H/¹H Homonuclear Shift Correlation by DQF-COSY Experiment

The double-quantum filtered (DQF)-COSY spectrum of an isoprenyl coumarin, along with ¹H-NMR spectrum and data are presented here. Determine the ¹H/¹H homonuclear interactions in the DQF-COSY spectrum.

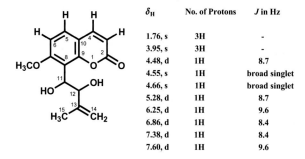

δ_H	No. of Protons	J in Hz
1.76, s	3H	-
3.95, s	3H	-
4.48, d	1H	8.7
4.55, s	1H	broad singlet
4.66, s	1H	broad singlet
5.28, d	1H	8.7
6.25, d	1H	9.6
6.86, d	1H	8.4
7.38, d	1H	8.4
7.60, d	1H	9.6

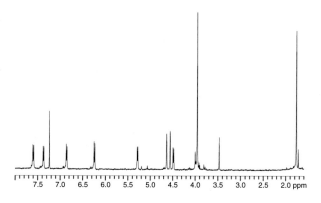

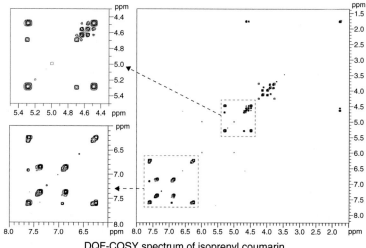

DQF-COSY spectrum of isoprenyl coumarin

Problem 7.17

¹H/¹H Homonuclear Shift Correlation by Long-Range COSY-45° Experiment
The LR COSY-45° spectrum, and ¹H-NMR chemical shifts of monoterpene (+)-5-hydroxyisopinocampheol, are given below. Determine the long-range ¹H/¹H homonuclear correlations based on the long-range COSY-45° spectrum. Demonstrate with reference to Fig. 7.10, how they can be helpful in interconnecting different spin systems?

δ_H	No. of Protons	J in Hz
4.05, ddd	1H	5.0, 5.0, 10.0
2.41, ddd	1H	3.0, 9.5, 13.0
2.33, ddd	1H	3.0, 7.0, 9.5
1.79, m	1H	-
1.77, m	1H	-
1.67, dd	1H	1.0, 7.5
1.47, d	1H	9.5
1.13, d	3H	7.0
1.13, s	3H	-
0.95, s	3H	-

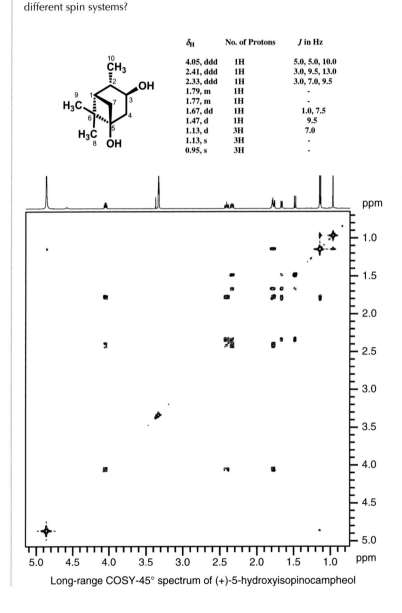

Long-range COSY-45° spectrum of (+)-5-hydroxyisopinocampheol

$^1H/^1H$ Homonuclear Shift Correlation by SECSY Experiment

The SECSY spectrum of 4′,5,7-trihydroxyflavone-7-rhamnoglucoside, isolated from *Potentilla evestita* L., along with the chemical shift assignments is given. The SECSY spectrum, a variant of COSY, allows the assignment of vicinal and geminal couplings between various protons. The display mode of SECSY spectra is different from that of normal COSY spectra. The spectrum comprises a horizontal line at δ 0.0 (unmodulated signals) and cross-peaks lying equidistant above or below it. When connecting such cross-peaks, the following considerations must be accounted for: (1) the cross-peaks to be connected should lie at equal distances above and below the horizontal line, and (2) all the connected cross-peaks should lie on *parallel* slanted lines. By dropping vertical lines from such connected cross-peaks to the horizontal line, we arrive at the chemical shifts of the coupled protons. By this method, the coupled protons can easily be identified. Interpret the spectrum, and determine the homonuclear correlations between various protons based on the SECSY spectrum.

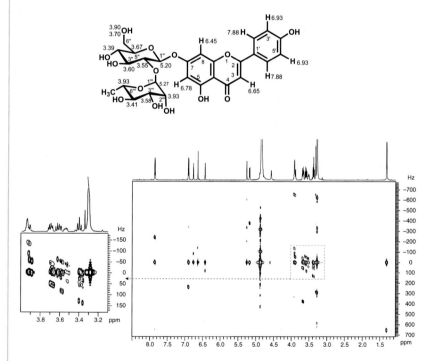

SECSY spectrum of 4′,5,7-trihydroxyflavone-7-rhamnoglucoside.

Problem 7.19

¹H/¹H Homonuclear Shift Correlation by SECSY Experiment

The SECSY spectrum of an isoprenyl coumarin, along with the ¹H-NMR chemical shifts are shown. Determine the homonuclear shift correlation between various protons based on the SECSY spectrum.

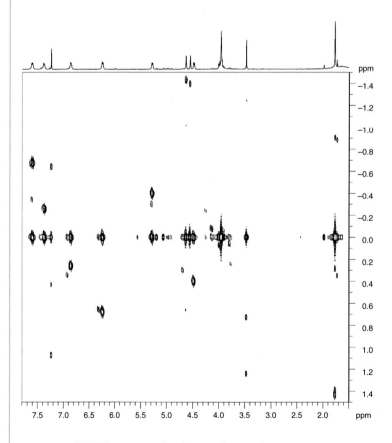

SECSY spectrum of an isoprenyl coumarin.

¹H/¹H Long-Range Total Homonuclear Shift Correlation by HOHAHA Experiment
The HOHAHA spectrum (100 ms) of podophyllotoxin is presented here. The HO-
HAHA, or *TO*tal *C*orrelation *S*pectroscop*Y* (TOCSY) spectrum (100 ms) shows
coupling interactions of all protons within a spin network, irrespective of whether
they are directly coupled to one another or not. As in COSY spectra, peaks on
the diagonal are ignored as they arise due to magnetization that is not modulated
by the coupling interactions. Podophyllotoxin has only one large spin system,
extending from the C-1 proton to the C-4 and C-15 protons. Identify all homonuclear correlations of protons within this spin system, based on the cross-peaks
in the spectrum.

δ_H	No. of Protons	*J* in Hz
2.76, m	1H	-
2.83, dd	1H	4.2, 13.8
3.73, s	6H	-
3.79, s	3H	-
4.07, t	1H	9.6
4.58, m	1H	-
4.59, m	1H	-
4.76, d	1H	9.0
5.95, d	1H	1.2
5.97, d	1H	1.2
6.35, s	2H	-
6.49, s	1H	-
7.09, s	1H	-

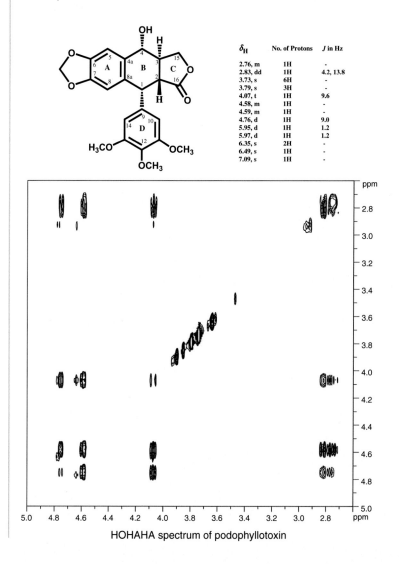

HOHAHA spectrum of podophyllotoxin

¹H/¹H Long-Range Total Homonuclear Shift Correlation by TOCSY Experiment
The TOCSY spectrum (100 ms) of 7-deoxyloganic acid, isolated from *Osyris wightiana* Wall. Ex. Wight., is shown. Interpret the spectrum, and determine the different spin systems from the cross-peaks in the spectrum.

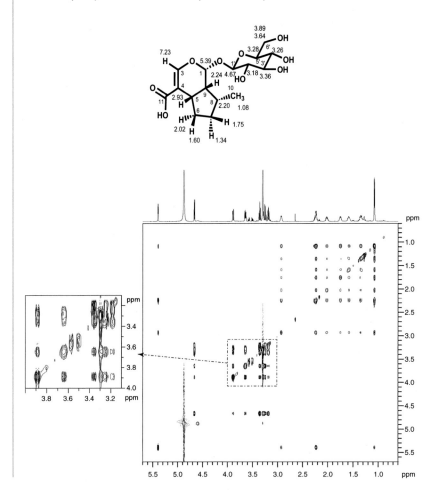

¹H/¹³C One-Bond Shift Correlation by Hetero-COSY (HETCOR) Experiment
The one-bond HETCOR spectrum, and ¹³C-NMR chemical shift data of podophyllotoxin, are shown. The one-bond heteronuclear shift correlations can readily be
identified from the HETCOR spectrum by locating the positions of the cross-peaks
and the corresponding δ_H and δ_C chemical shift values. The ¹H-NMR chemical

shifts are labeled on the structure. Assign the ^{13}C-NMR resonances to the various protonated carbons, based on the heteronuclear correlations in the HETCOR spectrum.

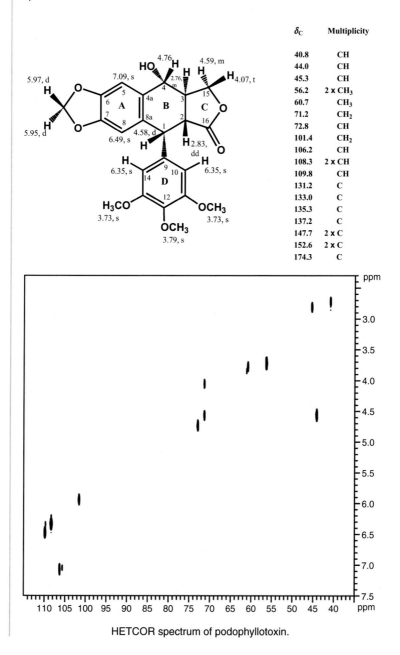

δ_C	Multiplicity
40.8	CH
44.0	CH
45.3	CH
56.2	2 x CH$_3$
60.7	CH$_3$
71.2	CH$_2$
72.8	CH
101.4	CH$_2$
106.2	CH
108.3	2 x CH
109.8	CH
131.2	C
133.0	C
135.3	C
137.2	C
147.7	2 x C
152.6	2 x C
174.3	C

HETCOR spectrum of podophyllotoxin.

$^1H/^{13}C$ One-Bond Shift Correlation by Hetero-COSY Experiment

The hetero-COSY spectrum of mangiferin, isolated from *Iris germanica* (Linn.) along with the ^{1}H-NMR assignments and ^{13}C-NMR data are shown. Assign the ^{13}C-NMR chemical nuclear shifts to various protonated carbons, based on the one-bond heteronuclear shift correlations.

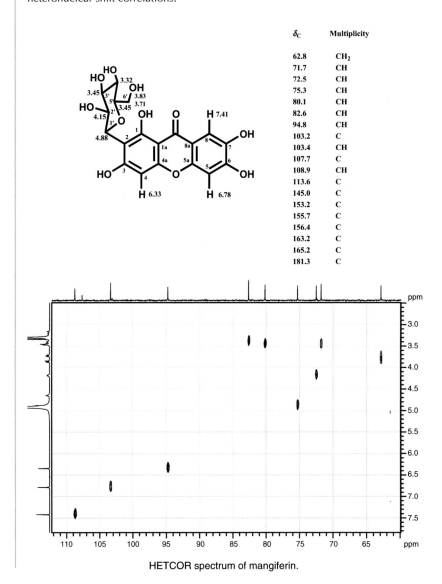

δ_C	Multiplicity
62.8	CH$_2$
71.7	CH
72.5	CH
75.3	CH
80.1	CH
82.6	CH
94.8	CH
103.2	C
103.4	CH
107.7	C
108.9	CH
113.6	C
145.0	C
153.2	C
155.7	C
156.4	C
163.2	C
165.2	C
181.3	C

HETCOR spectrum of mangiferin.

Problem 7.24

¹H/¹³C One-Bond Shift Correlation by Hetero-COSY Experiment

The hetero-COSY spectrum of (+)-catechin, isolated from *Osyris wightiana* Wall. Ex. Wight., along with the ¹H-NMR chemical shift data is shown. Assign the ¹³C-NMR shifts to various protonated carbons based on the cross-peaks in the hetero-COSY spectrum.

δ_C	Multiplicity
28.5	CH_2
68.8	CH
82.8	CH
95.5	CH
96.3	CH
100.8	C
115.1	CH
116.1	CH
120.0	CH
132.2	C
146.22	C
146.25	C
156.3	C
157.6	C

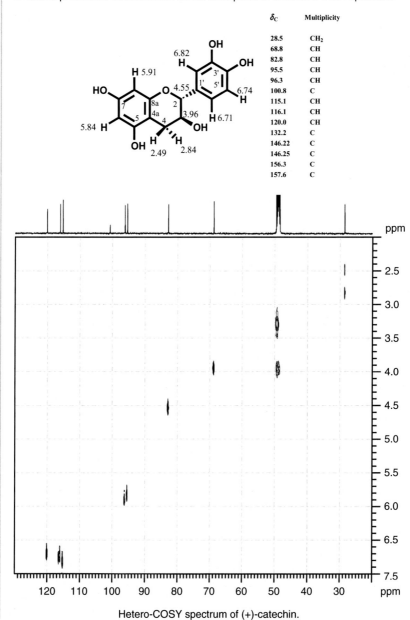

Hetero-COSY spectrum of (+)-catechin.

Problem 7.25

^{1}H/^{13}C-NMR Chemical Shift Correlation by HSQC Experiment

The heteronuclear single-quantum coherence (HSQC) spectrum, ^{1}H-NMR chemical shift assignments, and ^{13}C-NMR data of podophyllotoxin are shown. Determine the chemical shifts of various carbons and attached protons. The HSQC spectrum provide information about the one-bond correlations of protons and carbons. These spectra are fairly straightforward to interpret. The correlations are made by noting the position of each cross-peak and identifying the corresponding δ_H and δ_C values. Based on this technique, interpret the following spectrum.

δ_C	Multiplicity
40.8	CH
44.0	CH
45.3	CH
56.2	2 x CH$_3$
60.7	CH$_3$
71.2	CH$_2$
72.8	CH
101.4	CH$_2$
106.2	CH
108.3	2 x CH
109.8	CH
131.2	C
133.0	C
135.3	C
137.2	C
147.7	2 x C
152.6	2 x C
174.3	C

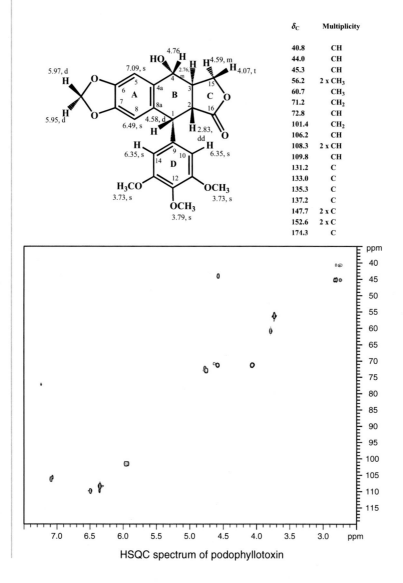

HSQC spectrum of podophyllotoxin

$^1H/^{13}C$-NMR Chemical Shift Correlation by HSQC Experiment
The HSQC spectrum, ^{1}H-NMR chemical shift assignments, and ^{13}C-NMR data of 5,7-dihydroxy-4'-methoxyisoflavone, isolated from *Potentilla evestita* L., are shown. Determine the carbon framework of the spin systems.

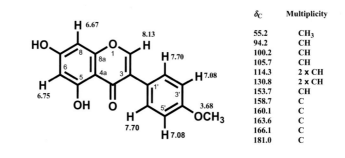

δ_C	Multiplicity
55.2	CH$_3$
94.2	CH
100.2	CH
105.7	CH
114.3	2 x CH
130.8	2 x CH
153.7	CH
158.7	C
160.1	C
163.6	C
166.1	C
181.0	C

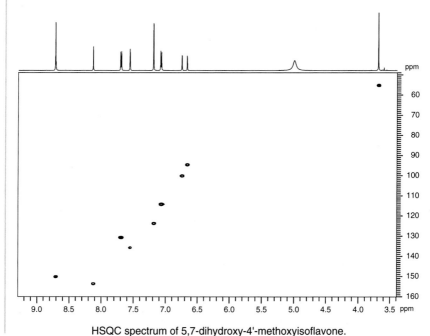

HSQC spectrum of 5,7-dihydroxy-4'-methoxyisoflavone.

$^1H/^{13}C$ Long Range Interactions from the HMBC Spectrum
The HMBC spectrum of podophyllotoxin is shown. The cross-peaks in the HMBC spectrum represent long-range heteronuclear ^{1}H/^{13}C interactions within the same sub-structure or between different sub-structures. Interpretation should start with

a readily assignable carbon (or proton), and then you should identify the proton/s (or carbon/s) with which it has coupling interactions. Then proceed from these protons, and look for the carbon two, three, or, occasionally, four bonds away. One-bond heteronuclear interactions may also appear in the HMBC spectrum.

The ^{1}H-NMR and ^{13}C-NMR chemical shifts have been assigned and substructures have been deduced on the basis of COSY-DQF (Problem 7.15), and other spectroscopic observations. Interpret the HMBC spectrum and identify the heteronuclear long-range coupling interactions between the ^{1}H and ^{13}C nuclei.

Carbon No.	δ_C	Multiplicity
C-1	44.0	CH
C-2	45.3	CH
C-3	40.8	CH
C-4	72.8	CH
C-4a	133.0	C
C-5	106.2	CH
C-6	147.7	C
C-7	147.7	C
C-8	109.8	CH
C-8a	131.2	C
C-9	135.3	C
C-10	108.3	CH
C-11	152.6	C
C-12	137.2	C
C-13	152.6	C
C-14	108.3	CH
C-15	71.2	CH$_2$
C-16	174.3	C
C-17	101.4	C

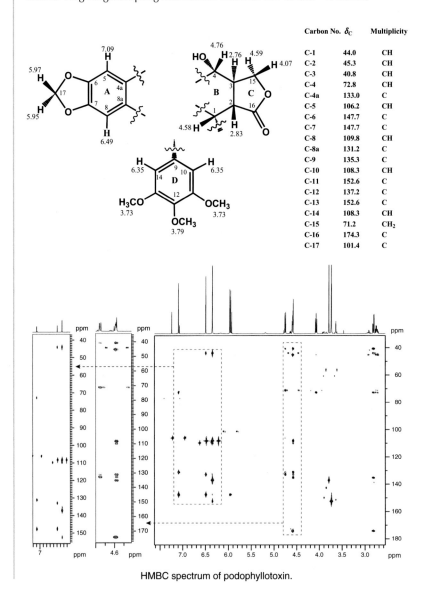

HMBC spectrum of podophyllotoxin.

¹H/¹³C Long-Range Interactions from the HMBC Spectrum

The HMBC spectrum of 5,7-dihydroxy-4′-methoxyisoflavone, isolated from *Potentilla evestita* L., along with the ¹H-NMR assignments is shown. Determine the ¹H/¹³C long-range heteronuclear shift correlations based on the HMBC experiment, and explain how HMBC correlations are useful in chemical shift assignments of nonprotonated quaternary carbons.

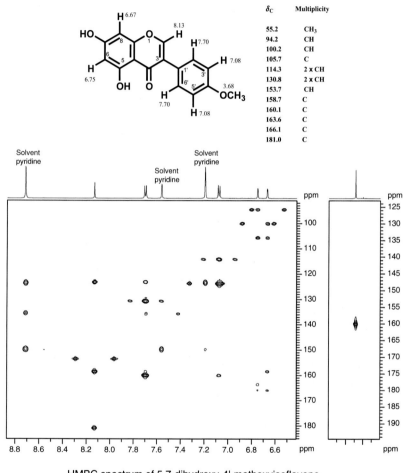

δ_C	Multiplicity
55.2	CH₃
94.2	CH
100.2	CH
105.7	C
114.3	2 x CH
130.8	2 x CH
153.7	CH
158.7	C
160.1	C
163.6	C
166.1	C
181.0	C

HMBC spectrum of 5,7-dihydroxy-4′-methoxyisoflavone

Problem 7.29

¹H/¹³C Long-Range Shift Correlation by COLOC Experiments
Like the HMBC, the COLOC experiment provides long-range heteronuclear
chemical shift correlations. The COLOC spectrum, ¹H-NMR, and ¹³C-NMR data
of 7-hyroxyfrullanolide are presented here. Use the data to assign the quaternary
carbons.

δ_H	No. ofprotons	δ_C	Multiplicity
0.94, s	3H	18.7	CH₃
1.18, m	1H	25.5	CH₃
1.31, m	1H	17.6	CH₂
₂1.33, m	1H	30.7	CH₂
1.39, m	1H	32.5	CH₂
1.45, m	1H	34.3	CH₂
1.48, m	1H	38.1	CH₂
1.56, m	1H	120.5	CH₂
1.64, s	3H	80.9	CH
1.72, m	1H	32.0	C
1.97, m	1H	75.1	C
1.99, m	1H	126.4	C
4.97, s	1H	139.5	C
5.71, s	1H	144.4	C
6.06, s	1H	169.4	C

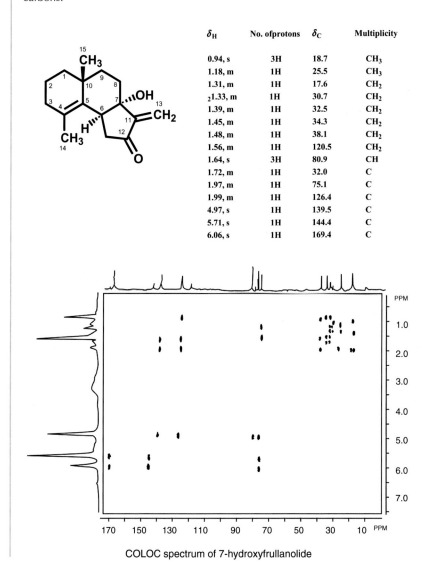

COLOC spectrum of 7-hydroxyfrullanolide

Stereochemical Correlations by nOe Experiment
The nOe difference and 1D ^{1}H-NMR spectra of podophyllotoxin are shown. Justify the stereochemical assignment of the C-2, C-3, and C-4 asymmetric centers based on the nOe data, assuming that the C-4 proton is β-oriented.

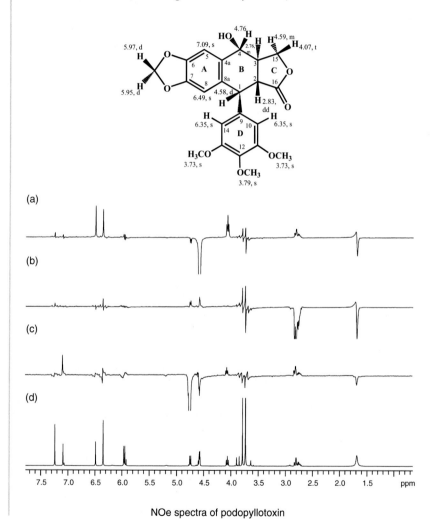

NOe spectra of podopyllotoxin

nOe Difference Spectrum and Its Use in Structure Elucidation
Two possible structures, A and B, were initially considered for a naturally occurring coumarin isolated from *Murraya paniculata* (L.) Jack. Its nOe difference and

^{1}H-NMR spectra are presented. The relevant ^{1}H chemical shift values are given around structure A. On the basis of the nOe data, identify the correct structure of the coumarin.

^{1}H-NMR and nOe spectra of coumarin

Structural Information from NOESY Experiment
The NOESY spectrum appears in a similar format as that of COSY spectra; i.e., un-modulated signals lie along the diagonal, whereas the chemical shift frequencies of a proton modulated by the chemical shift of another proton that lies spatially close to it (dipolar coupling) gives rise to an off-diagonal NOESY cross-peak. This cross-peak appears at the junction of the chemical shift frequencies of the two protons. Three possible structures, A, B, and C, were considered for buxatenone, $C_{22}H_3O_2$, a triterpene isolated from *Buxus papillosa* C. K. Schneid. The relevant ^{1}H-NMR chemical shifts have been assigned to the various protons. Assign the correct structure based on the NOESY interactions.

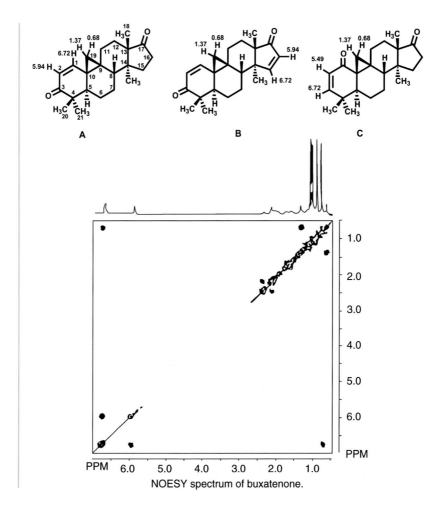

NOESY spectrum of buxatenone.

Structural Correlations by ROESY Experiment

The ROESY spectrum affords homonuclear *transverse* nOe interactions as cross-peaks between the various dipolarly coupled hydrogens. This may be compared with NOESY spectra, which arise due to a decrease in the *longitudinal magnetization* (z-magnetization) of one nucleus (the irradiated nucleus) that causes an increase in the z-magnetization of another nucleus that lies spatially close to the irradiated nucleus. ROESY spectra, on the other hand, involve an exchange of *transverse magnetization*, i.e., magnetization in the x'y'-plane. The ROESY spectrum and ¹H-NMR data of podophyllotoxin are shown. Deduce the stereochemistry and ¹H–¹H couplings in this compound.

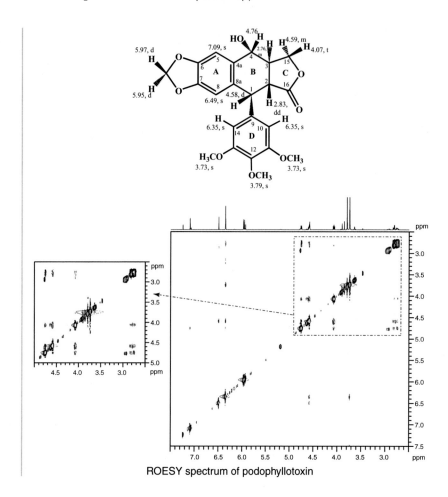

ROESY spectrum of podophyllotoxin

Problem 7.34

Homonuclear 2D J-Resolved Assignments

Homonuclear 2D J-resolved spectra are generally used to determine the coupling constants and multiplicity of complex proton signals. In these spectra, the chemical shifts of protons are displayed along one axis, and the coupling constants along the other axis, thereby eliminating the overlap between close-lying multiplets. The homonuclear 2D J-resolved spectrum of scopoletin is shown. The ^{1}H-NMR chemical shifts are assigned on the structure. Determine the approximate coupling constants and multiplicities for each signal.

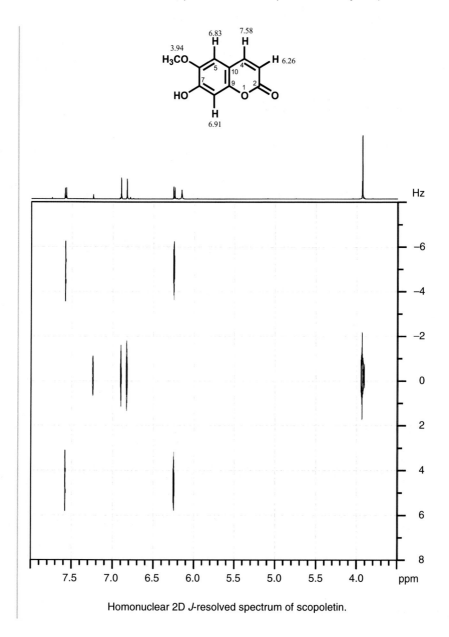

Homonuclear 2D *J*-resolved spectrum of scopoletin.

Heteronuclear 2D J-Resolved Assignments

Heteronuclear 2D J-resolved spectra are used to determine the coupling constants between carbon and proton signals. In these spectra, chemical shifts are presented on x-axis and coupling constants on y-axis. These spectra can be used to determine the α or β attachment of the sugar moieties to the aglycon. Heteronuclear 2D J-resolved spectrum of citroside B isolated from *Osyris wightiana* Wall. ex Wight is shown here. Determine the coupling constants for each signal.

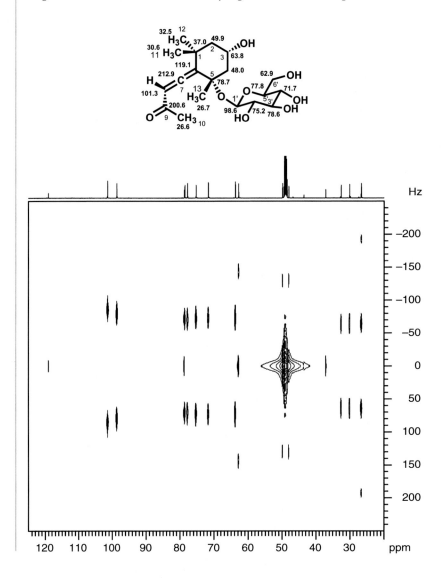

Carbon–Carbon Connectivity Assignment by 2D INADEQUATE Experiment
The 2D INADEQUATE spectrum contains satellite peaks representing direct coupling interactions between adjacent ^{13}C nuclei. The 2D INADEQUATE spectrum
and ^{13}C-NMR data of methyl terahydrofuran are shown. Assign the carbon–carbon
connectivities using the 2D INADEQUATE plot.

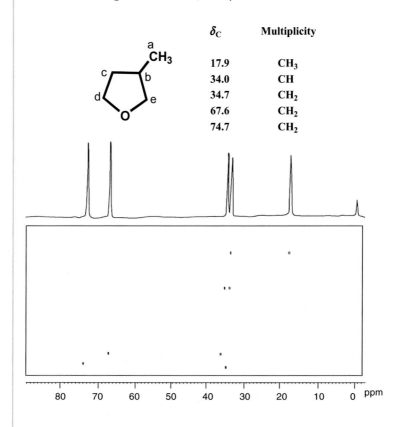

δ_C	Multiplicity
17.9	CH_3
34.0	CH
34.7	CH_2
67.6	CH_2
74.7	CH_2

2D INADEQUATE spectrum of 3-methyltetrahydrofuran.

Carbon–Carbon Connectivity Assignment by 2D INADEQUATE Experiment
The 2D INADEQUATE spectrum and ^{13}C-NMR data of 7-hydroxyfrullanolide are
given. Determine the carbon–carbon connectivities and subsequently build the
entire carbon skeleton based on the 2D INADEQUATE experiment.

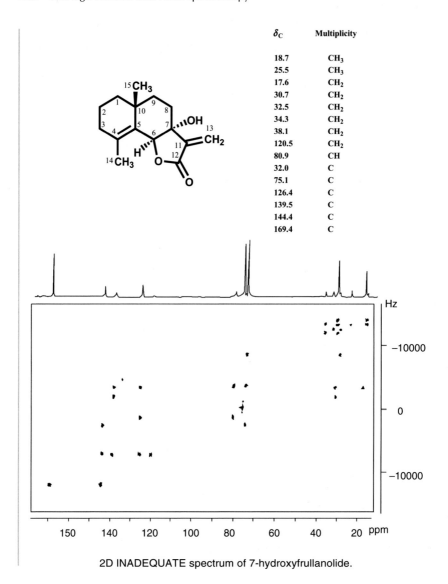

δ_C	Multiplicity
18.7	CH_3
25.5	CH_3
17.6	CH_2
30.7	CH_2
32.5	CH_2
34.3	CH_2
38.1	CH_2
120.5	CH_2
80.9	CH
32.0	C
75.1	C
126.4	C
139.5	C
144.4	C
169.4	C

2D INADEQUATE spectrum of 7-hydroxyfrullanolide.

SOLUTIONS TO PROBLEMS

7.1 Homonuclear shift-correlation spectroscopy (COSY) is a standard method for deducing proton coupling networks. Diagonal and off-diagonal peaks appear with respect to the two frequency dimensions, F_1 and F_2. The off-diagonal peaks (cross-peaks) represent the direct coupling interactions between protons. By working through cross-peaks, one can easily correlate protons that are coupled to each other. Several versions of the COSY experiment have been designed to get optimum performance in a variety of situations (such as DQF-COSY, COSY-45°, and COSY-60°).

7.2 The 2D shift-correlation technique is based on Jeener's original experiment and has now become the standard method for establishing proton–proton coupling networks and proton–carbon connectivity. When we excite protons with a 90° pulse and allow them to precess in the $x'y'$-plane, they acquire phases that differ according to the differences in their respective chemical shifts and J-coupling interactions. The second 90°_x pulse causes the mixing of spin states within a spin system. The $x'y'$-magnetization of one proton can thus be transferred to another coupled proton by a pulse applied along an axis in the $x'y'$-plane. The effect of this second 90°_x pulse on the magnetization vector will vary, depending on its position in the $x'y'$-plane. Fourier transformation of these data will then yield a spectrum with two frequency axes, F_1 and F_2. The "off-diagonal" peaks in the spectrum represent the spin–spin couplings between the protons. The peaks on the diagonal, however, represent unperturbed magnetization.

7.3 In the COSY-45° spectrum, diagonal peaks are not important. The off-diagonal cross-peaks are the ones that arise due to vicinal and geminal coupling interactions between coupled protons of a spin system. In the COSY spectrum of citroside B, the cross-peak A represents the coupling interactions between the signals at δ 4.52 and 3.13, assigned to the C-1' anomeric and C-2' methine protons of the sugar moiety. Cross-peak B is due to the COSY interaction of the proton at δ 4.31 (C-3H) with that at δ 2.48 (C-4H$_a$). The same proton (C-3H) is further coupled to δ 1.91 (C-2H$_a$) and δ 1.33/ 1.30 (δ C-4H$_b$ and C-2H$_b$), as is evident from cross-peaks C and D. Cross-peaks E and F represent the geminal interactions between C-4H$_a$ and C-4H$_b$, and C-2H$_a$ and C-2H$_b$. The vicinal couplings between glucose protons is indicated with cross-peaks G to K.

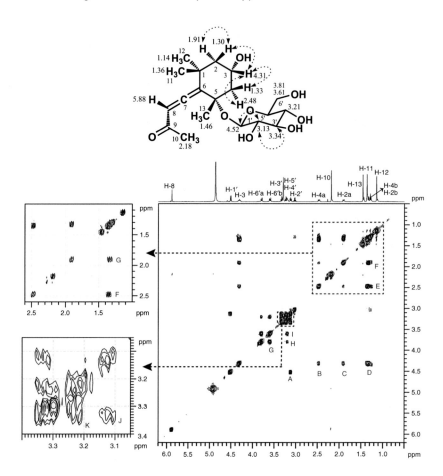

7.4 One-dimensional (1D) double-resonance or homonuclear spin–spin de-coupling experiments can be used to furnish information about the spin network. However, we have to irradiate each proton signal sequentially and to record a larger number of 1D ^{1}H-NMR spectra if we wish to determine all the coupling interactions. Selective irradiation (saturation) of an individual proton signal is often difficult if there are protons with close chemical shifts. Such information, however, is readily obtainable through a single COSY experiment.

7.5 The COSY-45° experiment has the following major advantages over the basic COSY-90° experiment.

1. The COSY-45° spectrum has reduced intensity of the component signals near the diagonal. This simplifies this region, thereby making it possible to identify correlations that would otherwise be hidden in the cluster of peaks close to the diagonal.

2. The cross-peaks visible in the COSY-45° spectrum represent essentially the couplings between directly connected transitions. This allows us to determine the relative signs of the coupling constants.

7.6 Decoupling in the v_1 dimension is carried out by fixing the interval between the two pulses to a value of t_dS and then applying a 180° refocusing pulse after a variable $t_{1/2}$ delay following the 90° pulse. Since the amplitude of the transferred magnetization depends on J_{td}, no J-modulation of the signal will occur during t_1. The signal will therefore be modulated by the chemical shift frequency. It will be at a maximum at $t_1 = 0$, zero at $t_1 = t_{d/2}$, and maximum again at $t_1 = t_d$. This results in the narrowing of cross-peaks along v_1, and can facilitate the identification and assignments of the coupled protons, especially in cases where the protons lie close to one another. Decoupled-COSY is the method of choice in molecules in which the cross-peaks are not distinguishable due to overlap in normal coupled COSY spectra, since in decoupled COSY spectra all the protons produce narrow cross-peaks at their chemical shifts on the v_1-axis.

7.7 The pulse sequence of the delayed-COSY experiment has a fixed delay at the end of the evolution period t_1 and before the acquisition period t_2. This results in the weakening and often the disappearance of stronger direct-coupling interactions (cross-peaks) and intensification of the weaker long-range interactions. This allows long-range couplings to be seen with greater clarity.

7.8 Exchange correlation spectroscopy (E. COSY), a modified form of COSY, is useful for measuring coupling constants. The pulse sequence of the E. COSY experiment has a mixing pulse β of variable angle. A number of experiments with different values of β are recorded that eliminate the multiplet components of unconnected transitions and leave only the multiplet components for connected transitions. This simplified 2D plot can then be used to measure coupling constants.

7.9 Spin-echo correlated spectroscopy (SECSY) is a modified form of the COSY experiment. The difference in the pulse sequence of the SECSY experiment is that the acquisition is delayed by time $t_{1/2}$ after the mixing pulse, while the mixing pulse in the SECSY sequence is placed in the middle of the t_1 period. The information content of the resulting SECSY spectrum is essentially the same as that in COSY, but the mode of presentation is different. If the individual nuclei in the SECSY spectrum are coupled to others with similar chemical shifts, then the F_1 spectral width may be less and the resolution consequently higher in SECSY spectra than in COSY spectra. However, for routine purposes the SECSY experiment does not offer any significant advantages over the COSY experiment.

7.10 COSY and HETCOR experiments are extremely useful in the structure elucidation of complex organic molecules. The geminal and vicinal protons and their one-bond C–H connectivities are first identified from the

HETCOR spectrum, and then the geminal couplings are eliminated from the COSY spectrum, leaving vicinal connectivities. By careful interpretation of the COSY and the one-bond HETCOR spectra, it is then possible to obtain information about the carbon–carbon connectivities of the *protonated* carbons ("pseudo-INADEQUATE" information). In this way, the carbon–carbon connectivity information of protonated carbons is obtainable through a combination of COSY and HETCOR experiments.

Long-range carbon–proton coupling interactions provide important information regarding coupling of protons with carbons across two, three, or four bonds ($^2J_{CH}$, $^3J_{CH}$, $^4J_{CH}$). Interaction of protons with quaternary carbons across two, three, or four bonds is particularly important in joining the various fragments together.

7.11 2D J-resolved NMR experiments are based on the elimination of chemical shift modulations during t_1, allowing the observation of J-coupling only in the F_1 domain. Dephasing in the $x'y'$-plane after the initial 90°_x pulse occurs due to both chemical shift effects and J-couplings in the system. The 180° refocusing pulse at the end of t_1 reverses only the dephasing of chemical shifts, whereas the effect of J-coupling continues to modulate the observed nucleus during t_2. This is then recorded as the second dimension in the 2D plot.

7.12 In both experiments, chemical shifts and coupling constants are separated along the two axes. In homonuclear 2D J-resolved spectra, the chemical shifts are along one axis and the couplings with the same type of nucleus lie along the other axis (e.g., ^1H chemical shifts along one axis and $^1H/^1H$ couplings along the other). In heteronuclear 2D J-resolved spectra, however, one axis would contain the chemical shifts of one type of nucleus, for example, ^{13}C, while the other axis would contain the coupling of the ^{13}C nuclei with another type of nucleus, for example, ^1H, i.e., δ_C along one axis and J_{CH} along the other. Both experiments are based on the same principle of J-modulation during the evolution period t_1, but the pulse sequences are substantially different. This can be understood by considering the behavior of the magnetization vectors in a spin-echo experiment. Let us first consider vector dephasing due to spin–spin coupling and chemical shifts in a homonuclear AX system. Application of the single 180° pulse at this stage will not only affect the nucleus (A) which is being observed, but will also affect the coupling partner nucleus (X). The vector components will therefore continue to diverge from each other at a rate governed by J, so the amplitudes of the signals detected during t_2 will be modulated according to their respective J values ("J-modulation"). In the case of heteronuclear AX spin coupling, application of simultaneous 180° pulses in the frequency range of both nuclei generates similar J_{AX} modulation effects. The chemical shifts will be refocused at the end of t_1 by the 180° pulse, while heteronuclear J-modulation effects are retained.

7.13 The NOESY and ROESY experiments essentially both provide the same information (internuclear distances, stereochemical information, etc.). However, they have several key differences. The NOESY experiments are based on longitudinal relaxation, while ROESY methods are based on a transverse relaxation process. Similarly, the NOESY experiments work well for molecules of very low and very high molecular weight, but they often do not work for medium sized molecules (molecular weights ~1000–3000 Da). In this molecular weight range, the nOes are very close to zero (so called nulling condition). A ROESY experiment, therefore, is used to get nOe information in intermediate sized molecular entities. For high molecular weight molecules, both NOESY and ROESY experiments give very similar results with the exception that the cross-peaks will be in phase with respect to the diagonals for a NOESY, and 180° out of phase with the diagonals in the case of a ROESY spectrum.

7.14 HSQC is a frequently recorded NMR spectrum in protein NMR spectroscopy. The most favorable characteristics of HSQC are its high resolution, and capacity to provide cross-peak as singlets for each proton directly connected to a ^{15}N or ^{13}C. Proteins are large molecules with many amino acids. With the exception of proline, all amino acids have "^1H–^{15}N" moieties (protons attached to nitrogen of the peptide linkage). They resonate between δ 6-8 region in the ^1H-NMR spectrum often as crowded and overlapping signals. The ^{15}N-HSQC spectrum provides a correlation between the ^{15}N and the amide proton, and each amide yields a cross-peak in the HSQC spectra. Each residue (except proline) can therefore produce an observable cross-peak in the HSQC spectra, cleanly separated based on its correlation with the attached ^{15}N. In practice, all the peaks are not always seen due to a number of factors, including line broadening, and chemical exchange of N-terminal residues of NH_3^+ with the solvent. In addition to the backbone amide resonances, side chain residues (such as NH of tryptophan, asparagine, and glutamine residues) also give rise to observable ^1H–^{15}N cross peaks in HSQC spectrum.

Although it is possible to record HSQC spectrum in natural abundance of nitrogen, but proteins under study are usually expressed with ^{15}N labeling.

The low resolution and multiplets in the cross-peaks due to ^1H–^1H coupling in direct dimension make HMQC unsuitable for protein NMR spectroscopy. Recently, a variant of the HMQC, i.e., fast-HMQC, has been used for the protein NMR spectroscopy.

7.15 The COSY-DQF spectrum of podophyllotoxin shows several cross-peaks (A-G). The obvious starting point in interpretation is the most downfield member of the spin network, i.e., the C-4β proton, geminal to the hydroxy function, resonating at δ 4.76, which shows the coupling interaction with the C-3 proton (δ 2.76) represented by cross-peak E in the spectrum. Cross-peaks A, B, and C are due to the correlations of the C-3 proton (δ 2.76) with the C-2 (δ 2.83), C-15β (δ 4.07), and C-15α (δ 4.59) protons. Cross-peak F establishes that the protons at δ 4.07 and 4.59 are coupled with each other. Cross-peak D corresponds to the coupling of C-2βH (δ 2.83) with the C-1 proton (δ 4.58). The chemical shift assignments, based on COSY cross-peaks, are presented around the structure.

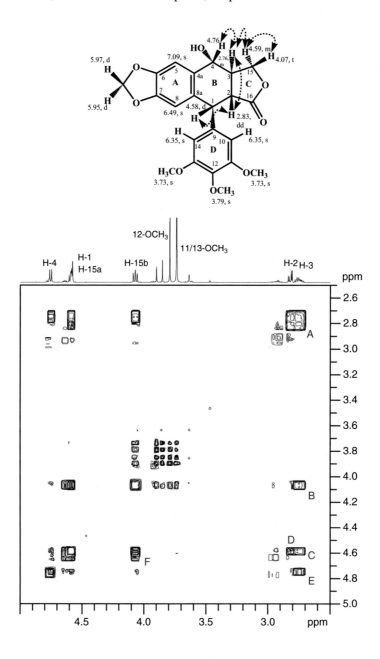

7.16 The DQF-COSY spectrum of isoprenyl coumarin displays clean and well-formed cross-peaks, in contrast to the COSY-45° spectrum. Two distinct spin systems are clearly visible in the spectrum. The upfield spin system comprises cross-peaks A and B; the downfield spin system contains cross-peaks C and D. The upfield spin system represents the oxygen-bearing methine protons and the vinylic protons. Cross-peak A represents the geminal interaction between the vinylic methylene protons, resonating at δ 4.55 and 4.66 (C-14 protons). Cross-peak B is due to the coupling between C-12H (δ 4.48) and C-11H (δ 5.28). The downfield spin system falls in the aromatic region. Cross-peak D represents coupling between the aromatic C-5 (δ 6.86) and C-6 (δ 7.38) protons, respectively; the coupling between the C-3 (δ 6.25) and C-4 (δ 7.60) vinylic protons is inferred from the cross-peak C in the DQF-COSY spectrum. Their downfield chemical shifts indicate their conjugation with lactone carbonyl group.

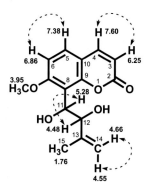

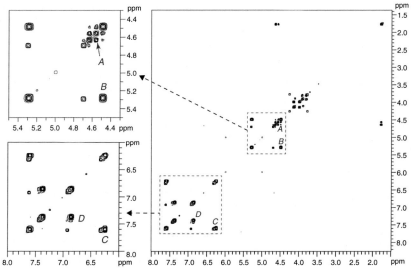

7.17 The LR COSY-45° spectrum of the monoterpene exhibits several long-range correlations that are easily distinguishable in comparison to the COSY-45° spectrum. There are some additional peaks when compared with the COSY-45° spectrum. Cross-peaks A, B, and C represent the long-range interactions; cross-peaks D, E, F, G, H, I, and J are due to direct interactions. This is a common phenomenon in the long-range COSY spectra; i.e., not only the long-range coupling interactions appear, but the stronger vicinal (or geminal) couplings also often appear. Cross-peak A is due to the interaction of the C-8 methyl protons (δ 0.95) with the C-9 methyl protons, resonating at δ 1.13. Cross-peak B represents the vicinal interaction of the C-7 methylene protons (δ 1.47) with the C-1 methine proton (δ 1.67). Cross-peak D represents coupling between the C-1 and C-2 protons, while cross-peak C is due to long-range interaction between the C-4a and C-7a protons. These long-range correlations between various protons may very well be observed due to "W" coupling often visible in such rigid ring systems. Cross-peaks E, G, I, and J represent the vicinal couplings between the H-10, H-1, H-3, H-4a with H-2, H-7, H-2, and H-3, respectively, in the molecule. Cross-peaks F and H show geminal couplings of C-7 and C-4 methylene protons. The long-range interactions are presented as broken arrows; the direct couplings are displayed as solid arrows around the structure.

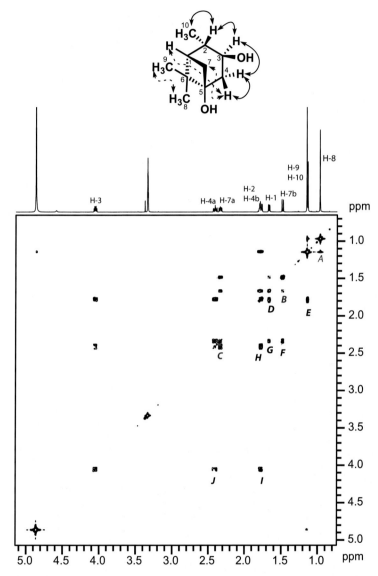

Long-range interactions are shown in red color while those cross-peaks that are similar to COSY 45° spectra are in black color.

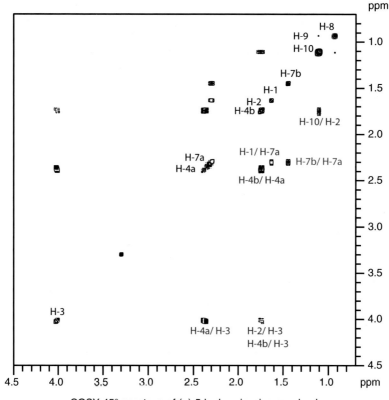

COSY-45° spectrum of (+)-5-hydroxyisopinocampheol, obtained by biotransformation of (+)-isopinocampheol with *Cunninghamella blakesleeana.*

7.18 The SECSY spectrum of 4′,5,7-trihydroxyflavone-7-rhamnoglucoside displays a number of cross-peaks. Pairs of interconnected cross-peaks can be identified by connecting cross-peaks that lie at equal distances above and below the horizontal line at δ 0.0 on parallel slanted lines. For example, cross-peak A connects to cross-peak A′ by the diagonal line, establishing the connectivity of the proton at δ 7.88 (C-2′H/ C-6′H) with the protons at δ 6.93 (C-3′H/ C-5′H). Another cross-peak, B of C-6 proton (δ 6.78) shows long-range connectivity *via* a line parallel to the first diagonal line to cross-peak B′ for C-8H at δ 6.45. Similarly, cross-peaks C and C′ represent vicinal connectivity between the anomeric C-1‴ proton (δ 5.27) and the C-2‴ (δ 3.53) proton of rhamnose sugar moiety. Cross-peaks D and D′ are due to the vicinal correlations of the anomeric C-1″ proton (δ 5.27) of glucose moiety with the C-2″ proton (δ 3.53). The cross-peaks E and E′, F and F′, G and G′, H and H′ also show vicinal interactions of sugar moieties. The C-6‴ proton (δ 1.32) shows cross-peaks I and I′, representing coupling with C-5‴ proton resonating at δ 3.93. These SECSY interactions establish the respective positions of the various protons and are presented around the structure and in the spectrum.

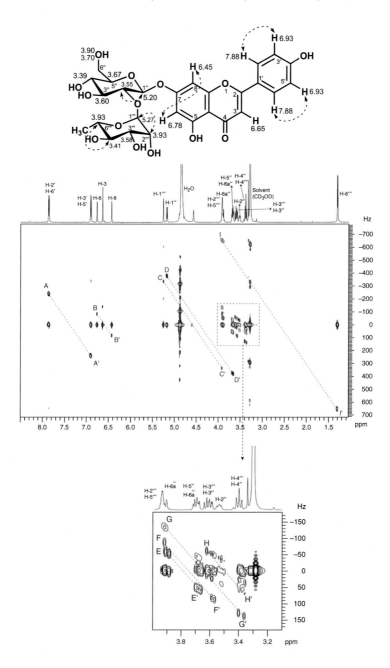

7.19 The SECSY spectrum of the coumarin presents cross-peaks for various coupled nuclei. These cross-peaks appear on diagonal lines that are parallel to one another. By reading the chemical shifts at such connected cross-peaks, we arrive at the chemical shifts of the coupled nuclei. For instance, cross-peaks A and A′ exhibit connectivity between the vinylic C-4 and C-3 protons resonating at δ 7.60 and 6.25, respectively. The C-4 methine proton appears downfield due to its β-disposition to the lactone carbonyl. Similarly, cross-peaks B and B′ show vicinal couplings between the C-5 and C-6 methine protons (δ 7.38 and 6.86, respectively) of the aromatic ring. The cross-peaks C and C′ represent the correlation between the oxygen-bearing C-11 (δ 5.28) and C-12 (δ 4.48) methine protons in the side chain. These interactions are presented around the structure.

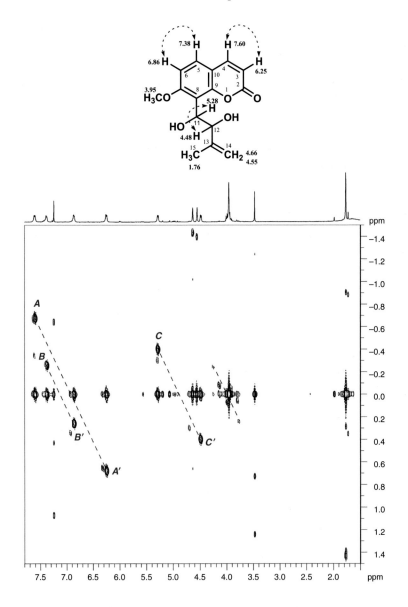

7.20 The HOHAHA spectrum of podophyllotoxin shows interactions between all protons of the spin systems in rings B and C of the molecule. Cross-peaks A, B, and C represent long-range interactions of the C-2 methine proton (δ 2.83) with the C-15 α and β protons (δ 4.59 and 4.07), as well as with the C-4β proton (δ 4.76). Cross-peak D represents geminal coupling between the C-15α and β protons (δ 4.59 and 4.07). Cross-peaks E and F are due to the interactions of the C-4 methine proton (δ 4.76) with the C-15β (δ 4.07) and C-15α (δ 4.59) protons, respectively. Coupling between C-3H (δ 2.76) and C-2H (δ 2.83) is not clearly visible due to closely lying chemical shifts. Other uncoupled protons only appear as diagonal peaks. The HOHAHA interactions between various protons of a spin system are shown as broken lines on the structure.

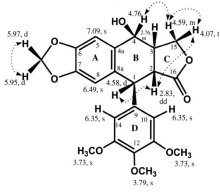

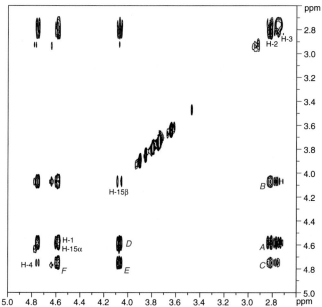

7.21 The presence of two very distinct spin systems (I, II) can easily be inferred from the TOCSY (100 ms) spectrum of 7-deoxyloganic acid isolated from *Osyris wightiana* Wall. *ex*. Wight. The cross-peaks between δ 2.93–1.07 represent an upfield spin system (I) which comprises 10 protons of the aglycone. The cross-peaks A, B, and C represent coupling of C-1 proton (δ 5.39) with C-5, C-9/ C-8, and C-10 protons, resonating at δ 2.93, 2.24/2.22, and 1.08. The cross-peaks of the second spin system (II) of glucose moiety appear between δ 4.67–3.17. The cross-peaks D, E, and F are due to vicinal and long-range coupling interactions of the anomeric proton (δ 4.67) with all the oxygenated methine protons δ 3.36 (H-3′), 3.28 (H-5′), 3.23 (H-4′), and 3.18 (H-2′) and with the oxygenated methylene protons of C-6′ (δ 3.89, 3.64), respectively, of the sugar moiety. Other protons of the sugar moiety also show interaction with each other. All protons in a spin system show strong correlations with each other, which is characteristic of the TOCSY (100 ms) spectrum. The TOCSY interactions are presented on the spectrum.

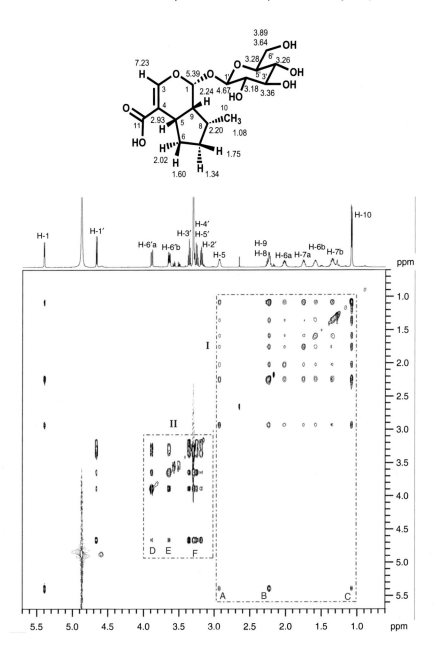

7.22 Twelve cross-peaks (A-L) are visible in the HETCOR spectrum of podo-phyllotoxin, representing the 13 protonated carbons (including two magnetically equivalent protonated carbons with the same chemical shifts) in the molecule.

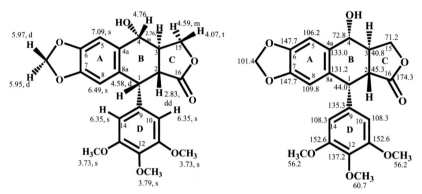

¹H-NMR Chemical shift assignments   ¹³C-NMR Chemical shift assignments

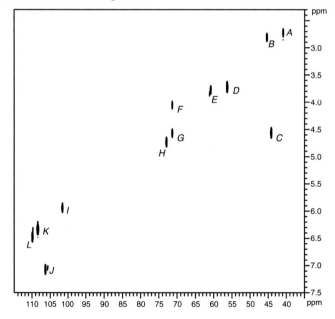

The most upfield cross-peak A is due to the interaction of the C-3 proton (δ 2.76) with the most upfield carbon at δ 40.8 (C-3). Cross-peaks B and C display heteronuclear couplings of the C-2 proton (δ 2.83) with the carbon at δ 45.3, and the C-1 proton (δ 4.58) with the carbon at δ 44.0. Cross-peaks F and G are due to the one-bond interactions of the C-15 methylene protons (δ 4.07 and 4.59) with the same carbon, resonating at δ 71.2. Cross-peak D represents heteronuclear interactions between protons of two magnetically equivalent methoxy groups (δ 3.73) with their respective carbons (δ 56.2), while cross-peak E displays heteronuclear coupling between the protons of the C-12 methoxy protons (δ 3.79) and its carbon, which resonates at δ 60.7. Coupling between the protons and carbon of the methylenedioxy group is represented by cross-peak I (linking δ 5.95/5.97 with δ 101.4). Cross-peak K is due to the coupling between the magnetically equivalent C-10 and C-14 methine protons and carbons; cross-peaks J and L represent direct heteronuclear couplings between H-5/C-5 and H-8/C-8, respectively. The ^{13}C-NMR chemical shift assignments, based on heteronuclear one-bond correlations visible in the HETCOR experiment, are displayed on the structure.

7.23 The HETCOR spectrum of mangiferin exhibits cross-peaks for all 10 protonated carbons.

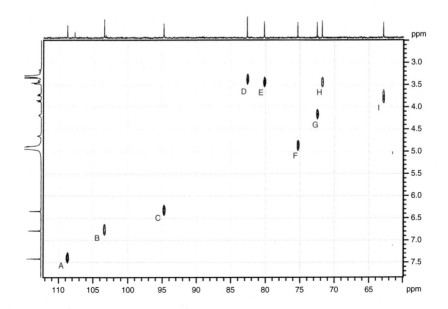

^{1}H-NMR Chemical shift assignments   ^{13}C-NMR Chemical shift assignments

Cross-peaks A–C, in the downfield region, represent the heteronuclear couplings of the aglycone moiety, while cross-peaks D–I are due to heteronuclear couplings between the sugar protons and their respective carbons. The downfield methine protons at δ 7.41, 6.78, and 6.33 show one-bond heteronuclear interactions (cross-peaks A, B, and C) with the carbons resonating at δ 108.9, 103.4, and 94.8, respectively. The cross-peaks D-I are representative of $^1H/^{13}C$ couplings of the six oxygen-bearing carbons of the hexapyranose moiety. Cross-peak F represents coupling between the anomeric proton (δ 4.88) and C-1' (δ 75.3). The upfield appearance (chemical shift) of the anomeric carbon indicates a C-linked sugar moiety. The ^{13}C-NMR assignments, based on the heteronuclear correlations, are summarized around the structure.

7.24 The one-bond hetero-COSY spectrum of (+)-catechin exhibits interactions for all eight protonated carbons. The most downfield cross-peaks A-C represent one-bond heteronuclear correlations of the ABX spin system on C ring protons, resonating at δ 6.82 (H-2'), 6.74 (H-5'), and 6.71 (H-6') with the C-2' (δ 115.2), C-5' (δ 116.0), and C-6' (δ 120.0). Similarly, cross-peak D and E display couplings of C-6, and C-8 methine protons (δ 5.91 and 5.84) of ring A with C-6 (δ 96.1) and C-8 (δ 95.3), respectively. The C-2 and C-3 protons, which resonate downfield at δ 4.55 and 3.96, due to the directly bonded oxygen atoms, display correlations with the carbons, resonating at δ 82.8 and 68.7 (cross-peaks F and G, respectively). The protons resonating at δ 2.84 and 2.49 both show one-bond heteronuclear interactions (cross-peaks H and I) with the carbon resonating at δ 28.5. This indicated that the carbon at δ 28.5 is a methylene carbon, since two different protons show one-bond coupling interactions with it. Various other one-bond heteronuclear interactions are delineated around the structure.

¹H-NMR Chemical shift assignments   ¹³C-NMR Chemical shift assignments

7.25 The HSQC spectrum of podophyllotoxin shows heteronuclear cross-peaks for all 13 protonated carbons (including two sets of protonated carbons). Each cross-peak represents a one-bond correlation between the ¹³C nucleus and the attached proton. It also allows us to identify the pairs of geminally coupled protons, since both protons display cross-peaks with the same carbon. For instance, peaks A and B represent the one-bond correlations of protons at δ 4.07 and 4.59 with the carbon at δ 71.2, and thus represent a methylene group (C-15). Cross-peak D is due to the heteronuclear correlation between the C-4 proton (δ 4.76) and the carbon at δ 72.8, assigned to the oxygen-bearing benzylic C-4. Heteronuclear shift correlations between the aromatic protons and carbons are easily distinguishable as cross-peaks J-L, while I represents ^{1}H/^{13}C interactions between the methylenedioxy protons (δ 5.95/ 5.97) and the carbon at δ 101.4. The ^{13}C-NMR and ^{1}H-NMR chemical shift assignments, based on the HMQC cross-peaks, are summarized on the structure.

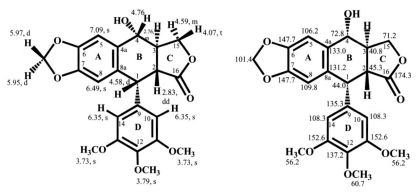

^{1}H-NMR Chemical shift assignments ^{13}C-NMR Chemical shift assignments

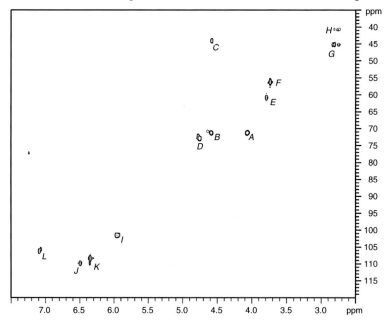

7.26 The HSQC spectrum of 5,7-dihydroxy-4′-methoxyisoflavone shows six cross-peaks of protonated carbons. The most downfield methine proton (δ 8.13) shows one-bond heteronuclear correlation with the carbon resonating at δ 153.7 (C-2) (cross-peak A). The downfield shift is due to the electron withdrawing effect of carbonyl carbon. Cross-peaks B and C represent heteronuclear interactions between protons of two pairs of magnetically equivalent methine (δ 7.70 and 7.08) with their respective carbons, which resonated at δ 130.8 and 114.3, respectively. The cross-peaks D and E indicate interactions of the methine protons of ring A, appearing upfield at δ 6.75 and 6.67, with C-6 (δ 100.2) and C-8 (δ 94.2), respectively. The coupling between protons and carbon of the C-4′ methoxy groups is represented by cross-peak F.

^{1}H-NMR Chemical shift assignments

^{13}C-NMR Chemical shift assignments

7.27 The HMBC spectrum of podophyllotoxin exhibits a number of cross-peaks representing long-range heteronuclear interactions between the various carbons and protons. For instance, the downfield oxygen-bearing aromatic carbon(s) (δ 147.7) can be recognized as the oxygen-bearing aromatic carbon(s), i.e., C-6 and C-7, and they exhibit HMBC interactions with the protons resonating at δ 7.09 and 6.49 (cross-peaks A and D, respectively). This establishes that the aromatic methines C-5 and C-8 are bound to the oxygen-bearing quaternary C-6 and C-7, respectively. Cross-peak B is due to the heteronuclear coupling interaction between the C-5 proton at δ 7.09 and the quaternary C-8a (δ 131.2). The methylenedioxy protons (δ 5.95 and 5.97) show long-range correlation with C-6/C-7 (δ 147.7) (cross-peak K) indicating its attachment with ring A. Similarly, long-range heteronuclear relationships between the C-8 proton (δ 6.49) and C-4a (cross-peak E) and between the C-1 proton (δ 4.58) and C-8a (cross-peak O') are also visible in the HMBC spectrum. The C-4 proton geminal to the oxygen atom, which is also the terminus of the largest substructure (i.e., spin system), exhibits long-range heteronuclear coupling interaction with C-3 (δ 40.8; cross-peak L), establishing correlation in the spin system. Cross-peaks O, P, P', Q, and R represent multiple-bond correlations of the C-1 proton (δ 4.58) with C-9, C-8, C-10, C-2, and C-3, respectively. Cross-peak M represents long-range interaction between the oxygen-bearing C-4 proton (δ 4.76) and C-15 (δ 71.2), while cross-peak N is due to the long-range interaction of the C-4 proton with the quaternary C-4a. The C-2 carbon shows long-range heteronuclear interaction with the C-16 (δ 174.3) (cross-peak U), and with the C-9 (δ 135.3) and C-4 (δ 72.8), respectively, as displayed by cross-peaks V and W. The heteronuclear multiple-bond correlations in the HMBC plot are presented around the structure.

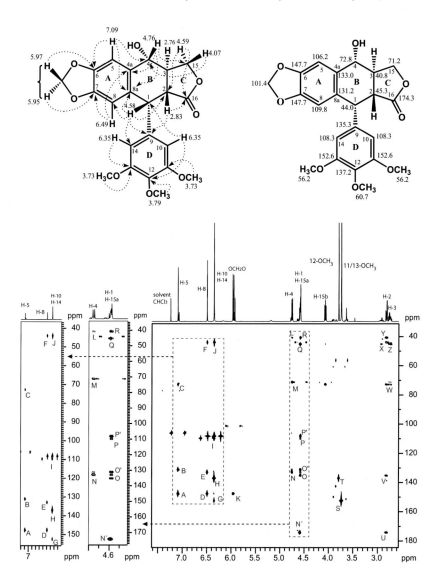

7.28 The HMBC spectrum of 5,7-dihydroxy-4′-methoxyisoflavone displays long-range heteronuclear shift correlations between the various ^{1}H/^{13}C nuclei. These correlations are very helpful to determine the ^{13}C-NMR chemical shifts of quaternary carbons and allow the interlinking of the different substructures obtained.

Cross-peaks A, B, and C represent interactions of the C-2 proton resonating at δ 8.13 with the C-4, C-9, and C-1′ (δ 181.0, 158.7, and 123.5, respectively). Cross-peaks D, E, and F correspond to the coupling between the C-2′/C-6′ protons (δ 7.70) with the C-4′ (δ 160.1), C-6′/C-2′ (δ 130.8), and C-1′ (δ 123.5). Cross-peaks G, H, and I represent long-range correlations between the C-3′/C-5′ protons (δ 7.08) and the C-4 (δ 160.1), C-2 (δ 123.9), and the C-5′/C-3′ methine (δ 114.3) carbons, respectively. Cross-peaks J, K, L, and M are due to the coupling of C-6 (δ 6.75) with the carbons of ring A (i.e., C-7 (δ 166.1), C-5 (δ 163.6), C-10 (δ 105.7), and C-8 (δ 194.2)). The C-8 methine proton (δ 6.67) with C-7 (δ 166.1), C-9 (δ 158.7), C-10 (δ 105.7), and C-6 (δ 100.2) appears as cross-peaks N–Q, respectively, in the HMBC plot. Various long-range heteronuclear shift correlations based on the HMBC experiment are shown on the structure.

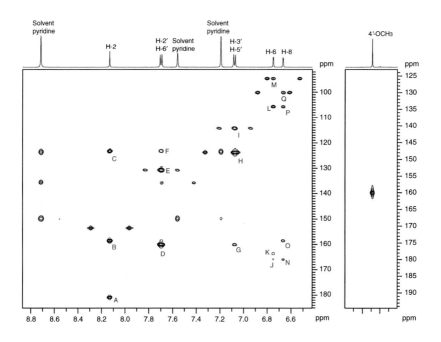

7.29 The COLOC spectrum of 7-hydroxyfrullanolide shows long-range (^2J, ^3J, and ^4J) heteronuclear interactions of various carbons with protons that are two, three, or four bonds away and thus provide a tool to identify the signals of nonprotonated (quaternary) carbons. The ^{1}H-NMR and ^{13}C-NMR data along with the multiplicities of the signals (determined by DEPT and broad-band experiments) are essential for the interpretation of the COLOC plot. Cross-peaks A–C represent ^3J interactions of the C-1 methylene protons (δ 1.33 and 1.31) with the carbons at δ 30.7 (CH$_2$), 34.3 (CH$_2$), and 25.6 (CH$_3$) assigned to C-8, C-9, and C-15, respectively. The C-2 protons (δ 1.56 and 1.45) exhibit cross-peaks D and E with the carbon resonances at δ 32.5 (CH$_2$) and 32.0 ($-$C$-$). These carbon resonances are therefore assigned to the C-3 methylene carbon and the C-10 quaternary carbon, respectively. Cross-peak F is due to the heteronuclear interaction of the C-6 proton (δ 4.97) with a downfield quaternary carbon at δ 126.3, assigned to the olefinic C-5. The exomethylenic protons at δ 5.71 and 6.06 exhibit three sets of cross-peaks, G, H, and I, displaying their respective heteronuclear interactions with the quaternary carbons resonated at δ 75.1, 144.4, and 169.4. These downfield quaternary signals are assigned to the oxygen-bearing C-7, olefinic C-11, and ester carbonyl C-12, respectively, based on their chemical shifts. The C-6 proton (δ 4.97) also shows a ^3J interaction (cross-peak) with a quaternary carbon at δ 139.5, which may be ascribed to the C-4 olefinic carbon. The chemical shift assignments to various quaternary carbons based on long-range heteronuclear COLOC interactions are presented around the structure.

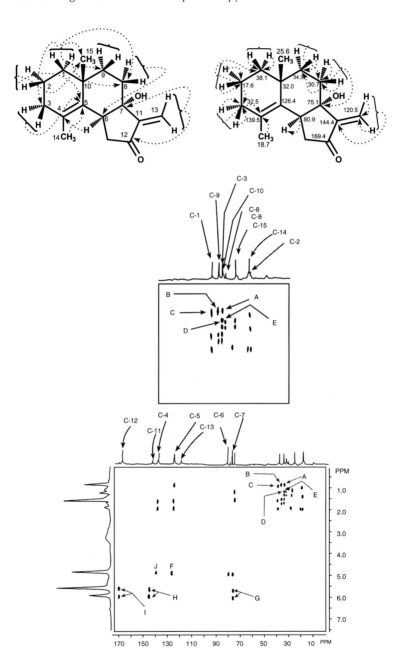

7.30 Assuming a β-orientation of the C-1 methine proton (δ 4.58), the stereochemistry at other centers can be assigned. For instance, β-orientation of
the C-2 methine proton (δ 2.83) is confirmed by nOe interaction between
C-2 and C-1 protons. Selective irradiation of C-1 proton (δ 4.58) results
in an nOe on C-2 proton (δ 2.83) and C-15β proton (δ 4.07) (spectrum
a). When C-2 protons (δ 2.83) were irradiated (spectrum b), nOe effects
on C-4βH (δ 4.76), C-1βH (δ 4.58), and C-15βH (δ 4.07) were observed.
The nOe between the C-4 proton (δ 4.76) and to C-2H (δ 2.83) and C-
15βH (δ 4.07) indicates that these protons are β-oriented (spectrum c).

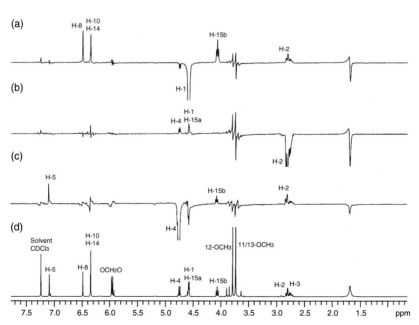

7.31 The nOe difference measurements not only help in stereochemical assignments but also provide connectivity information. A large nOe at δ 6.83 (C-5H), resulting from the irradiation of the methyl singlet of the methoxy group (δ 3.94), confirms their proximity in space. This nOe result is consistent with structure A for the coumarin.

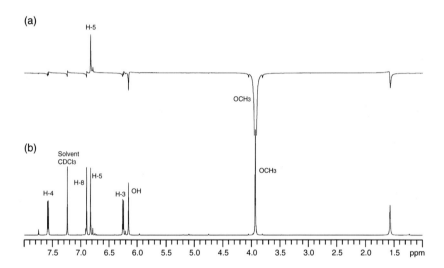

7.32 The NOESY spectrum of buxatenone shows four cross-peaks, A–D. Cross-peak B represents the dipolar coupling between the most upfield C-19 cyclopropyl proton (δ 0.68) with the most downfield olefinic proton (δ 6.72). This could be possible only when the double bond is located either between C-1 and C-2 or between C-11 and C-12. The possibility of placing a double bond between C-11 and C-12 can be excluded on the basis of chemical shift considerations, since conjugation with the carbonyl group leads to a significant downfield shift of the olefenic protons β- to the carbonyl group. Cross-peaks A, C, and D are due to couplings between C-1H/C-2H, C-16αH/C-16βH, and C-19αH/C-19βH, respectively.

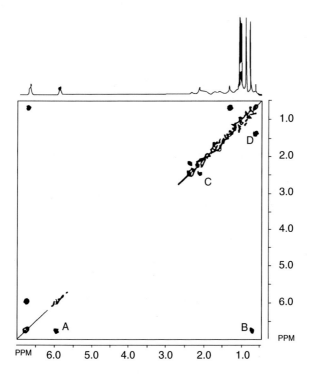

7.33 The ROESY spectrum of podophyllotoxin exhibits a number of cross-peaks (A–D) representing interactions between dipolarly coupled (through space coupling) hydrogens, which can be helpful to determine the stereochemistry at different asymmetric centers. For example, based on the assumption that the C-1 proton (δ 4.58) is β-oriented, we can trace out the stereochemistry of other asymmetric centers. Cross-peak "B" represents dipolar coupling between the C-1 proton (δ 4.58) and the C-2 proton (δ 2.83), thereby confirming that the C-2 proton is also β-oriented. The C-4 proton resonating at δ 4.76 also exhibits cross-peaks "C" and "E" with the C-2 proton (δ 2.83) and C-15β proton (δ 4.07), thereby indicating its β-stereochemistry in space. The β-position of the C-1 proton establishes the α-stereochemistry of the attached substituted aromatic ring D. The C-15α proton resonating at δ 4.59 shows a cross-peak with the C-15β proton (cross-peak D) as well as vicinal interactions with the C-3α proton (δ 2.76) (cross-peak A). These interactions could only be possible if the hydroxyl group at the C-4 position possesses α-orientation. The homonuclear transient nOe interactions based on the ROESY spectrum are shown on the structure.

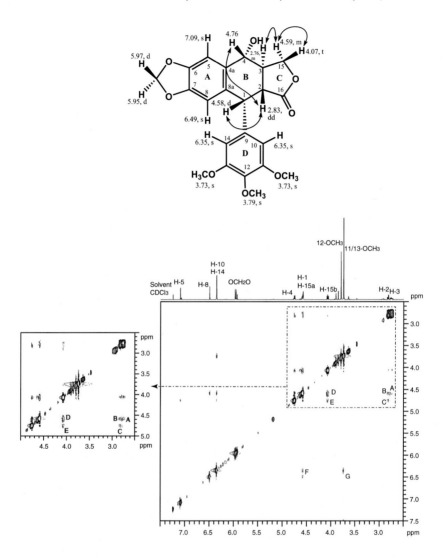

7.34 The homonuclear 2D *J*-resolved spectrum of scopoletin shows signals for all five types of magnetically distinct protons. The most downfield doublet "A" represents C-4 proton, which is coupled to C-3 proton, and this resonates as a doublet at δ 7.58 with the coupling constants *J* = 9.6 Hz. The coupling constants are determined by measuring the distance between the centers of different components of the doublet. This is more accurately done after plotting the corresponding projections so that the centers of the peaks can easily be located. The vertical axis has a *J* scale in hertz. Similarly, the coupling of the C-3 with C-4 proton can be measured by doublet "D." Signals B, C, and E appear as a singlets for the methine (δ 6.83 and 6.91) and methoxy (δ 3.94) protons.

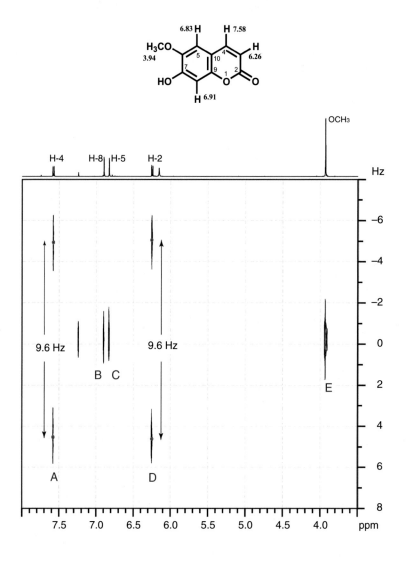

7.35 The heteronuclear 2D J-resolved spectrum of citroside B shows signals for all carbons. Similar to the homonuclear J-resolved spectrum (Problem 7.34), the ^{13}C–^1H coupling is calculated by measuring the distances between different components of the signal (doublet, triplet, etc.). The bands A, E, and M represent C-6, C-5, and C-1 carbons and they appear as singlets since there is no proton attached to them. The bands B–D, and F–I appearing as doublets represent C-8, C-1', C-3', C-5', C-2', C-4', and C-3 carbons as methines, respectively. The C-3 and C-8 methine carbons appear at δ 63.8 and 101.3 with J_{CH} coupling constants 140, and 176 Hz, respectively, while the coupling constants (J_{CH}) of the methine carbons of the sugar moiety are between 140.0 and 144.0 Hz. The band C represents the anomeric carbon, resonating at δ 98.6 (J_{CH} = 160.0 Hz). This indicates the β-orientation of the anomeric carbon of the D-glucose moiety. The methylene carbon of the sugar appears as a triplet at δ 62.9 (J_{CH} = 280.0 Hz), whereas the C-2 and C-4 methylene carbons of the aglycon appear at δ 49.9 and 48.0 with J_{CH} = 264.0 Hz. The C-11 to C-13 methyl carbons appear as quartets with J_{CH} = 380.0 Hz.

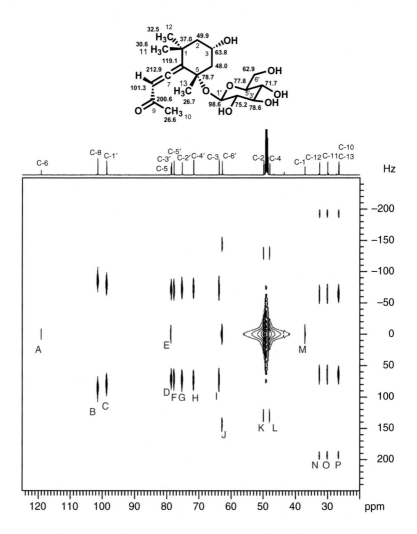

7.36 In order to establish the carbon–carbon connectivities, a horizontal line is drawn between the satellite peaks that lie at equal distances on either side of the diagonal. For example, a horizontal line is drawn from satellite peak A, which corresponds to the carbon resonating at δ 17.9 (signal "a") to another cross-peak, B, at δ 34.0 (signal "b"), indicating that the two carbons are connected with each other by a C–C bond. A vertical line is next drawn from B downward to the satellite B_1. B_1 has another pair of satellites, C, on the same horizontal axis, thereby establishing the connectivity between the carbons at δ 34.0 and δ 34.7 (signal "C"). Similarly, satellite peak B_2 is connected to another satellite, E, thereby establishing the connectivity between the carbons resonating at δ 34.0 and 74.7 (signal "e"). By dropping a vertical line from satellite peak C, we reach C_1. Its "mirror image" partner satellite on the other side of the diagonal line is D, thereby establishing the connectivity between the carbon at δ 34.7 with the carbon at δ 67.6 (signal "d"). These ^{13}C–^{13}C connectivities are presented around the structure of methyl tetrahydrofuran.

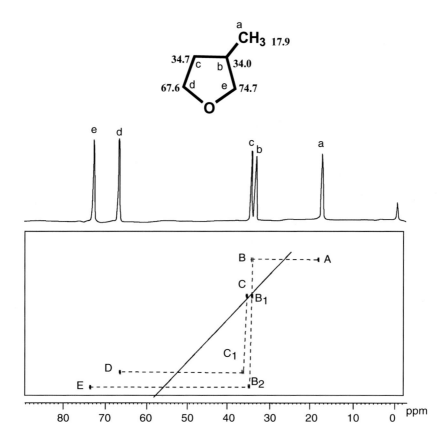

7.37 The 2D INADEQUATE spectrum provides carbon–carbon connectivity information and allows the entire carbon framework to be built up. The best strategy for the interpretation of a complex INADEQUATE spectrum is to start with the most downfield satellite carbon resonance and to trace the subsequent $^{13}C-^{13}C$ connectivities. Using this strategy, the signal at δ 169.4 for the carbonyl carbon, which appeared as satellite peak A in the spectrum, was chosen as the starting point. A horizontal line from satellite peak A to another peak B (δ 144.4), connects the C-12 and C-11 carbons. A vertical line is next drawn upward to the satellite peaks B' and B'', which have other pairs of satellites peaks C and D, at mirror image positions across the diagonal on the horizontal axis, thereby establishing the connectivity of the C-11 olefinic quaternary carbon (δ 144.4) with C-13 and C-7 (δ 120.5 and 75.1), respectively. Similarly, satellite peaks D' and D'' on the same vertical axis show connectivities with other satellites E and K, respectively, which correspond to the C-6 and C-8 carbons (δ 80.9 and 30.7), respectively. By drawing a line vertically upward from peak K to K' and then a horizontal line across the diagonal connecting the cross-peaks K' and L, the interaction between the C-8 and C-9 carbons (δ 30.7 and 34.3, respectively) can be traced out. The C-9 methine carbon does not show any coupling interaction with the C-10 carbon (δ 32.0) in the spectrum. This serves to illustrate the point while the presence of a double-quantum peak in the 2D INADEQUATE spectrum provides concrete evidence of a bond, the absence of such a peak does not necessarily prove that a bond is absent. By considering these interactions, fragment **1** could be obtained.

A horizontal line between satellite peaks E' and F establishes the connectivity between the C-6 and C-5 carbons (δ 80.9 and 126.4), respectively. The mirror image partners of satellites F' and F'' appearing on the same horizontal axis are M and G, thereby establishing the connectivity of the C-5 carbon with the C-10 and C-4 carbons resonating at δ 32.0 and 139.5, respectively. Signals G' and G'', which correspond to the C-4 carbon, can be similarly connected to satellites H and I, so connectivities of the C-4 carbon with the C-3 and C-14 carbons resonating at δ 32.5 and 18.7, respectively, can be deduced. The two pairs of satellites ($J-W$) and ($J'-N$) establish connectivities of the C-2 carbon (δ 17.6) with the C-3 and C-1 carbons (δ 32.5 and 38.1, respectively). The satellites M' and M'', lying vertically, indicate connectivities of the C-10 carbon (δ 32.0) with the C-1 (M'-N') and C-15 (M''-Q) carbons (δ 38.1 and 25.5, respectively). These connectivities lead to a fragment **2**.

By joining fragments **1** and **2** together, the following structure for 7-hydroxyfrullanolide was deduced.

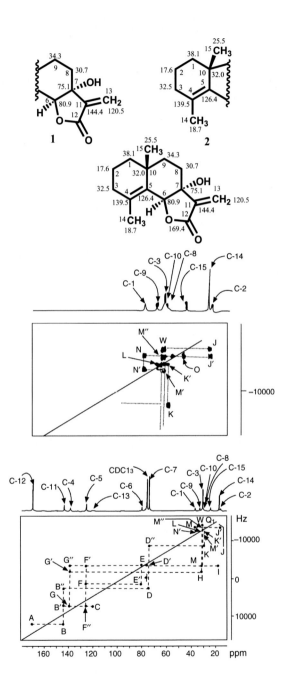

REFERENCES

Bauer, C., Freeman, R., Wimperis, S., 1984. Long-range carbon-proton coupling constant. J. Magn. Reson. 58 (3), 526–532.

Bax, A., Freeman, R., 1981. Investigation of complex networks of spin-spin coupling by two-dimensional NMR. J. Magn. Reson. 44 (3), 542–561.

Bax, A., 1983. Broadband homonuclear decoupling in heteronuclear shift correlation NMR spectroscopy. J. Magn. Reson. 53 (3), 517–520.

Bax, A., Lerner, L., 1986. Two-dimensional nuclear magnetic resonance spectroscopy. Science 232, 960–967.

Becker, E.D., Ferretti, J.A., Farrar, T.C., 1969. Driven equilibrium Fourier transform spectroscopy. A new method for nuclear magnetic resonance signal enhancement. J. Am. Chem. Soc. 91 (27), 7784–7785.

Bodenhausen, G., Ruben, D.J., 1980. Natural abundance nitrogen-15 NMR by enhanced heteronuclear spectroscopy. Chem. Phys. Lett. 69 (1), 185–189.

Garbow, J.R., Weitekamp, D.P., Pines, A., 1982. Bilinear rotation decoupling of homonuclear scalar interactions. Chem. Phys. Lett. 93 (5), 504–509.

Giraudeau, P., Akoka, S., 2007. A new detection scheme for ultrafast 2D J-resolved spectroscopy. J. Mag. Reson. 186 (2), 352–357.

Jin-shan, W., De-Zheng, Z., Tao, J., Xiu-Wen, H., Guo-Bao, C., 1982. A rapid method for heteronuclear two-dimensional J spectroscopy. J. Magn. Reson. 48 (2), 216–224.

Kessler, H., Bermel, W., Griesinger, C., 1985. Recognition of NMR proton spin systems of cyclosporin A via heteronuclear proton-carbon long-range couplings. J. Am. Chem. Soc. 107 (4), 1083–1084.

Kessler, H., Bermel, W., Griesinger, C., Kolar, C., 1986. The elucidation of the constitution of glycopeptides by the NMR spectroscopic COLOC technique. Angew. Chem. Int. Ed. Engl. 25 (4), 342–344.

Kessler, H., Griesinger, C., Lautz, J., 1984a. Determination of connectivities via small proton-carbon couplings with a new two-dimensional NMR technique. Angew. Chem. Int. Ed. Engl. 23 (6), 444–445.

Kessler, H., Griesinger, C., Zarbock, J., Lossli, H.R., 1984b. Assignment of carbonyl carbons and sequence analysis in peptides by heteronuclear shift correlation via small coupling constants with broadband decoupling in t_1 (COLOC). J. Magn. Reson. 57 (2), 331–336.

Lupulescu, A., Olsen, G.L., Frydman, L., 2012. Toward single-shot pure-shift solution ^{1}H NMR by trains of BIRD-based homonuclear decoupling. J. Mag. Reson. 218, 141–146.

Müller, L., 1979. Sensitivity enhanced detection of weak nuclei using heteronuclear multiple quantum coherence. J. Am. Chem. Soc. 101 (16), 4481–4484.

Neuhaus, D., Williamson, M., 1989. The Nuclear Overhauser Effect in Structural and Conformational Analysis (second edition). VCH Pubs, New York, p 278.

Paudel, L., Adams, R.W., Kiraly, P., Aguilar, J.A., Foroozandeh, M., Cliff, M.J., Nilsson, M., Sandor, P., Waltho, J.P., Morris, G.A., 2013. Simultaneously enhancing spectral resolution and sensitivity in heteronuclear correlation NMR spectroscopy. Angew. Chem. Int. 52 (44), 11616–11619.

Quast, M.J., Zektzer, A.S., Martin, G.E., Castle, R.N., 1987. Response-intensity modulation in long-range heteronuclear two-dimensional NMR chemical-shift correlation. J. Magn. Reson. 71 (3), 554–560.

Reynolds, W.F., McLean, S., Tay, L.L., Yu, M., Enriquez, R.G., Estwick, D.M., Pascoe, K.O., 1997. Comparison of ^{13}C resolution and sensitivity of HSQC and HMQC sequences and application of HSQC-Based sequences to the total ^{1}H and ^{13}C spectral assignment of Clionasterol. Magn. Reson. Chem. 35 (7), 455–462.

Sakhaii, P., Haase, B., Bermel, W., 2013. Broadband homodecoupled heteronuclear multiple bond correlation spectroscopy. J. Mag. Reson. 228, 125–129.

Schenker, K.V., Philipsborn, von W., 1985. Optimization of INEPT and DEPT experiments for spin systems with hetero- and homonuclear couplings. J. Magn. Reson. 61 (2), 294–305.

States, D.J., Habackorn, R.A., Ruben, D.J., 1982. A two-dimensional nuclear Overhauser experiment with pure absorption phase in four quadrants. J. Magn. Reson. 48 (2), 286–292.

Thrippleton, M., Edden, R.A.E., Keeler, J., 2005. Suppression of strong coupling artefacts in J-spectra. J. Mag. Reson. 174 (1), 97–109.

Wang, J.S., Wong, T.C., 1985. An improved rapid method for heteronuclear two-dimensional J spectroscopy. J. Magn. Reson. 61 (1), 59–66.

Welti, D.H., 1985. Revision of the ^{13}C NMR signal assignment of Harman, based on one- and two-dimensional INADEQUATE spectroscopy. Magn. Reson. Chem. 23 (10), 872–874.

Wernly, J., Lauterwin, J., 1985. Assignment of the quaternary olefinic carbon atoms of β-carotene by 2D ^{1}H, ^{13}C-chemical shift correlation *via* long-range couplings. J. Chem. Soc. Chem. Commun. (18), 1221–1222.

Wimperis, S., Freeman, R., 1984. An excitation sequence which discriminates between direct and long-range CH coupling. J. Magn. Reson. 58 (2), 348–353.

Wimperis, S., Freeman, R., 1985. Sequences which discriminate between direct and long-range CH couplings: compensation for a range of $^1J_{CH}$ values. J. Magn. Reson. 62 (1), 147–152.

Zektzer, A.S., Quast, M.J., Linz, G.S., Martin, G.E., Mackenny, J.D., Johnston, M.D., Castle, R.N., 1986. New pulse sequence for long-range two-dimensional heteronuclear NMR chemical shift correlation. J. Magn. Reson. 24 (12), 1083–1088.

Zektzer, A.S., John, B.K., Martin, G.E., 1987a. Repression of one-bond modulations in long-range heteronuclear 2D NMR spectra using a modified long-range optimized heteronuclear chemical shift correlation pulse sequence. J. Magn. Reson. 25 (9), 752–756.

Zektzer, A.S., John, B.K., Castle, R.N., Martin, G.E., 1987b. "Decoupling" modulations due to one-bond heteronuclear spin couplings in long-range heteronuclear chemical-shift correlation spectra. J. Magn. Reson. 72 (3), 556–561.

Chapter 8

Playing with Dimensions in NMR Spectroscopy

8.1 THREE-DIMENSIONAL SPECTROSCOPY

8.1.1 Basic Philosophy

The basic philosophy of three-dimensional (3D) NMR spectroscopy is inherent in triple-resonance experiments, in which a system under perturbation by two different frequencies is subjected to further perturbation by a third frequency (Cohen et al., 1964). With the development of two-dimensional (2D) NMR spectroscopy (Aue et al., 1976a; Chandrakumar and Subramanian, 1987), it was only a matter of time before 3D NMR experiments were introduced, particularly for resolving heavily overlapping regions of complex molecules. The earlier 3D experiments included 3D *J*-resolved COSY (Plant et al., 1986), 3D correlation experiments (Griesinger et al., 1987b,c), and 3D combinations of shift-correlation and cross-relaxation experiments (Griesinger et al., 1987c; Oschkinat et al., 1988). These experiments illustrated the potential and power of 3D methods for structure elucidation. 3D NMR experiments can be directed at two main objectives: (1) unraveling complex signals that overlap in 1D and 2D spectra (Ernst et al., 1987), and (2) establishing connectivity of nuclei *via* *J*-couplings or their spatial proximity *via* dipolar couplings or cross-relaxation effects.

Solving Problems with NMR Spectroscopy. http://dx.doi.org/10.1016/B978-0-12-411589-7.00008-5
387

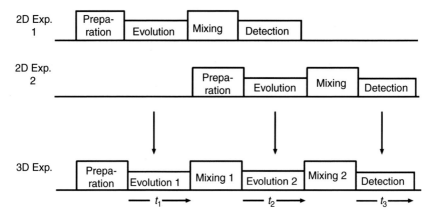

FIGURE 8.1 Pulse sequence for 3D experiments obtained by merging two different 2D sequences. *(Reprinted from Griesinger et al., (1989) J. Magn. Reson., 84, 14–63, with permission from Academic Press, Inc.)*

2D NMR spectroscopy has two broad classes of experiments: (1) 2D *J*-resolved spectra (Müller et al., 1975; Aue et al., 1976b), in which no coherence transfer or mixing process normally occurs, and chemical shift and coupling constant frequencies are spread along two different axes, ω_1 and ω_2, and (2) 2D shift-correlation spectra, involving either coherent transfer of magnetization [e.g., COSY, hetero-COSY (Maudsley and Ernst, 1977), relayed COSY (Eich et al., 1982), TOCSY (Braunschweiler and Ernst, 1983), 2D multiple-quantum spectra (Braunschweiler et al., 1983), etc.] or incoherent transfer of magnetization (Kumar et al., 1980; Macura and Ernst, 1980; Bothner-By et al., 1984) [e.g., 2D cross relaxation experiments, such as NOESY, ROESY, 2D chemical-exchange spectroscopy (EXSY) (Jeener et al., 1979; Meier and Ernst, 1979), and 2D spin-diffusion spectroscopy (Caravatti et al., 1985)].

In 3D experiments (Misiak and Kozminski, 2007), two different 2D experiments are combined so that three frequency coordinates are involved. In general, the 3D experiment may be made up of the preparation, evolution (t_1), and mixing periods of the first 2D experiment, combined with the evolution (t_2), mixing, and detection (t_3) periods of the second 2D experiment. The 3D signals are, therefore, recorded as a function of *two* variable evolution times, t_1 and t_2, and the detection time t_3. This is illustrated in Fig. 8.1.

3D NMR spectra, like 2D NMR spectra, may be broadly classified into three major types: (1) 3D *J-resolved spectra* (in which the chemical shift frequencies and homonuclear or heteronuclear coupling frequencies are resolved in three different dimensions, no coherence transfer or mixing process being normally involved); (2) *3D shift-correlated spectra*, in which three resonance frequencies are correlated by two coherent or incoherent transfer processes. The two transfer processes may be homonuclear or heteronuclear and may involve transfer of antiphase coherence (e.g., COSY) or in-phase coherence (e.g., TOCSY), leading to such 3D experiments as COSY-COSY, COSY-TOCSY, and hetero-COSY-TOCSY. It is possible to have combinations of *J*-resolved and shift-correlated spectra, for example, 3D *J*-resolved COSY in which the 2D COSY spectrum is spread into a third dimension through scalar coupling (Plant et al., 1986; Vuister and Boelens, 1987); (3) *3D exchange spectra*, in which two successive exchanges are recorded. The exchange processes may involve either chemical exchange (EXSY) or cross-relaxation in the laboratory frame (NOESY) or in the rotating frame (ROESY), e.g., NOESY-ROESY or EXSY-EXSY spectra.

In some of the most useful 3D experiments, the coherent transfer of magnetization (e.g., COSY) may be combined with incoherent magnetization transfer (e.g., NOESY) to give such 3D experiments as NOESY-COSY or NOESY-TOCSY (Griesinger et al., 1987c; Oschkinat et al., 1988; Vuister et al., 1988). Such 3D experiments are finding increasing use in the study of biological molecules.

Since there are two time variables, t_1 and t_2, to be incremented in 3D experiment (in comparison to one time variable to increment in the 2D experiment), such experiments require a considerable data storage space in the computer and also consume much time. It is, therefore, practical to limit such experiments to certain limited frequency domains of interest. Some common pulse sequences used in 3D time-domain NMR spectroscopy are shown in Fig. 8.2.

8.1.2 Types and Positions of Peaks in 3D Spectra

Five different types of peaks can occur in 3D spectra. These are illustrated in an ABC spin system, in which the Larmor frequencies of the three nuclei are v_A, v_B, and v_C, and the coherence transfers are associated with two different mixing processes, M_1 and M_2:

	M_1	M_2
Cross-peaks	v_A----------v_B	----------v_C
Cross-diagonal peaks ($\omega_1 = \omega_2$)	v_A----------v_A	----------v_B
Cross-diagonal peaks ($\omega_2 = \omega_3$)	v_A----------v_B	----------v_B
Back-transfer peaks	v_A----------v_B	----------v_A
Diagonal peaks	v_A----------v_A	----------v_A

The *cross-peaks* arise when two different mixing processes lead to transfer of magnetization. Cross-peaks correlate three Larmor frequencies and hence provide information on connectivity or spatial proximity of sets of three nuclei. In a COSY-COSY spectrum, for instance, they provide information on three nuclei coupled with one another. In the case of *cross-diagonal peaks*, only one mixing process is involved in the magnetization transfer, while *diagonal peaks* involve no magnetization transfer at all. *Back-transfer peaks* arise by the retransfer of magnetization to the original nucleus, rather than its being passed onto the

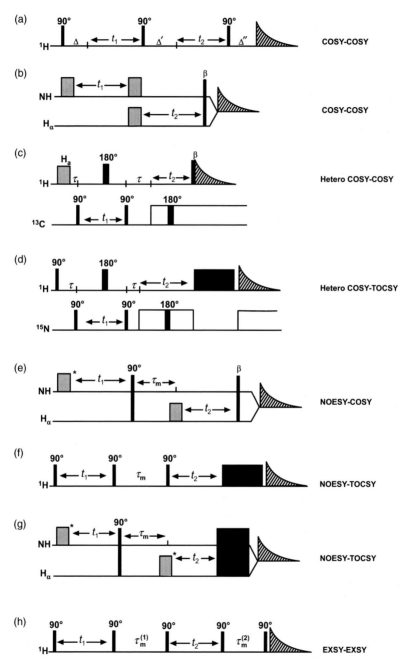

FIGURE 8.2   Pulse sequences for some common 3D time-domain NMR techniques. Nonselective pulses are indicated by filled bars (width represents π and $\pi/2$). Nonselective pulses of variable flip angle are shown by the flip angle β. Frequency-selective pulses ($\pi/2$) are drawn with grey color bars. *(Reprinted from Griesinger et al., (1989) J. Magn. Reson., 84, 14–63, with permission from Academic Press, Inc.)*

third nucleus; hence, two magnetization transfer processes are again involved. Back-transfer peaks provide interesting information in spectra in which unequal mixing processes are involved. For instance, in a NOESY-TOCSY spectrum they would allow us to identify protons that are spatially close and part of the same spin system (Oschkinat et al., 1989).

Increasing the number of dimensions from two to three may result in a reduction in the signal/noise (S/N) ratio. This may be due either to the distribution of the intensity of the multiplet lines over three dimensions or to some of the coherence transfer steps being inefficient, resulting in weak 3D cross-peaks.

The major advantage of 3D spectra is that they can enhance resolution of overlapping cross-peaks in 2D spectra. This is presented in schematic form in Fig. 8.3. Let us consider two pairs of protons attached to four different carbons, and assume that each pair has the same resonance frequency. Let us also assume that proton A is spatially close to proton B and gives rise to a corresponding

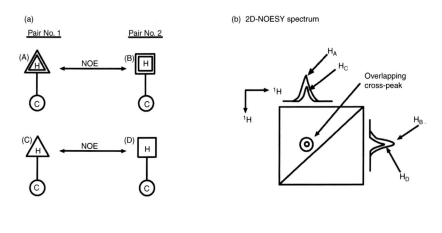

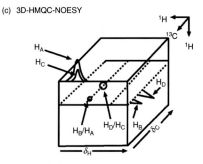

FIGURE 8.3 Schematic representation of the resolution advantages of 3D NMR spectroscopy. (a) Both pairs of protons have the same resonance frequency. (b) Due to the same resonance frequency, both pairs exhibit overlapping cross-peaks in the 2D NOESY spectrum. (c) When the frequency of the carbon atoms is plotted as the third dimension, the problem of overlapping is solved, since their carbon resonance frequencies are different. The NOESY cross-peaks are thus distributed in different planes.

cross-peak in the 2D NOESY spectrum. Similarly, proton C is spatially close to proton D and also gives rise to a second cross-peak in the 2D NOESY spectrum. However, since protons A and C have the same resonance frequencies, as do protons B and D, the two cross-peaks would overlap and only a single cross-peak would actually be observed. If we could somehow introduce a third dimension, say, the ^{13}C chemical shift, then the two overlapping cross-peaks would be separated, since the carbon atoms to which protons B and D are attached have differing chemical shifts. Thus, the 3D HMQC-NOESY spectrum would contain two well-separated cross-peaks that had overlapped in the NOESY spectrum.

Figure 8.4 shows the general features of a 3D spectrum, which can be visualized as peaks appearing in a 3D space, such as a cube. The cube can be subdivided by two different cross-diagonal planes ($\omega_1 = \omega_2$ and $\omega_2 = \omega_3$) and one back-transfer plane. A diagonal line connecting two opposite corners of the cube is the "space

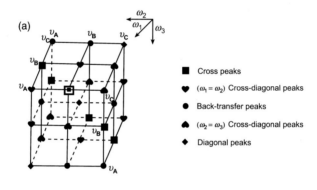

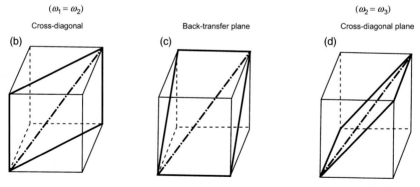

FIGURE 8.4 3D spectrum of a three-spin system showing peak types appearing in a 3D space. Three diagonal peaks, six ($\omega_1 = \omega_2$) and six ($\omega_2 = \omega_3$) cross-signal peaks, six back-transfer peaks, and six cross-peaks are present in the cube. (a) The cubes (b–d) represent three planes in which cross-diagonal peaks and the back-transfer peaks appear on their respective ($\omega_1 = \omega_2$), ($\omega_2 = \omega_3$), and ($\omega_1 = \omega_3$) planes. (*Reprinted from Griesinger et al., (1989) J. Magn. Reson., 84, 14–63, with permission from Academic Press, Inc.*)

diagonal"; diagonal peaks ($v_A \rightarrow v_A \rightarrow v_A$) would appear on this line. The cross-diagonal peaks and the back-transfer peaks appear on their respective planes.

A 3D spectrum of a three-spin subsystem in which all the nuclei are coupled to one another, such as $C(H_A)(H_B)-C(H_C)$, will lead to 27 peaks, comprising six cross-peaks, 12 cross-diagonal peaks (six at $\omega_1 = \omega_2$ and the other six at $\omega_2 = \omega_3$), six back-transfer peaks, and three diagonal peaks. However, in the case of a linear three-spin network (e.g., $CH_A-CH_B-CH_C$), the number of peaks will depend on whether two equal (e.g., COSY-COSY or NOESY-NOESY) or unequal (e.g., COSY-NOESY) mixing processes are involved. If the mixing processes are equal, then the forward and reverse transfers, such as:

$$v_A \rightleftharpoons v_B \rightleftharpoons v_C$$
$$v_C \rightleftharpoons v_B \rightleftharpoons v_C$$

produce 17 peaks distributed as two cross-peaks, eight cross-diagonal peaks (four at $v_1 = v_2$ and four at $v_2 = v_3$), four back-transfer peaks, and three diagonal peaks (Fig. 8.5a).

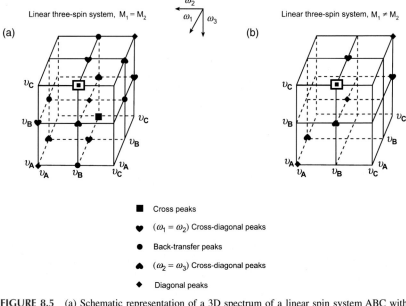

FIGURE 8.5 (a) Schematic representation of a 3D spectrum of a linear spin system ABC with identical mixing processes M_1 and M_2. In a "linear" spin system, the transfer of magnetization between A and C is forbidden for both M_1 and M_2. (b) Schematic representation of a 3D spectrum of a linear spin system ABC, where transfer *via* M_1 is possible only between A and B and transfer *via* M_2 occurs only between B and C. *(Reprinted from Griesinger et al., (1989) J. Magn. Reson., 84, 14–63, with permission from Academic Press, Inc.)*

In the case of a COSY-NOESY spectrum having an unequal mixing, let us consider two nuclei (say, A and B) that are spatially close but belong to different coupling networks, and nuclei B and C, which have scalar coupling with each other but are spatially distant. The only transfers allowed in this situation are:

$$v_A \xrightleftharpoons{M_1} v_B \xrightleftharpoons{M_2} v_C$$

So a simplified 3D spectrum is obtained having eight peaks, i.e., one cross-peak, four cross-diagonal peaks (two at $v_1 = v_2$ and two at $v_2 = v_3$), three diagonal peaks, and no back-transfer peaks (Fig. 8.5b).

3D spectra can be conveniently represented by 2D cross-sections taken at appropriate points and perpendicular to one of the three axes. In such cross-sections, the cross-diagonal planes and back-transfer planes appear as lines running across the 2D spectrum. 2D cross-sections may be taken so that they cut across the cross-peak due to the magnetization transfer $v_A \rightarrow v_B \rightarrow v_C$ in a three-spin system. Three such cross-sections are illustrated in Fig. 8.6, taken perpendicular to the three different axes v_1, v_2, and v_3. The cross-section (v_2, v_3), taken perpendicular to v_1 at the chemical shift frequency of nucleus A $(v_1 = v_A)$, shows those cross-peaks that arise from the transfer of magnetization from nucleus A to the other spins during the two transfers. The 2D cross-section (v_1, v_2), taken perpendicular to v_2 at $v_2 = v_B$, represents the magnetization reaching and leaving nucleus B. This cross-section may contain not only the cross-peak (v_A, v_B, v_C) but also its symmetry equivalent peak (v_C, v_B, v_A) (hollow circle) if such a pathway is allowed. The third cross-section (v_1, v_2), taken perpendicular to v_3 at $v_3 = v_C$, provides information about all the magnetization arriving at v_C.

The peak shapes in 3D spectra can be obtained from the phases of the corresponding signals in the two 2D experiments from which the 3D spectrum is

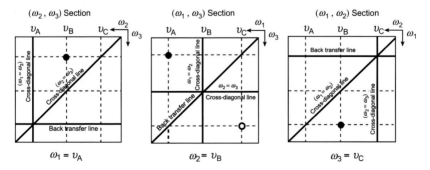

FIGURE 8.6 Three 2D cross-section through the A→B→C cross-peaks of a 3D spectrum taken at constant ω_1, ω_2, and ω_3 frequencies, respectively. The open circle indicates a symmetry-related peak with respect to the back-transfer diagonal line. *(Reprinted from Griesinger et al., (1989) J. Magn. Reson., 84, 14–63, with permission from Academic Press, Inc.)*

derived. Thus, if the two 2D spectra have pure phases, e.g., absorption signals, then the 3D cross-peaks will also be in the pure-absorption mode, and if the two 2D spectra from which the 3D spectrum is constituted have dispersive-mode signals, then the 3D spectrum will also have dispersive peak shapes. Mixed peak shapes in 2D spectra lead to mixed peak shapes in the corresponding 3D spectrum.

Symmetric 2D spectra result from a mixing sequence that is time- and phase-reversal (TPR) symmetric (Griesinger et al., 1987a). 3D correlation spectra constituted from two identical symmetric mixing sequences, such as COSY-COSY or ROESY-ROESY, are themselves symmetric with respect to the plane $\omega_1 = \omega_3$. If unequal mixing processes are involved in the creation of 3D spectra, e.g., COSY-NOESY or NOESY-TOCSY, then such spectra lack global symmetry. The 3D EXSY-EXSY spectrum of heptamethyl benzenonium sulfate in sulfuric acid shown in Fig. 8.7 serves as an example of a 3D spectrum having $\omega_1 = \omega_3$ reflection symmetry. Peaks lying in the same plane ($\omega_1 = \omega_3$) are shown connected by dotted lines and have $\omega_1 = \omega_3$ reflection symmetry (i.e., symmetry

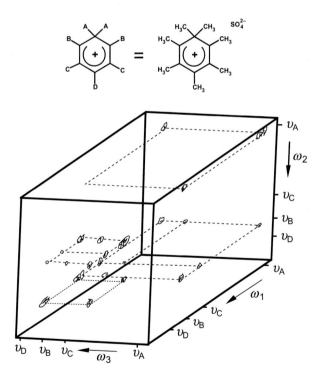

FIGURE 8.7 The 3D EXSY-EXSY spectrum of heptamethyl benzenonium sulfate in H_2SO_4 representing $\omega_1 = \omega_3$ reflection symmetry. Cross-peaks lying on identical planes (ω_1, ω_3) are connected by lines. *(Reprinted from Griesinger et al., (1989) J. Magn. Reson., 84, 14–63, with permission from Academic Press, Inc.)*

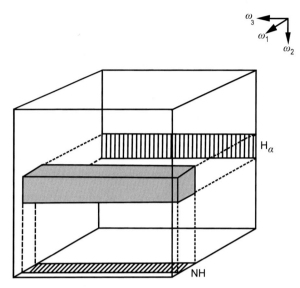

FIGURE 8.8 Volume chosen for a selective 3D NOESY-TOCSY spectrum of a peptide with NH region in ω_1, Hα in ω_2, and all protons in ω_3, *(Reprinted from Griesinger et al., (1989) J. Magn. Reson., 84, 14–63, with permission from Academic Press, Inc.)*

about the diagonal line connecting the two opposite crosses, or in other words, drawn with $\omega_1 = \omega_3$). The exchange processes giving rise to the EXSY-EXSY spectrum are due to the 1,2-shifts of the methyl groups (A ↔ B, B ↔ C, C ↔ D).

To record a 3D spectrum that covers the entire spectral range in all three dimensions with a resolution comparable to that obtained in 2D spectra, an enormous amount of data must be processed, which may not only be beyond the storage capabilities of most computer systems, but may also take weeks to accumulate. It is, therefore, usual to record partial 3D spectra in only a certain region of interest. For instance, in Fig. 8.8, the volume chosen for the 3D NOESY-TOCSY spectrum of a peptide covers the NH region in ω_1, H$_\alpha$ in ω_2, and all protons in ω_3.

3D NMR experiments are proving to be of particular use in assignments to amino acids in polypeptides. Some of the 3D and 4D NMR techniques used for the structure determination of proteins are presented in Table 8.1. The NMR experiments and the types of labeling suited for protein structure determination are given in Table 8.2. The development of 3D NMR techniques (Griesinger et al., 1987a,b,c; Vuister et al., 1988; Fesik and Zuiderweg, 1988; Oschkinat et al., 1988; Kay et al., 1989; Zuiderweg and Fesik, 1989) and 4D NMR techniques (Kay et al., 1990) has allowed protein structures of molecular weights of up to 35 kDa to be investigated. 3D experiments that combine

TABLE 8.1 A Selection of 3D and 4D NMR Techniques Used in Protein Structure Determination*

Experiments	Information/Linkage Obtained
(a) Unlabeled Proteins—Homonuclear Experiments	
3D-TOCSY-TOCSY	^{1}H spin system/amino acids
3D-TOCSY-NOESY	Sequential assignment of the spin systems, identification of nOes
3D-NOESY-NOESY	Identification of nOes
(b) ^{15}N Labeled Proteins	
3D-TOCSY-^{1}H$-^{15}$N-HMQC/HSQC	^{1}H spin system, ^{15}N shifts
3D-NOESY-^{1}H$-^{15}$N-HMQC/HSQC	Sequential assignment of the spin systems, identification of nOes with amide protons
3D-HNHB	$H_N(i)-N(i)-H_\beta(i)$, stereospecific assignment of methyl protons
(c) ^{13}C-^{15}N Labeled Proteins	
3D-TOCSY-^{1}H$-^{13}$C-HMQC/HSQC	^{1}H spin system, ^{13}C shifts
3D-NOESY-^{1}H$-^{13}$C-HMQC/HSQC	Sequential assignment of the spin systems, in particular, identification of nOes between side chains
3D-HCCH-COSY	^{1}H spin systems, ^{13}C shifts
3D-HCCH-TOCSY	^{1}H spin systems, ^{13}C shifts
3D-HC(C)NH-TOCSY	Connection of H_N and side-chain ^{1}H and ^{13}C
3D-H(CCO)NH	Correlation of all side-chain ^{1}H and ^{13}C of amino acid i with $H_N(i+1)$
H(CA)NH	$H_\alpha(i)-N(i)-H_N(i)$
3D-HCACO	$H_\alpha(i)-C_\alpha(i)-CO(i)$
3D-HNCA	$H_N(i)-N(i)-C_\alpha(i)/C\alpha(i-1)$
3D-HNCACB	$H_N(i)-N(i)-C_\alpha(i)/C_\alpha(i-1)/C_\beta(i)/C_\beta(i-1)$
3D-HN(CA)NNH	$HN(i)-N(i+1)-H_N(i+1)$
3D-HA(CA)NNH	$H_\alpha(i)/H_\alpha(i-1)-N(i)H_N(i)$
3D-HN(NCA)NNH	$H_N(i)-N(i+1)-H_N(i+1)$
3D-HCA(CO)N	$H_\alpha(i)-C_\alpha(i)-N(i+1)$
3D-HNCO	$H_N(i)-N(i)-CO(i-1)$
3D-HN(CA)HA	$H_N(i)-N(i)-H_\alpha(i)/H_\alpha(i-1)$
3D-HN(CO)CA	$H_N(i)-N(i)-C_\alpha(i-1)$
3D-HN(COCA)HA	$H_N(i)-N(i)-H_\alpha(i-1)$
3D-HNHA	$H_N(i)-N(i)-H_\alpha(i)$
3D-CBCA(CO)NH	$H_\beta(i-1)-C_\beta(i-1)-C_\alpha(i-1)-N(i)-H_N(i)$
3D-HN(CA)CO	$HN(i)-N(i)-CO(i)/CO(i-1)$

(Continued)

TABLE 8.1 A Selection of 3D and 4D NMR Techniques Used in Protein Structure Determination* *(cont.)*

Experiments	Information/Linkage Obtained
3D-HN(CA)CB	$H_N(i)-N(i)-C_\beta(i)/C_\beta(i-1)$
3D-H(N)CACO	$H_N-C_\alpha(i)/C_\alpha(i-1)\text{-}CO(i)/CO(i-1)$
3D H(C)(CO)NH TOCSY	Correlation of side-chain ^1H and ^{13}C with $H_N(i,i-1)$ and $N(i,i-1)$
4D HCANNH	$H_\alpha(i)-C_\alpha(i)-N(i)-H_N(i)$
4D HCA(CO)NNH	$H\alpha(i)-C_\alpha(i)-N(i+1)-H_N(i+1)$
4D HNCAHA	$H_N(i)-N(i)-C_\alpha(i)-H\alpha(i)$
4D HCCH TOCSY	^1H spin systems, ^{13}C shift
4D $^1H-^{13}C$ HMQC-NOESY $^1H-^{13}C$ HMQC	Identification of nOes between side chain
4D $^1H-^{15}N$ HMQC-NOESY $^1H-^{13}C$ HMQC	Identification of nOes with amide protons
$^2J_{COH\alpha}$—E. COSY	$^2J(CO, H_\alpha)$
3D HNCA	$^3J(HN, H_\alpha)$, ϕ angle
3D HMQC J_{HH}TOCSY	$^3J(HN, H_\alpha)$, ϕ angle
3D H,C COSY ct-C,C COSY-β-INEPT	Vicinal $^1H-^1H$ couplings
Soft HCCH COSY	$^3J(H_\beta,CO)$
Soft HCCH F. COSY	$^3J(H_\alpha,H_\beta)$

*Reproduced with permission from Oschkinat et al., (1994). Angew. Chem. Int. Ed. Engl. **33**(3), 277–293.*

TABLE 8.2 NMR Experiments and the Types of Labeling Suited for Protein Structure Determination*

	Normal (without isotopic labeling	With ^{15}N labeling	With ^{15}N and ^{13}C labeling
Assignment of spin system	TOCSY-TOCSY	TOCSY-HMQC TOCSY-TOCSY	3D/4D HCCH-TOCSY HCCH-COSY HC(C)NH-TOCSY
Sequential assignments	TOCSY-NOESY	NOESY-HMQC TOCSY-NOESY	HNCACB HCACO HNCA HNCO HCA(CO)N CBCA(CO)NH HNHA HN(CA)CO
Identification of nOes	NOESY-NOESY TOCSY-NOESY	NOESY-HMQC NOESY-NOESY TOCSY-NOESY	3D NOESY-HMQC 4D HMQC-NOESY HMQC

*Reproduced with permission from Oschkinat et al. (1994) Angrew. Chem. Int. Ed. Engl. **33**, 277–293.*

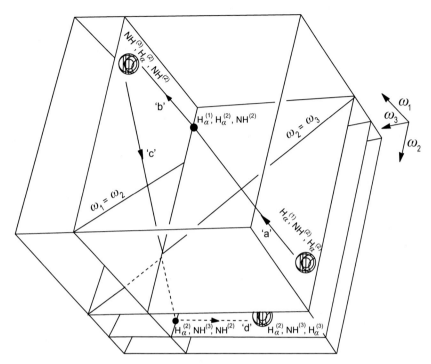

FIGURE 8.9 NOESY-TOCSY spectrum and its sequence analysis: ω_2, ω_3 planes intersect ω_1, ω_2 planes. Cross-peak, ($H_\alpha^{(1)}$, $HN^{(2)}$, $H_\alpha^{(2)}$), when reflected across the ω_2, ω_3 diagonal, yields cross-peak ($H_\alpha^{(1)}$, $H_\alpha^{(2)}$, $NH^{(2)}$). Along the line $\omega_1^{(2)}$, cross-peak $HN^{(3)}$, $H_\alpha^{(2)}$, $NH^{(2)}$ is found. By reflection at line $\omega_1 = \omega_2$ in the ω_1, ω_2 section, the latter peak is converted into point $H_\alpha^{(2)}$, $NH^{(3)}$, $NH^{(2)}$, from which cross-peak $H_\alpha^{(2)}$, $NH^{(3)}$, $H_\alpha^{(3)}$ is found by a search paralleled to ω_3. *(Reprinted from Griesinger et al., (1989) J. Magn. Reson., **84**, 14–63, with permission from Academic Press, Inc.)*

incoherent and coherent transfers of magnetization (NOESY-COSY, NOESY-TOCSY, hetero-COSY-NOESY, and ROESY-TOCSY) would give cross-peaks in which each cross-peak correlates three different nuclei. Two of these nuclei would belong to the same amino acid and will be correlated through a coherent magnetization transfer step (e.g., through COSY or TOCSY); the third nucleus would belong to an adjacent amino acid, thereby establishing the sequential connectivity of the two amino acids. Once a cross-peak has been assigned, we proceed to the next cross-peak by keeping two of the frequency coordinates fixed and searching for the next cross-peak along the third frequency domain of the 3D space.

Different assignment strategies can be employed, depending on whether selective or nonselective pulses have been used in recording 3D spectra. A homonuclear 3D NOESY-TOCSY spectrum in which the NH/H_α region has been recorded is presented in a schematic form in Fig. 8.9.

The superscripts in brackets indicate the number of the amino acid in a given sequence. For example, the cross-peak due to ($H_\alpha^{(1)}$, $NH^{(2)}$, $H_\alpha^{(2)}$) represents the TOCSY interaction between the α protons of the first amino acid (H_α) and NH proton of the second amino acid (NH) and the NOESY interactions between this $NH^{(2)}$ proton and the α proton of the second amino acid, $H_\alpha^{(3)}$. Having located the first cross-peak due, say, to $H_\alpha^{(1)}$, $NH^{(2)}$, $H_\alpha^{(3)}$, we reflect this cross-peak across the diagonal line (the $\omega_2 = \omega_3$ diagonal) within the (ω_2, ω_3) section containing this cross-peak (line "a" in Fig. 8.9). This takes us to a position in the (ω_2, ω_3) section that is determined by $\omega_3 = \delta_{NH}^{(2)}$ and $\omega_2 = H_\alpha^{(2)}$.

The peak at this position may be very weak or even absent due to the weak nOe expected between $H_\alpha^{(1)}$ and $H_\alpha^{(2)}$. Next, we pass a line "b" through this peak and parallel to the ω_1-axis. This takes us to the ($NH^{(3)}$, $H_\alpha^{(2)}$, $NH^{(2)}$) cross-peak. Reflection of this cross-peak across the $\omega_1 = \omega_2$ diagonal takes us along line "c" to position ($H_\alpha^{(1)}$, $NH^{(3)}$, $NH^{(2)}$). This cross-peak will not appear, since there is no J-coupling between $NH^{(2)}$ and $NH^{(3)}$. However, by following line "d" from this point parallel to the ω_3-axis, we arrive at the ($H_\alpha^{(2)}$, $NH^{(3)}$, $H_\alpha^{(3)}$) cross-peak. By repeating this process we can assign the amino acids sequentially.

The pulse sequences for HMQC-COSY and HMQC-NOESY experiments are presented in Fig. 8.10. The 3D HMQC-COSY spectrum of a ^{15}N labeled tripeptide is shown in Fig. 8.11. Since the coherence transfer involved in this experiment is ^{15}N (t_1) $\rightarrow$ ^{15}N$\underline{H}$ (t_2) $\rightarrow$ ^{15}NH (t_3) + C$_\alpha$$\underline{H}$ (t_3), it is expected that the ^{15}N resonance will appear in ω_1, that the ^{15}N$\underline{H}$ resonance will appear in ω_2, and that ^{15}NH and C$_\alpha$H resonance will appear in ω_3. The HMQC-NOESY spectrum shown in Fig. 8.12 has been edited with respect to ^{15}N frequencies in ω_1. The spectrum shows that the amide protons of residues 1 and 3 have close chemical shifts, but that the ^{15}N nuclei to which they are coupled have well-separated chemical shifts (90 and 119 ppm). Similarly, the near overlap in the ^{15}N dimension (Leu-2 at 120 ppm and Leu-3 at 119 ppm) is well resolved in the spectrum. Readers are directed to following references for a more detailed discussion on the theory and applications of various types of 3D NMR experiments (Griesinger et al., 1989; Oschkinat et al., 1989, 1994; Rule and Hitchens, 2005; Cavanagh et al., 2007).

8.2 CONCATENATED NMR TECHNIQUES

With the basic understanding of the concepts and applications of 1D, 2D, and 3D NMR techniques, we are now well placed to discuss another innovation in NMR spectroscopy, *vis-a-vis* combined or "concatenated" NMR experiments. Concatenation "*is the action of linking things together in a series.*" The discussion on concatenated techniques here is an extension of the knowledge of 2D NMR methods. We explain here with the help of illustrations and problem-solving exercises, the benefits and applications of concatenated NMR techniques in structure determination of small organic molecules.

(a)

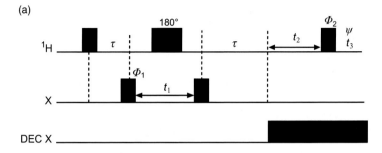

(b)

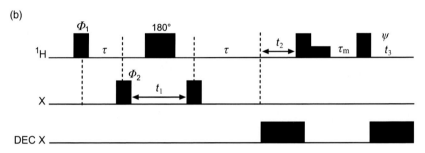

FIGURE 8.10 Pulse sequences for (a) the HMQC-COSY experiments and (b) the HMQC-NOESY experiments. *(Reprinted from Fesik and Zuiderweg (1988) J. Magn. Reson. 78, 588–593, with permission from Academic Press, Inc.)*

Concatenated or hybrid NMR techniques are very powerful tools for chemists which are not only easily accessible through standard NMR spectrometers, but are also capable of providing useful information for structure elucidation, such as assignment of multiplicities to diastereomeric protons, simplifying the problems associated with the overlapping of signals in protons, multiplicity assignments in proton spectrum, as well as for conformation and configuration analysis.

One such experiment is DEPT-HSQC or *multiplicity edited* ^{1}H-X HSQC. This is a combination of DEPT and HSQC (2D one-bond heteronuclear ^{1}H-X, say ^{1}H-^{13}C secondary quantum coherence) experiments. In this concatenated experiment, *DEPT-derived multiplicity information* (CH, CH$_2$, CH$_3$) is detected in the 2D square plot of HSQC and illustrated in different colors. In other words, the DEPT-HSQC plot shows only those HSQC interactions in which the carbon multiplicities have been separated by DEPT. This is a very robust method to extract information about the multiplicity of carbon signals, particularly CH$_2$ signals, which cannot be obtained from HSQC alone (Section 7.6.1.2). Such combinations of NMR methods have enhanced the effectiveness NMR spectroscopy in solving complex structural problem.

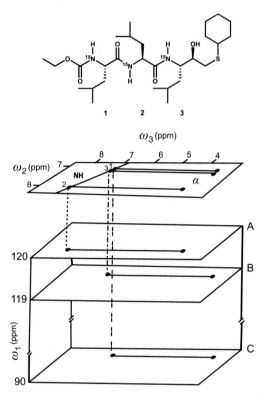

FIGURE 8.11   A 3D heteronuclear HMQC-COSY spectrum of a tripeptide. The ω_1-axis represents
^{15}N chemical shifts, whereas the ω_2- and ω_3-axes exhibit proton chemical shifts. The assignment
pathways are indicated in the top spectrum for reference purposes, not as part of the 3D experiment.
*(Reprinted from Fesik and Zuiderweg (1988) J. Magn. Reson. 78, 588–593, with permission from
Academic Press, Inc.)*

Many of the combined NMR techniques, such as HSQC-NOESY, HSQC/
HMQC-TOCSY, NOESY-TOCSY, etc., are, in a strict sense, 3D NMR techniques, extensively used in the structure elucidation of macromolecules
(Cong et al., 2006; Blinov et al., 2007). A detailed description of such techniques as 3D NMR spectroscopy, and their applications in structural biology
is beyond the scope of this book. However, *"2D analogs"* of some common
3D NMR techniques are now extensively used in organic chemistry research,
very much the way the 2D NMR experiments, such as TOCSY and NOESY,
are used as their 1D analogs.

Following is a brief conceptual framework of dimensionality in concatenated
NMR techniques:

1. Concatenated techniques involve the combination of two 2D NMR techniques,
    such as DEPT-HSQC (1D + 2D NMR techniques), HSQC-TOCSY, etc.

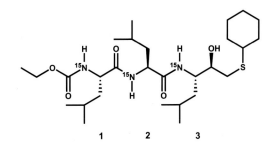

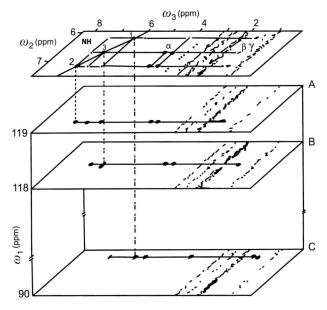

FIGURE 8.12 A 3D heteronuclear HMQC-NOESY spectrum of a tripeptide. The ω_1-axis represents ^{15}N chemical shifts, whereas ω_2- and ω_3-axes exhibit proton chemical shifts. The assignment pathways are indicated in the top spectrum for reference purposes, not as part of the 3D experiment. *(Reprinted from Fesik and Zuiderweg (1988) J. Magn. Reson. 78, 588–593, with permission from Academic Press, Inc.)*

2. In the case of such combinations of two 2D NMR techniques, three time periods exist in the NMR pulse sequence (variable evolution times t_1 and t_2, as well as the acquisition time period t_3).

3. Variable evolution times t_1 and t_2 are the incremented delays, giving rise to the 2nd (F_2) and 3rd (F_3) dimensions, while t_3 provides the F_1 axis.

4. Increment of both t_1 and t_2 leads to the full 3D spectrum (F_1, F_2, and F_3).

5. Increment in one of the time periods, either t_1 or t_2, gives rise to 2D planes (equivalent to a ^{1}H, ^{1}H-TOCSY, ^{1}H-^{1}H-NOESY, or ^{1}H,^{15}N HSQC).
6. If there are no increments in t_1 and t_2 (i.e., if t_1 and t_2 are fixed), then it yields a 1D *type* spectrum.
7. This means that concatenated NMR spectra can be designed (by appropriate pulse sequences), and displayed in a variety of ways (ranging from a full 3D spectrum to a 2D spectrum or a 1D spectrum) with varying degrees of information content, and ease of interpretation.

There are a large number of concatenated NMR techniques available, but this chapter is focused on the most commonly used 2D concatenated techniques (or 2D analogs of 3D NMR techniques). These hybrid or concatenated NMR techniques are now playing an increasingly prominent role in structural studies of complex organic molecules (Williamson et al., 2001). These techniques can be broadly divided into following two categories, based on their underlying principles and applications.

Problem 8.1

What is the key difference between 3D NMR and 2D concatenated NMR techniques?

8.2.1 Two-Dimensional Multiplicity Edited Spectra

Multiplicity information is an essential requirement for structure determination. However, many of the most useful NMR experiments do not provide multiplicity information. Heteronuclear NMR experiments, such as HSQC and HMQC, do not provide information about the diastereomeric XH$_2$ (CH$_2$, NH$_2$) groups.

8.2.1.1   DEPT-HMQC or DEPT-HSQC

The hybrid NMR experiments, **D**istortionless **E**nhancement by **P**olarization **T**ransfer- **H**eteronuclear **M**ultiple **Q**uantum **C**oherence (DEPT-HMQC) or DEPT-**H**eteronuclear **S**ingle **Q**uantum **C**oherence (DEPT-HSQC) are also called *multiplicity edited* HMQC/HSQC techniques. These are ideal techniques for carbon multiplicity assignments by inverse polarization transfer.

Such experiments are especially suitable, when small quantities of samples do not allow the recording of X-spin edited spectra (GASPE, APT, DEPT, INEPT, etc.). This added advantage of increasing sensitivity due to inverse polarization transfer makes even small quantities of samples accessible for structural studies.

In the pulse sequence (Fig. 8.13), the DEPT sequence is combined with the HMQC sequence to generate a combined/concatenated DEPT-HMQC sequence. The DEPT pulse sequence generates the heteronuclear multiple

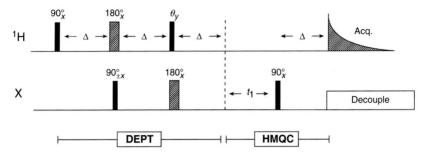

FIGURE 8.13 Pulse sequence for DEPT- HMQC spectrum.

quantum coherence that eventually evolves during t_1 in the HMQC pulse sequence. Depending on the θy pulse in the DEPT sequence, one can produce HMQC spectra with different multiplicities highlighted. For example, a DEPT pulse sequence with $\theta = 180°$ produces a HMQC spectrum with XH_2 (CH_2 or NH_2) correlation inverted, relative to other multiplicities (CH and CH_3) (Fig. 8.14).

8.2.2 Utilization of X Spin Dispersion

One of the major bottlenecks in the interpretation of various homonuclear correlation spectra (^1H or ^1H-detected spectra) is over-crowding which results in overlapping peaks. For example, the TOCSY spectrum, with very useful information about the J-couplings within various spin systems, often fails because of overlapping signals which are too difficult to interpret. Similarly, the NOESY spectrum too can have many overlapping signals which complicate the interpretation of spatial or through space couplings between protons. In such cases, a greater dispersion of chemical shifts of the attached heteroatom (^{13}C or ^{15}N) can be used as a means to separate the overlapping proton–proton correlations. Inherently protons can suffer serious overlapping of signals as they resonate within a narrow chemical shift range (0–12 ppm), whereas carbon signals are dispersed in a larger chemical shift range (0–210 ppm). Following are two examples of concatenated techniques, where X spin dispersion is used to solve the problem of overlapping in correlation spectra (TOCSY and NOESY), thereby greatly facilitating their interpretation.

8.2.2.1 HSQC-TOCSY

The concatenated *H*eteronuclear *S*ingle *Q*uantum *C*oherence-*TO*tal *C*orrelation *S*pectroscop*Y* (HSQC/HMQC-TOCSY) technique utilizes X (^{13}C, ^{15}N, etc.) dispersion to aid in the interpretation of the proton spectrum. This technique has been developed for the measurement of long-range heteronuclear coupling

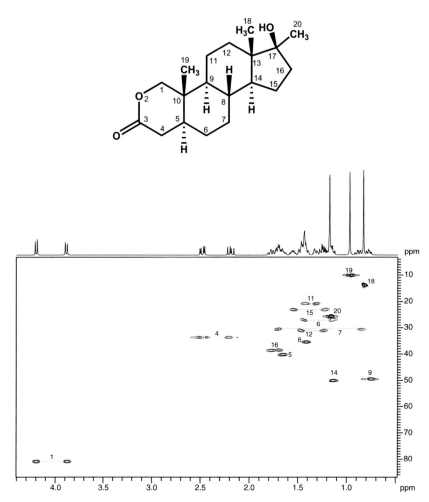

FIGURE 8.14 The DEPT-HSQC spectrum of an anabolic drug oxandrolone. Apart from the usual one-bond heteronuclear correlations, multiplicities of carbon signals are also shown. Positive phase CH$_3$, and CH signals, are presented in black, while negative phase CH$_2$ is shown as red. Sensitivity advantage due to inverse polarization transfer (^{1}H-detection) has clearly demonstrated 5 mg of sample was used.

constants. It is a powerful tool for the structural studies of complex organic molecules, such as saponins, polysaccharides, and peptides, where TOCSY experiment (or HOHAHA) may fail due to serious overlapping of proton signals (Bassarello et al., 2001). For example saponins have one or more sugar moieties attached to the aglycon part (steroidal or triterpenoidal skeleton) and sugar protons generally resonate in the region of 3–5 ppm. Similarly, in the case of cyclic peptides, the α-protons of individual amino acids are likely to appear in the region of δ 3–4.5 ppm. Such molecules can be handled by

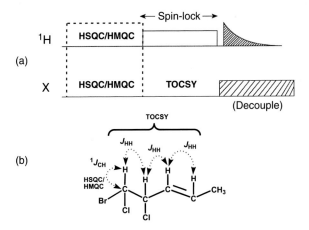

FIGURE 8.15 (a) The pulse sequence for HSQC-TOCSY spectrum. (b) The correlations thus obtained. Both one-bond heterounuclear interactions (^{1}H–^{13}C) and J-coupled homonuclear correlations (^{1}H–^{1}H within a spin system) are produced.

concatenated techniques. HSQC-TOCSY spectrum provides not only the one-bond heteronuclear correlations, but also clearly maps the homonuclear correlations (J-coupled) between the protons of a spin system. This hybrid technique can, therefore, be considered to be a 2D TOCSY spectrum in which the proton overlapping can be resolved by a spread across the carbon dimension. The 2D HSQC-TOCSY experiment is a combination of an inverse experiment involving an initial HSQC pulse sequence, followed by a TOCSY sequence. The HSQC-TOCSY experiment benefits from the greater spectral dispersion in the carbon range. It is especially suited when overlap in the proton spectrum prevents assignments of protons while the corresponding carbons are well resolved. Cross-peaks are seen between all the J-coupled protons within a spin system, as well as with each carbon in that spin system. Similar information can be obtained from the analogous 2D HMQC-TOCSY experiment.

The pulse sequence of the combined HSQC-TOCSY NMR technique is presented in Fig. 8.15. It involves the initial excitation of all protons. During the subsequent evolution time (t_1), their chemical shifts are labeled and total correlation transfer (TOCSY spin-lock mixing causing relay of magnetization along the proton network) takes place between all the protons within each spins system. In the ^{13}C domain, magnetization is transferred from the ^{13}C-attached protons to ^{13}C nuclei *via* INEPT. (This means that only ^{1}H–^{13}C protons are selected, and at the same time sensitivity is enhanced by direct polarization transfer from the more sensitive ^{1}H nuclei to the less sensitive ^{13}C nuclei). At this point, ^{13}C chemical shift labeling takes place, followed by reverse inverse polarization transfer (INEPT) from ^{13}C to the attached ^{1}H, the HSQC sequence. Finally, during the acquisition time t_3, FIDs of ^{1}H (attached to ^{13}C) are collected. The resulting spectrum contains information about the heteronuclear correlations between

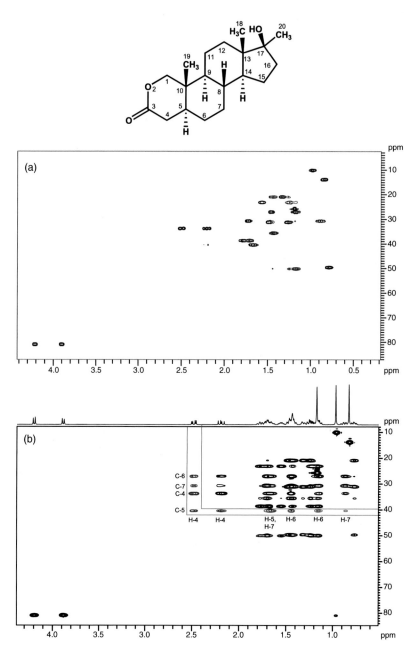

FIGURE 8.16 (a) HSQC spectrum of oxandrolone. (b) HSQC-TOCSY spectrum showing all *J*-coupled protons.

^{1}H–^{13}C (HSQC), further extended to homonuclear correlations of *the selected proton* with its partners within the ^{1}H–^{1}H spin systems (TOCSY).

The concatenated HSQC-TOCSY technique thus fully solves the signal overlapping problem by displaying the protons of individual spin systems separately (Fig. 8.16), aligned to the corresponding carbon chemical shift (since overlapping is not a major problem in the carbon chemical shift range).

8.2.2.2 HSQC-NOESY

The concatenated HSQC-NOESY experiment is designed to obtain X-edited NOESY spectra of molecules (both small and large biomolecules). The 2D HSQC-NOESY experiment is a hybrid inverse experiment consisting of an initial HSQC pulse sequence, followed by a NOESY sequence. In the first step, magnetization is transferred from the X nucleus (^{13}C or ^{15}N) to the directly bonded proton nuclei *via* $^1J_{(CH)}$. In the second step, polarization transfer is achieved *via* dipolar coupling (NOE). Similar results are obtained from the analog 2D HMQC-NOESY experiment.

The basic pulse sequence of the 2D HSQC-NOESY experiment (Fig. 8.17) is based on the 2D HSQC experiment in which a NOESY mixing sequence is inserted, followed by acquisition of FIDs. Carbon decoupling is optionally applied during the proton acquisition. The first step of this pulse sequence is to record an HSQC experiment, in which the ^{1}H magnetization is transferred to the directly-bonded ^{13}C ($^1J_{XH}$). ^{13}C magnetization evolves during t_1 and then is

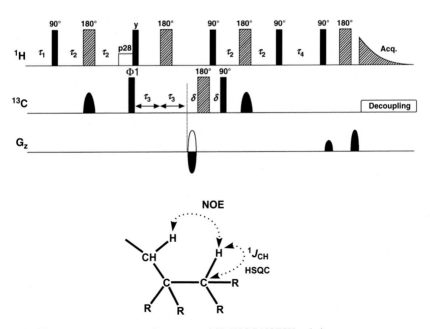

FIGURE 8.17 Pulse sequence of concatenated 2D HSQC-NOESY technique.

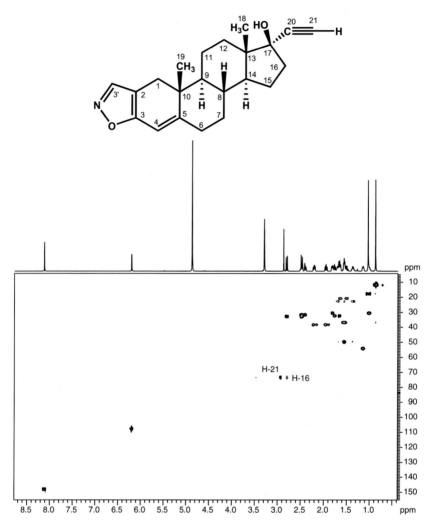

FIGURE 8.18 HSQC-NOESY spectrum of danazol. The NOE peaks (red) have an opposite sign as compared to the directly bonded proton peaks (black). Only protonated carbons appear in the spectrum.

transferred back to the directly bonded proton(s). The completion of the sequence is a NOESY mixing period, in which magnetization is transferred from the "carbon spin-labeled" proton to other protons *via* dipolar coupling (nOe).

Improved results can be obtained from the gradient enhanced version of the same experiment (see ge-2D HSQC-NOESY experiment) (Fig. 8.17).

The HSQC-NOESY (Fig. 8.18) spectrum correlates chemical shifts of heteronucleus X (^{13}C or ^{15}N) in F_1 dimension with protons (F_2 dimension) *via* the direct heteronuclear coupling $^1J_{(XH)}$. Two different cross-peaks are present for each X resonance: directly bonded ^1H–^{13}C appeared along the vertical ^{13}C-axis (so-called HSQC). These protons then show cross-peaks with the dipolarly coupled ^1H partners as NOESY. These two types of cross-peaks are generally shown in two different colors for clarity. The same NOE effects can also be detected by a 2D HMQC-NOESY, 2D HMQC-ROESY, and 2D HSQC-ROESY experiments. These are generally very insensitive experiments, as NOESY effects are very weak to begin with. Such experiments require several milligrams of samples for acceptable spectral quality.

SOLUTION TO PROBLEM

8.1 The key difference between a 3D NMR experiment and a 2D concatenated NMR technique lies in the variable evolution times. In 3D experiments, both t_1 and t_2 are incremented, whereas in a 2D concatenated NMR experiment, only one of these evolution times (either t_1 or t_2) is incremented. The resulting spectrum is a square plot, just like the 2D TOCSY, NOESY, or HSQC.

REFERENCES

Aue, W.P., Bartholdi, E., Ernst, R.R., 1976a. Two-dimensional spectroscopy. Application to nuclear magnetic resonance. J. Chem. Phys. 64 (5), 2229–2246.

Aue, W.P., Karhan, J., Ernst, R.R., 1976b. Homonuclear broad band decoupling and two-dimensional J-resolved NMR spectroscopy. J. Chem. Phys. 64 (10), 4226–4227.

Bassarello, C., Bifulco, G., Zampella, A., D'Auria, M.V., Riccio, R., Gomez-Paloma, L., 2001. Stereochemical studies on sphinxolide: advances in the J-based NMR determination of the relative configuration of flexible systems. Eur. J. Org. Chem. 2001 (1), 39–44.

Blinov, K.A., Williams, A.J., Hilton, B.D., Irish, P.A., Martin, G.E., 2007. The use of unsymmetrical indirect covariance NMR methods to obtain the equivalent of HSQC-NOESY data. Magn. Reson. Chem. 45 (7), 544–546.

Bothner-By, A.A., Stephens, R.L., Lee, J., Warren, C.D., Jeanloz, R.W., 1984. Structure determination of a tetrasaccharide: transient nuclear Overhauser effects in the rotating frame. J. Am. Chem. Soc. 106 (3), 811–815.

Braunschweiler, I., Ernst, R.R., 1983. Coherence transfer by isotropic mixing: application to proton correlation spectroscopy. J. Magn. Reson. (1969) 53 (3), 521–528.

Braunschweiler, L., Bodenhausen, G., Ernst, R.R., 1983. Analysis of networks of coupled spins by multiple quantum N.M.R. Mol. Phys. 48 (3), 535–560.

Caravatti, P., Neuenschwander, P., Ernst, R.R., 1985. Characterization of heterogeneous polymer blends by two-dimensional proton spin diffusion spectroscopy. Macromolecules 18 (1), 119–122.

Cavanagh, J., Fairbrother, W.J., Palmer III, A.G., Rance, M., Skelton, N. J., 2007. Protein NMR Spectroscopy: Principles and Practice (Second Edition), Elsevier Academic Press, MA 01803, USA, 535–673.

Chandrakumar, N., Subramanian, S., 1987. Modern Techniques in High Resolution FT-NMR. First edition, Springer, New York, 67-225.

Cohen, A.D., Freeman, R., McLauchlan, K.A., Whiffen, D.H., 1964. Determination of the signs of long-range proton spin coupling constants by nuclear magnetic triple resonance. Mol. Phys. 7 (1), 45–56.

Cong, Y., Li, W., Liu, C., Wang, J., Li, X., 2006. Application of HSQC-TOCSY to the analysis of saponins. Asian J. Tradit. Med. 1 (1), 20–24.

Eich, G.W., Bodenhausen, R., Ernst, R.R., 1982. Exploring nuclear spin systems by relayed magnetization transfer. J. Am. Chem. Soc. 104 (13), 3731–3732.

Ernst, R.R., Bodenhausen, G., Wokaun, A., 1987. Principles of NMR in One and Two Dimensions. Clarendon, Oxford.

Fesik, S.W., Zuiderweg, E.R.P., 1988. Heteronuclear three-dimensional nmr spectroscopy. A strategy for the simplification of homonuclear two-dimensional NMR spectra. J. Magn. Reson. (1969) 78 (3), 588–593.

Griesinger, C., Gemperle, C., Sørensen, O.W., 1987a. Symmetry in coherence transfer: Application to two-dimensional NMR. Mol. Phys. 62 (2), 295–332.

Griesinger, C., Sørensen, O.W., Ernst, R.R., 1987b. A practical approach to three-dimensional NMR spectroscopy. J. Magn. Reson. (1969) 73 (3), 574–579.

Griesinger, C., Sørensen, O.W., Ernst, R.R., 1987c. Novel three-dimensional NMR techniques for studies of peptides and biological macromolecules. J. Am. Chem. Soc. 109 (23), 7227–7228.

Griesinger, C., Sørensen, O.W., Ernst, R.R., 1989. Three-dimensional Fourier spectroscopy: Application to high-resolution NMR. J. Magn. Reson. (1969) 84 (1), 14–63.

Jeener, J., Meier, B.H., Bachmann, P., Ernst, R.R., 1979. Investigation of exchange processes by two-dimensional NMR spectroscopy. J. Chem. Phys. 71 (11), 4546–4553.

Kay, L.E., Clore, G.M., Bax, A., Gronenborn, A.M., 1990. Four-dimensional heteronuclear triple-resonance NMR spectroscopy of interleukin-1 beta in solution. Science 249 (4967), 411–414.

Kay, L.E., Marion, D., Bax, A., 1989. Practical aspects of 3D heteronuclear NMR of proteins. J. Magn. Reson. (1969) 84, 72–84.

Kumar, A., Ernst, R.R., Wüthrich, K., 1980. A two-dimensional nuclear Overhauser enhancement (2D NOE) experiment for the elucidation of complete proton-proton cross-relaxation networks in biological macromolecules. Biochem. Biophys. Res. Commun. 95 (1), 1–6.

Macura, S., Ernst, R.R., 1980. Elucidation of cross relaxation in liquids by two-dimensional N.M.R. spectroscopy. Mol. Phys. 41 (1), 95–117.

Maudsley, A.A., Ernst, R.R., 1977. Indirect detection of magnetic resonance by heteronuclear two-dimensional spectroscopy. Chem. Phys. Lett. 50 (3), 368–372.

Meier, B.H., Ernst, R.R., 1979. Elucidation of chemical exchange networks by two-dimensional NMR spectroscopy: the heptamethylbenzenonium ion. J. Am. Chem. Soc. 101 (21), 6441–6442.

Misiak, M., Koźmiński, W., 2007. Three-dimensional NMR spectroscopy of organic molecules by random sampling of evolution time space and multidimensional Fourier transformation. Magn. Reson. Chem. 45 (2), 171–174.

Müller, L., Kumar, A., Ernst, R.R., 1975. Two-dimensional carbon-13 NMR spectroscopy. J. Chem. Phys. 63 (12), 5490–5491.

Oschkinat, H., Cieslar, C., Holak, T.A., Clore, G.M., Gronenborn, M., 1989. Practical and theoretical aspects of three-dimensional homonuclear Hartmann–Hahn - nuclear Overhauser enhancement spectroscopy of proteins. J. Magn. Reson. (1969) 83 (3), 450–472.

Oschkinat, H., Griesinger, C., Kraulis, P.J., Sørensen, O.W., Ernst, R.R., Gronenborn, A.M., Clore, G.M., 1988. Three-dimensional NMR spectroscopy of a protein in solution. Nature (London) 332 (6162), 374–376.

Oschkinat, H., Muller, T., Diekmann, T., 1994. Protein structure determination with three-and four-dimensional NMR spectroscopy. Angew Chem. Int. Ed. Engl. 33 (3), 277–293.

Plant, H.D., Mareci, T.H., Cockman, M.D., Brey, W.S., 1986. 27th Experimental NMR Conference, Baltimore, Maryland, April 13–17.

Rule, G.S., Hitchens, T. K., 2005. Fundamentals of Protein NMR Spectroscopy, Springer, Netherlands, 239–351.

Vuister, G.W., Boelens, R., 1987. Three-dimensional J-resolved NMR spectroscopy. J. Magn. Reson. (1969) 73 (2), 328–333.

Vuister, G.W., Boelens, R., Kaptein, R., 1988. Nonselective three-dimensional NMR spectroscopy. The 3D NOE-HOHAHA experiment. J. Magn. Reson. (1969) 80 (1), 176–185.

Williamson, R.T., Marquez, B.L., Gerwick, W.H., Martin, G.E., Krishnamurthy, V.V., 2001. J-IMPEACH-MBC: a new concatenated NMR experiment for F_1 scalable, J-resolved HMBC. Magn. Reson. Chem. 39 (3), 127–132.

Zuiderweg, E.R.P., Fesik, S.W., 1989. Heteronuclear three-dimensional NMR spectroscopy of the inflammatory protein C5a. Biochemistry 28 (6), 2387–2391.

Chapter 9

Some Key Developments in NMR Spectroscopy

Chapter Outline

NMR spectroscopy continues to evolve, in terms of improvements in hardware, as well as in designing of innovative pulse sequences for new applications. A dazzling array of recent publications, referred to in this chapter, provide an overview of the key developments in this field. Some of the topics, such as DOSY, which were not included in the earlier chapters, are also presented here. The discussion is divided into three broad categories, i.e., key progress made in solid-state NMR spectroscopy, developments in NMR hardware, and advent of innovative pulse sequences for new applications (Edger, 2013).

9.1 DEVELOPMENT OF NEW HARDWARE

9.1.1 Bench-Top NMR Spectrometers

NMR spectroscopy is routinely used to support and manage a variety of academic and research activities in chemistry laboratories. Conventional NMR spectrometers are large, expensive, and often difficult to maintain. The advent of low-field bench-top NMR spectrometers, therefore, is changing the way NMR spectroscopy has been practiced for many years. With a small bench top NMR,

Solving Problems with NMR Spectroscopy. http://dx.doi.org/10.1016/B978-0-12-411589-7.00009-7
415

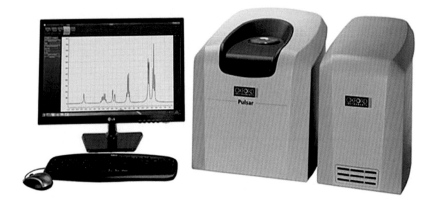

FIGURE 9.1   Oxford Instrument's Pulsar™ - 60 MHz (1.4 T) NMR spectrometer, fitted with cryogen-free permanent magnet. *(Reprinted with the permission of Oxford Instruments).*

there is no longer any need to line up at the core NMR facility, for instance to run a quick check on the purity of a natural product or determine the status of a synthesis. Whether it is for identifying a compound quantitation, reaction monitoring, structure elucidation, or kinetics measurements, bench-top NMR enables researchers, within limits, to carry out the tasks at hand. The 45, 60, 80, and 90 MHz (~1–2 T) bench top NMR spectrometers can provide a quick look at the NMR data, with the option to do a detailed analysis on a more powerful instrument later, if required (Fig. 9.1). Generally, a resolution of between 1 and 3 Hz can be achieved by these robust, easy to handle, and affordable spectrometers (Metz and Mäder, 2008).

Bench-top NMR spectrometers are equipped with cryogen-free stable permanent magnets, made from some rare earth metals. Samarium–cobalt and neodymium magnets are particularly suitable for instruments up to 90 MHz. They operate at temperatures from room temperature to 45°C and their designs allow the instruments to be made small enough to fit on a laboratory bench, and safe to be transported for on-site analysis. They require only single phase local electrical power, fitted with a UPS system, making them especially suitable for small research laboratories in developing countries. These spectrometers are also ideal for teaching and training of NMR spectroscopy in places where funding is a major constraint.

Bench-top NMR spectrometers normally use traditional 3–5 mm NMR sample tubes. The samples are dissolved in deuterated solvents, and the spectrometers can be shimmed and locked at ambient room temperature without any hassle. Because of the lower field magnets and corresponding low sensitivities, the sample requirements are relatively high, that is often not an issue in medicinal and synthetic chemistry studies. With a concentrated sample, a good signal-to-noise ratio can be obtained, and ^{1}H-NMR spectra may be recorded in

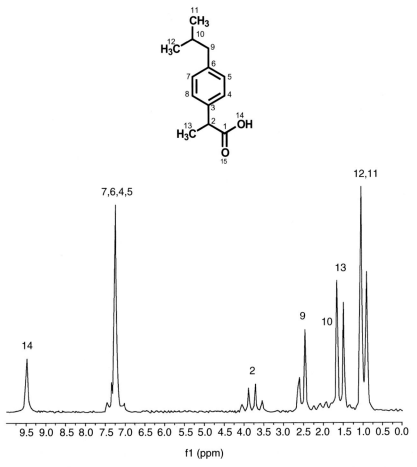

FIGURE 9.2 ^{1}H-NMR spectrum of ibuprofen (200 mM in CDCl$_3$), recorded in 10 seconds on a 43 MHz bench-top NMR spectrometer *(Reprinted with the permission of Magritek Limited).*

a few minutes (Fig. 9.2). Recently, a large number of on-site NMR studies on archeological samples have been reported with such magnets in the literature (Capitani et al., 2012).

9.1.2   Earth's Field NMR Spectrometer

The earth's field NMR spectrometers (EF-NMR) (Fig. 9.3) are *ultra*-low field NMR instruments. These spectrometers use the globally available homogenous earth's magnetic field (0.05 T) for the detection of NMR active nuclei. The resulting Larmor frequency of ^{1}H in the earth's field magnet is only a couple of kHz, as compared to 60 MHz in a 1.41 T low field 60 MHz NMR spectrometer. The EF-NMR signals are also affected by both the magnetic noises in the

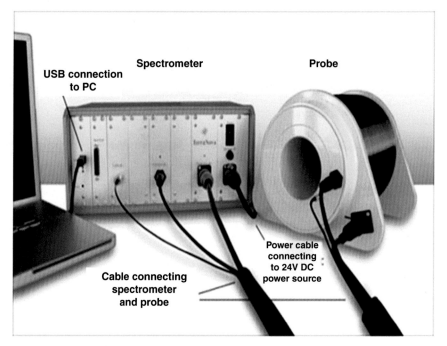

FIGURE 9.3   An earth's field NMR spectrometer (Terranova EF-NMR system by Magritek) *(Reprinted with the permission of Magritek Limited).*

laboratory environment, and the natural variations in the earth's field. The very low sensitivity, the dominance of scalar-couplings, and the lack of chemical-shift dispersion have been the major bottlenecks in the use of EF-NMR spectrometers.

However, such disadvantages have been overcome, to an extent, with the introduction several new methods including:

1. Introduction of electronic equipment which can compensate for the changes in ambient magnetic fields.
2. Use of a pulse sequence with only DC fields.
3. Pre-polarization of the sample by the use of a crude electromagnet with the field 350 times larger than the initial magnetization. The Pre-polarization methodology significantly increases the sensitivity and makes it possible to perform a large range of modern and sophisticated NMR experiments.

While chemical shifts are important in NMR, they are normally not of significance in the earth's field NMR instrument because of very low dispersion which results in most of them, if not all, overlapping together as one broad signal. In the absence of chemical shifts, several other spectral features, such as spin–spin multiplets (that are separated by high fields) are observed in the EF-NMR. The EF-NMR spectra are, therefore, dominated by spin–spin coupling

(*J*-coupling) effects. Software optimized for analyzing these spectra can provide useful information about the structure of the molecules in the sample. Efforts are being made to develop methods for chemical shift based NMR signals in the EF-NMR (Katz et al., 2012).

The EF-NMR spectrometer has certain advantages, including the portability of the machine, the ability to analyze samples on-site, and its low cost. In high-field NMR spectrometers, the magnetic field inhomogeneity is compensated in many ways but this is not an issue in EF-NMR spectrometers due to the excellent inherent homogeneity of the earth's magnetic field (at a particular location). EF-NMR spectrometers are good educational tools due to their relatively low cost and simplicity, but their research applications are severely limited because of their poor sensitivity and dispersion. Compounds containing hydrogen nuclei such as water, hydrocarbons (natural gas and petroleum), and carbohydrates (plants and animals) may be analyzed by ^{1}H-EF-NMR.

9.1.3   Ultra-High Field NMR Spectrometers

Efforts to build ultra-high field NMR spectrometers (over 1000 MHz) are underway. These include the development of a 1.2 GHz (1200 MHz) NMR spectrometer through the financial support (Euro 10 million) of the Netherlands government. Many other laboratories in US and Europe are also involved in the development of ultra-high field NMR spectrometers of 1.2 GHz or beyond. Some laboratories already have such machines, mostly custom made at site for specific purposes. Magnetic field stability with very low field drift rates, suppression of external magnetic field disturbances, and minimization of stray fields are the key challenges in the development of ultra-high field NMR spectrometers (Harrison et al., 2008).

9.1.4   New Liquid Nitrogen Cooled Cryogenic Probe

The first-generation cryogenically cooled probes, as discussed in Chapters 1 and 3 (Sections 1.2.2 and 3.3.2.1), require gaseous helium as the cryogen, and they are often difficult to maintain. A new range of cryogenically cooled probes have been recently developed which use liquid nitrogen to cool the *Rf* coils and the preamplifier, and they provide a sensitivity gain of a factor of 2 or more over equivalent room temperature (RT) probes. The probe assembly comprises, in addition to the probe, a control unit, an automatic tuning accessory, and a liquid nitrogen vessel. These probes are now available in various settings, such as broad-band and triple resonance excitation. Their simple design, the requirement of minimum infrastructure, ease of handling and maintenance, low running cost, and most importantly affordable price make them an excellent choice for routine analysis (Ramaswamy et al., 2013).

The liquid nitrogen cooled probes (Fig. 9.4) deliver a significant signal-to-noise (sensitivity) enhancement over the room temperature (RT) probes of a

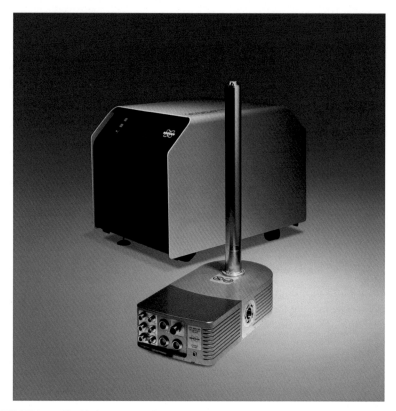

FIGURE 9.4 Liquid nitrogen cooled cryogenic Prodigy probe by Bruker Spectrospin *(Reprinted with the permission of Bruker corporation).*

factor of 2–3 for X-nuclei (such as ^{13}C, ^{15}N, ^{31}P, etc.), although it is not as much as that obtainable from helium cooled probes that provide a sensitivity enhancement by a factor of 4. Liquid nitrogen cooled probes enable time-demanding heteronuclear correlated NMR experiments to be performed up to 10 times faster than with RT probes. They are specially suited for the study of small molecules in academia, chemical industries, and pharmaceutical sector, and can enable such laboratories to substantially increase their sample throughput.

9.1.5   Multiple Receiver Technology

The use of dual receivers has been in practice in solid state NMR spectroscopy since long, as it allows rapid *parallel* data acquisition in each dimension. In conventional high resolution solution state NMR spectroscopy, the direct detection of only *single nucleus at a time* is commonly carried out. For example in COSY and TOCSY, as well as in HSQC and HMBC, ^{1}H is detected, while in INADEQUATE, ^{13}C nucleus is detected. As a result, NMR experiments are performed in sequence, *one after the other*. If we have the capacity to detect two or more nuclei at the same time, then we can record many experiments, parallel to each other. This has now become possible with spectrometers that incorporate two or more receivers so that they can tune into different nuclei at the same time.

The availability of NMR spectrometers that incorporate multiple receivers, tuned to different nuclei, has led to the development of a new class of NMR experiments, commonly called *P*arallel *A*cquisition *N*MR *S*pectroscop*Y* (PANSY), and *P*arallel *A*cquisition of *N*MR, an *A*ll-in-one *C*ombination of *E*xperimental *A*pplications (PANACEA).

PANSY is now used to simultaneously detect signals from up to four nuclear species, such as ^1H, ^2H, ^{13}C, ^{15}N, ^{19}F, and ^{31}P. Conventional COSY, TOCSY, HSQC, HMQC, and modified HMBC pulse sequences have been modified for the PANSY applications (Kupče et al., 2006). PANACEA, however, combines several standard NMR pulse sequences into a single entity. For example, PANACEA experiments with INADEQUATE, HSQC, and HMBC are designed for getting information about ^{13}C–^{13}C couplings, as well as direct ^1H–^{13}C and long-range $^1H/^{13}C$ correlations. Such parallel experiments offer a direct route to molecular structure (Kupče and Freeman, 2010). Figure 9.5 shows the PANACEA spectra, consisting of (1) INADEQUATE, and (2) multiplicity-edited HSQC spectra of cholesterol, recorded in parallel in 23-min measurements.

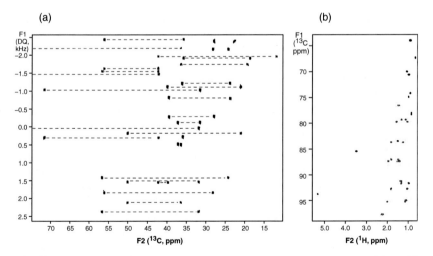

(a)   (b)

FIGURE 9.5 A PANACEA experiment exhibiting (a) INADEQUATE spectrum, and (b) multiplicity-edited HSQC spectrum (concatenated DEPT-HSQC) in $CDCl_3$ (1 M), recorded in parallel in 23 minutes. Two responses in (a) are correlated with signals that lie well outside this spectral window. (b) Multiplicity-edited HSQC spectrum shows red peaks that belong to CH_2 groups, while the black peaks represent CH and CH_3 groups. Four-fold aliasing was used to reduce the experimental duration. Sweep width = 20 kHz; sweep width (1) = 5.2 kHz with 4 k data points and two scans per increment *(Reprinted from Kupče and Freeman (2010). J. Magn. Reson. **206**, 147–153, with permission from Academic Press, Inc.).*

The PANSY and PANACEA experiments allow unambiguous structure elucidation of small organic molecules from a single measurement, and include an internal field/frequency correction routine. These experiments do not require the conventional NMR lock system, and can be obtained in pure liquids. Furthermore, long-range spin–spin couplings can be extracted from the PANACEA spectra, and used for three-dimensional structure refinement. These quick NMR measurements are especially suitable for samples where long-term stability is an issue. Similarly, the recording of several NMR spectra in parallel, rather than in series, leads to a major saving in instrument time.

The PANSY and PANACEA experiments are currently not optimized for structure determination of large molecules. However, this is one area in which these innovative pulse sequences are likely to have impact in the future, simply because conventional biomolecular NMR spectroscopic measurements in series, *one by one*, can take days if not weeks of NMR instrument time.

9.2 KEY DEVELOPMENTS IN SOLID-STATE NMR SPECTROSCOPY

Major recent developments in NMR spectroscopy have largely taken place in solid-state NMR and in NMR-based imaging. Both subjects are beyond the scope of this book. However, some key developments in the field of solid-state NMR spectroscopy are presented here as we feel that they will soon find extensive uses in structural chemistry, structural biology, and pharmaceutical industrial applications.

9.2.1 Solid-State Cryogenic MAS Probes

Magic angle spinning (MAS) probes have long been in use in solid-state NMR spectroscopy. Most of these probes function at room temperature. Recently, cryogenic MAS solid-state probes have been constructed which provide a sensitivity enhancement by *cooling the sample* to a temperature of 30 K. They also have excellent spinning stability which allows the recording of two-dimensional (2D) NMR spectra. This development should not be confused with the cryogenically cooled probes in solution state NMR spectroscopy, as in those probes, the sensitivity gain is based on cooling of the detection coil and circuitry, and not of the sample itself.

In cryogenic MAS probes, the solid sample is cooled with liquid gas, while room temperature nitrogen is used as the MAS bearing and drive gas (Thurber and Tycko, 2008). This probe can provide only medium speed magic angle spinning due to the relatively large sample volume (40–80 mL). The gain in sensitivity in these probes is based on the fact that by lowering the temperature of the sample, one can increase the spin polarization. At a temperature 1 K, the nuclear spin polarization at thermal equilibrium is proportional to $1/T$, where T is the temperature of the sample. Thus if we cool down the sample to

25 K, from the room temperature ($\sim$295 K), a 12-fold increase in spin polarization is expected, leading to a 12-fold increase in sensitivity (Fernandez and Pruski, 2012).

Substantial lowering of temperature in solution-state NMR spectroscopy for sensitivity enhancement is not practical due to a number of factors, including temperature gradient issues, precipitation of solute particles, and solidification or *icing* of deuterated solvents. It is, therefore, more logical to use other methods, including lowering of the temperature of the detection circuits to reduce the noise level and thus increase the *S/N* ratio. However, in solid-state NMR spectroscopy, modest lowering of the temperature of the sample in MAS probe is an excellent approach for the enhancement of *S/N* ratio. Figure 9.6 shows ^{13}C-NMR (CP-MAS) proton decoupled spectra of powdered ^{13}C-labeled valine at 300 and 31 K, recorded using a cryogenic MAS probe (revolution NMR LLC, USA). Over 10-fold enhancement in sensitivity is clearly seen (Polenova et al., 2015; Thurber and Tycko, 2008).

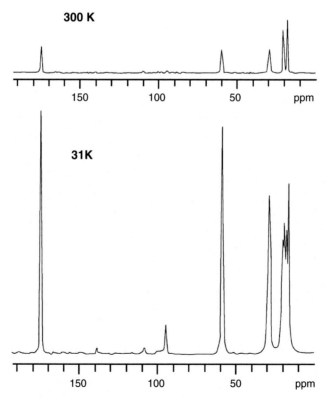

FIGURE 9.6 ^{13}C CP-MAS spectra of powdered ^{13}C-labeled valine powder (9 mg) in one scan, (a) ^{13}C CP-MAS NMR spectrum at 300 K, and (b) ^{13}C CP-MAS NMR spectrum at 31 K *(Reprinted with the permission of Revolution NMR LLC, USA).*

9.2.2 DNP-MAS NMR Spectroscopy

The development of dynamic-MAS NMR spectrometers represents one of the most important advances in the field of NMR spectroscopy. The process of *D*ynamic *N*uclear *P*olarization (DNP) provides a dramatic enhancement in sensitivity, both in liquid- and solid-state NMR studies. However, it is in the solid-state where DNP has found the greatest use.

As described in Section 3.6.1, the large electron polarization of radical molecules can be transferred to a nuclear spin by using *in situ* microwaves for low-temperature state NMR spectroscopy. These polarization transfers between electrons and coupled nuclei can occur spontaneously through electron-nuclear cross relaxation, and/or spin-state mixing among the electrons and nuclei. The DNP-MAS NMR spectrometers are now available in the range of 200–800 MHz, and they involve major developments in probe design and microwave technologies (gyrotron) (Barnes et al., 2012).

The DNP-enhanced solid-state NMR experiments at high field can yield sensitivity enhancements of up to 80-fold (Fig. 9.7). These systems are equipped with unique high power microwave sources, i.e., an easy-to-use software-controlled high-power gyrotron (a vacuum electronic device, capable of generating

FIGURE 9.7 A DNP solid-state NMR spectrometer *(Reprinted with the permission of Bruker Corporation).*

high-power, high-frequency THz electromagnetic radiations, which serves as the microwave source). Optimum beam propagation to the sample is ensured by microwave transmission lines. Low-temperature (cryogenic) MAS probe technology (~ 100 K), with cold spinning gas and built-in waveguide, are now available for a range of NMR systems, such as for 400, 600, and 800 MHz spectrometers.

9.3 INNOVATIVE PULSE SEQUENCES AND NOVEL APPLICATIONS OF NMR SPECTROSCOPY

9.3.1 Diffusion Order Spectroscopy (DOSY)

The introduction of second frequency dimension gave birth to 2D NMR spectroscopy, as described previously. A logical extension of the idea of the second dimension in NMR spectroscopy was the advent of additional NMR dimensions that depend on other molecular properties, such as size, shape, mass, and charge. In DOSY spectroscopy, the mixtures of organic molecules are analyzed based on molecular diffusion of each component.

Very often chemists need to handle mixtures of organic compounds. They can be mixtures of compounds in extracts of medicinal plants, reaction mixtures containing several products, etc. Historically, mixture analysis in organic chemistry was generally performed in two steps. In the first step, the various constituents are separated using chromatographic techniques, including high performance liquid chromatography or gas chromatography, while in the second step, these constituents are individually characterized by using various spectroscopic techniques. The advent of so-called hyphenated techniques such as LC-MS and GC-MS as well as LC-NMR made it possible to separate and analyze the individual constituents of complex mixtures in one step in a time-efficient manner. The LC-NMR technique suffers from a number of limitations, such as the need of setting up of ideal separation conditions for HPLC and issues of solubility in the appropriate deuterated solvent.

In the last decade, the $\underline{D}$iffusion $\underline{O}$rder $\underline{S}$pectroscop$\underline{Y}$ (DOSY) NMR technique has been developed for the analysis of mixtures of organic compounds. The method relies on the different rates of chemicals through a solution. These rates are directly related to the physical properties of the various components making up the mixture. DOSY NMR does not require the physical separation of the components in a mixture, and provides chemical shift information *in situ*. This has made DOSY NMR a powerful technique with wide applications.

The DOSY spectrum (Fig. 9.9) separates the NMR signals of different compounds in a mixture according to their diffusion coefficients. It is a 2D NMR experiment in which the information is spread in two dimensions. A series of spin echo spectra is recorded with different pulsed field gradient strengths, and the resulting signal decays are analyzed to extract a set of diffusion coefficients with which to synthesize the diffusion domain (diffusion dimension) of the DOSY spectrum.

The molecules, as a liquid or in the solution state, are in constant motion. This includes translational motion (Brownian motion) which provides a mechanism by which the diffusion process takes place. Brownian motion is that random motion of molecules that occurs as a consequence of their absorption of heat. Molecules will diffuse from volumes of high concentration to low concentration; this happens because the molecules are in constant random (Brownian) motion, and they tend to bump into each other more if they are in more concentrated areas. Hence, they tend to move away from each other through diffusion. The diffusion depends on many physical parameters, including molecular weight, size and shape of the molecule, and the temperature and viscosity of solution. Assuming a spherical size of the molecule, the diffusion coefficient D is described by Eq. 9.1.

$$D = \frac{kT}{6\pi\eta r_{\mathrm{s}}} \tag{9.1}$$

where k is the Boltzmann constant, T is the temperature, η is the viscosity of the liquid, and r_{s} is the (hydrodynamic) radius of the molecule.

Pulse gradient NMR spectroscopy is used to measure the translational diffusion of molecules. By the use of gradients, the molecules can be spatially labeled (marked depending on their position in the sample tube). If they move, after the initial labeling, during the diffusion time Δ, their new position is again decoded by using the second gradient. This difference in diffusion time directly affects the NMR signal intensity or signal strength. This simple theory forms the basis of differentiating different solutes in a mixture.

The basic pulse sequence of DOSY involves the characterization of molecular diffusion by pulsed field gradient spin-echo (PGSE). The PGSE constructs a 2D plot where the x-axis contains the chemical shifts of ^{1}H (in ppm), while the y-axis gives the diffusion coefficient D of each signal. Since diffusion coefficients of various components are different, one can record separate ^{1}H-NMR spectra for each compound in a mixture.

There are several basic sequences developed to map the diffusion rates of individual components in a complex mixture. In its simplest form, the magnetization is excited with a 90° Rf pulse, and then dispersed using a magnetic field gradient pulse. After a period of $\Delta/2$, an 180° Rf pulse is applied to invert the dispersed magnetization. After a period Δ, a second gradient pulse is applied to refocus the signal. The PGSE pulse sequence is presented in Fig. 9.8. The DOSY NMR experiments are also used for the measurement of diffusion rates of different constituents in a mixture. Figure 9.9 shows the DOSY spectrum of a mixture of three components, cholesterol, aspirin, and phenol. Each line, horizontal to the diffusion axis, presents a separate ^{1}H-NMR spectrum of each component, which is also shown in 1D slices.

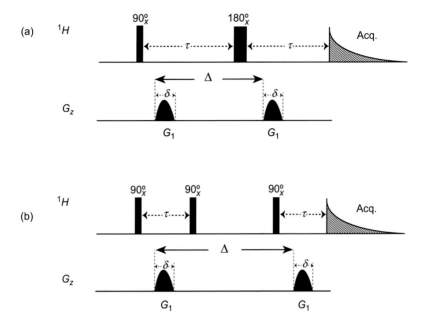

FIGURE 9.8 Basic pulse sequences for the measurement of molecular diffusion based on spin-echo. (a) Pulse field gradient spin-echo, and (b) PFG stimulated-echo. The molecular diffusions are measured during the delay period Δ with increasing gradient strengths.

9.3.2 Other Key Developments

In addition to developments mentioned above, there are several other interesting techniques that have been developed which are presented below:

1. Development of multiple-sample solid-state magic-angle spinning (MAS) probes which can simultaneously acquire NMR spectra of seven or more samples. They can substantially increase throughput of solid-state NMR spectroscopy (Nelson et al., 2006).
2. A "moving-tube" method has been developed to circumvent problems associated with long T_1 relaxation times. This utilizes a 5 mm NMR tube (1.5 m long), filled with a large quantity of sample solution. A motor physically moves this long NMR tube vertically after each pulse-acquisition (scan), and thus removes (moves away) the excited spins from the coil, and introduce relaxed spins (so called "fresh sample") in front of the detection coil. This method has been applied in recording of variant COSY spectrum (Donovan et al., 2012).
3. The methodology for quantitative ^{1}H-NMR (qHNMR) is continuously evolving, expanding its applications to various fields. A major challenge is to accurately and most comprehensively analyze complex mixtures (natural

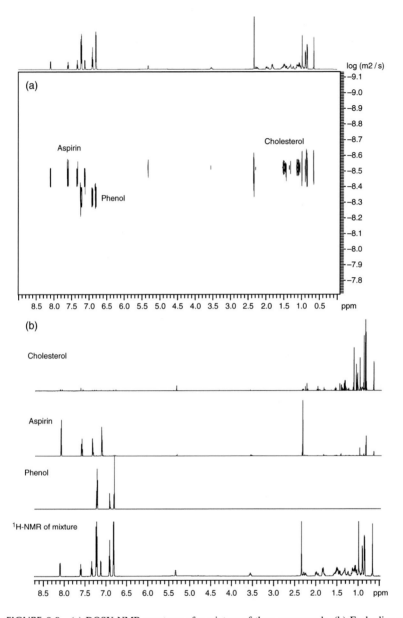

FIGURE 9.9 (a) DOSY NMR spectrum of a mixture of three compounds. (b) Each slice corresponds to the regular ¹H-NMR spectrum of one component of the mixture (cholesterol, aspirin, and phenol in this case). The diffusion rate of various constituents can be obtained from the DOSY spectrum.

products, environmental pollutants, biological fluids, toxins, etc.) containing low-level analytes. For this purpose, standardization, calibration, rapid identification, and computational methods are regularly reported in a large number of publications (Pauli et al., 2012).

REFERENCES

Barnes, A.B., Markhasin, E., Daviso, E., Michaelis, V.K., Nanni, E.A., Jawla, S.K., Mena, E.L., DeRocher, R., Thakkar, A., Woskov, P.P., Herzfeld, J., Temkin, R.J., Griffin, R.G., 2012. Dynamic nuclear polarization at 700 MHz/460 GHz. J. Magn. Reson. 224, 1–7.

Capitani, D., Di Tullio, V.D., Porietti, N., 2012. Nuclear magnetic resonance to characterize and monitor cultural heritage. Prog. Nucl. Magn. Reson. Spectrosc. 64, 29–69.

Donovan, K.J., Allen, M., Martin, R.W., Shaka, A.J., 2012. Improving the double quantum filtered COSY experiment by "Moving Tube" NMR. J. Magn. Reson. 219, 41–45.

Edger, M., 2013. Physical methods and techniques: NMR spectroscopy. Ann. Rep. Prog. Chem. Sect. B: Org. Chem. 109, 256–274.

Fernandez, C., Pruski, M., 2012. Solid State NMR. Probing quadrupolar nuclei by solid-state NMR spectroscopy: recent advances. Top. Curr. Chem. 306, 119–188.

Harrison, R., Bateman, R., Brown, J., Domptail, F., Friend, C.M., Ghoshal, P., King, C., Van der Linden, A., Melhem, Z., Noonan, P., Twin, A., Field, M., Hong, S., Parrell, J., Zhang, Y., 2008. Development trends in high field magnet technology. IEEE. Trans. Appl. Supercond. 18 (2), 540–543.

Katz, I., Shtirberg, L., Shakour, G., Black, A., 2012. Earth field NMR with chemical shift spectral resolution: Theory and proof of concept. J. Magn. Reson. 219, 13–24.

Kupče, E., Freeman, R., 2010a. High-resolution NMR correlation experiments in a single measurement (HR-PANACEA). Magn. Reson. Chem. 48 (5), 333–336.

Kupče, E., Freeman, R., 2010b. Molecular structure from a single NMR sequence (fast-PANACEA). J. Magn. Reson. 206 (1), 147–153.

Kupče, E., Freeman, R., John, B.K., 2006. Parallel acquisition of two-dimensional NMR spectra of several nuclear species. J. Am. Chem. Soc. 128 (30), 9606–9607.

Metz, H., Mäder, K., 2008. Benchtop-NMR and MRI- a new analytical tool in drug delivery research. Int. J. Pharm. 364 (2), 170–175.

Nelson, B.N., Schieber, L.J., Barich, D.H., Lubach, J.W., Offerdahl, T.J., Lewis, D.H., Heinrich, J.P., Munson, E.J., 2006. Multiple-sample probe for solid-state NMR studies of pharmaceuticals. Solid State Nucl. Magn. Reson. 29, 204–213.

Pauli, G.F., Godecke, T., Jaki, B.U., Lankin, D.C., 2012. Quantitative ^{1}H NMR. Development and potential of an analytical methods: an update. J. Nat. Prod. 75 (4), 834–851.

Polenova, T., Gupta, R., Goldbourt, A., 2015. Magic angle spinning NMR spectroscopy: a versatile technique for structural and dynamic analysis of solid-phase systems. Anal. Chem. 87 (11), 5458–5469.

Ramaswamy, V., Hooker, J.W., Withers, R.S., Nast, R.E., Edison, A.S., Brey, W.W., 2013. Microsample cryogenic probes: technology and applications. eMagRes. 2 (2).

Thurber, K.R., Tycko, R., 2008. Biomolecular solid state NMR with magic-angle spinning at 25 K. J. Magn. Reson. 195 (2), 179–186.

Chapter 10

Logical Approach for Solving Structural Problems

Chapter Outline

10.1 STRUCTURE ELUCIDATION STEP BY STEP

10.1.1 Spectroscopic Techniques

Determination of structure of an unknown compound is an intellectually demanding process. Information is obtained from various spectroscopic techniques, which are then used to assemble a *proposed structure*, which then goes through a rigorous process of *self-criticism*. Finally the structure *must agree* with all the spectral observations. Table 10.1 lists key structural information obtained from various spectroscopic techniques.

10.1.2 How to Prepare Samples for Spectroscopic Studies?

Chemists often have to deal with complex mixtures containing many constituents. These may be plant extracts containing thousands of primary and secondary natural products, environmental samples comprising dozens of organic pollutants, or biological fluids with many small and large metabolites. Many of them are closely related in their chemical and physical properties and it is often challenging to purify them using chromatographic techniques. Similarly

Solving Problems with NMR Spectroscopy. http://dx.doi.org/10.1016/B978-0-12-411589-7.00010-3
431

TABLE 10.1 Spectroscopic Techniques and Their Applications in Structure Elucidation of Organic Molecules

Spectroscopic Techniques	Radiation Absorbed	Effect on the Molecule	Structural Information Deduced
Ultraviolet/Visible spectrophotometry	Ultraviolet-visible λ, 190–400 nm and 400–800 nm	Changes in electronic energy levels within the molecule	Extent of π-electron systems. Presence of conjugated unsaturation, and conjugation with nonbonding electrons
Infrared spectrophotometry	Infrared (mid infrared) λ, 2.5–25 mm v, 400–4000 cm^{-1}	Changes in the vibrational and rotational movements of the molecule	Detection of functional groups, which have specific vibration frequencies, for example, $C{=}O$, NH_2, OH, etc.
NMR spectroscopy	Radiofrequency λ, 25 cm	Nuclei placed under the static magnetic field change their spins after absorption of radiofrequency radiations	The electronic environment of nuclei, their numbers and number of neighboring atoms
Circular dichroism (CD)/Optical rotatory dispersion (ORD)	Ultraviolet/visible λ 200–600 nm	Changes in the electronic energy level of a molecule	Identification of absolute configurations of small molecules in solution
Mass spectrometry	High-speed electron beam and other ionization sources	Effects of ionization and fragmentation of charged particles	Relative masses of molecular ions and fragments

pure compounds, thus isolated, can have chemical structures that are novel and unprecedented in the scientific literature. It is therefore essential that chemists develop a rigorous approach towards structure elucidation.

The development of deductive skills involves the effective use of spectroscopic data, incisive reasoning, careful search of the relevant literature, and critical thinking. After obtaining a pure compound from a mixture, the following steps need to be carried out:

1. Careful checking of the purity of the compounds by TLC, preferably precoated high performance thin layer chromatography (HPTLC), and if possible by HPLC. It is vital to have a pure compound before one begins the process of structure elucidation. Impurities are genuine compounds, and

appear as signals in various spectroscopic techniques, leading to incorrect conclusions. It is well worth the effort to ensure that the compound in hand has above 99% purity, as spurious signals will otherwise lead to wrong conclusions regarding the structure.

2. Measurement of the correct quantity through a properly calibrated analytical balance is important. A fundamental question regarding whether to proceed further is based on the available quantity. The choice of different spectroscopic techniques largely depends on the quantity of the sample. It is often a waste of time and effort to process a minor sample (less than 0.1 mg) as one may not be able to go beyond recording partial data, such as HREI–MS, UV, and IR spectra, and 1D ^{1}H-NMR spectrum. Even with the most modern and sensitive NMR spectrometers, it is often not easy to record ^{13}C-NMR and 2D-NMR experiment on very minor constituents. It is therefore advisable to make the extra effort and collect a large enough quantity of a minor compound of interest (preferable >5 mg), if serious structural work is to be carried out. You may, however, be able to recognize a known compound by recording its mass spectrum and confirming by comparison with the reported mass spectrum. This may need only microgram quantities. Of course the quantity needed will depend on the molecular weight of the compound and the experiments to be conducted.

3. It is also essential to check the solubility of the compound before attempting spectroscopic measurements.

Once a pure compound is available in adequate quantity and with good solubility in one of the many common solvents (you should make sure that the corresponding fully deuterated solvent is available in your NMR laboratory), you are now ready to proceed further for NMR and other spectroscopic studies. Figure 10.1 presents the various steps involved in the preparation of samples for spectroscopic analysis.

Solubility is an important physical property, which is based on the nature of the solute and the solvent. Figure 10.2 correlates the solubility in various solvents with the likely nature and class of the compound under study.

10.1.3 Which Spectroscopic Technique Is Used First?

Among the various spectroscopic techniques available, which one should be used first is a topic of debate. There are basically two options and the choice is largely based on the quantity of the sample available:

1. Non-destructive spectroscopic techniques (optical rotation, UV, IR, NMR spectroscopy, and single-crystal X-ray diffraction)
2. Destructive spectroscopic/analytical techniques (mass spectrometry, elemental analysis)

The choice of the above mentioned techniques is mainly dictated by the quantity of the sample available. In general non-destructive techniques should

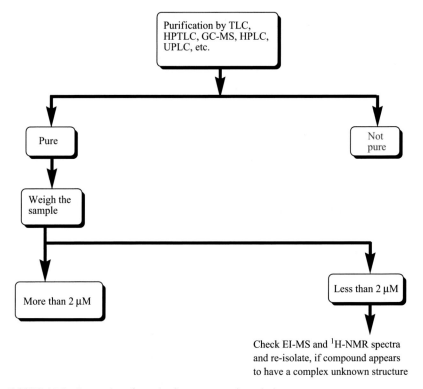

FIGURE 10.1   Preparation of samples for spectroscopic analysis.

be used first in the case of a low quantity of sample material being available. Once this critical decision is made, the process of spectroscopic analysis is initiated. A very small amount of the sample is needed to record a mass spectrum, and a wealth of information can be obtained quickly including the molecular formula. The possible structure may also be derived through a data search in an MS library with other known compounds. This could therefore be a preferred first measurement.

10.1.4   Getting Started – Preliminary Spectroscopic Analysis

It is imperative to obtain the molecular formula of the unknown compound by high-resolution electron impact mass spectrometry (HREI–MS) or elemental analysis (if sufficient compound is available) as a first step (Figs 10.3 and 10.4). The 1D ^{1}H- and ^{13}C-NMR spectra should also be inspected to verify the number of protons and carbons, as obtained by integration, and matched with the expected number as obtained from the molecular formula obtained by high resolution mass spectrometry. This allows the "degree of unsaturation" (DoU), or "index of hydrogen deficiency" to be calculated.

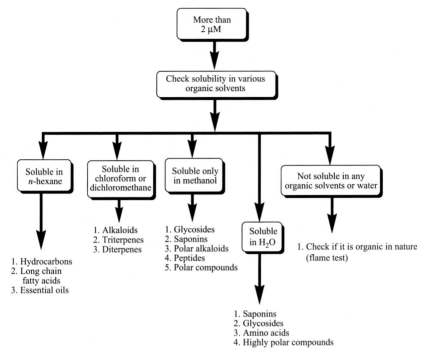

FIGURE 10.2   Preparation of samples for structure elucidation. Only pure (>99%) compounds which are in sufficient quantities (2 μM) are processed further.

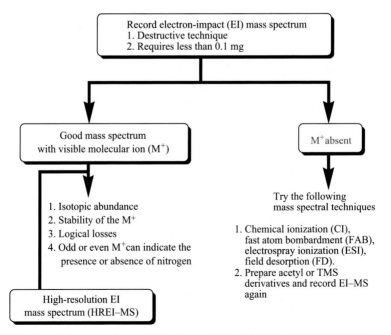

FIGURE 10.3   Determination of molecular mass and formula by using mass spectrometry.

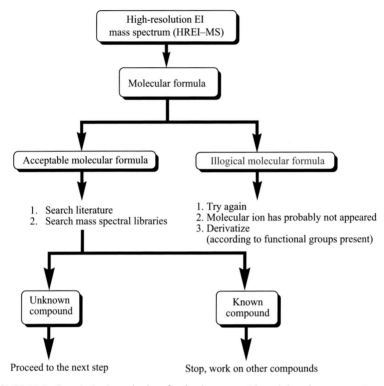

FIGURE 10.4 Steps in the determination of molecular mass and formula by using mass spectrometry.

The DoU is generally calculated by using the following equation:

$$DoU = \frac{2C + 2 + N - X - H}{2} \tag{10.1}$$

where C is the number of carbon atoms, N is number of nitrogen atoms and X is the number of hydrogen or halogen atoms.

Further information about functionalities and unsaturation sites can be obtained from the IR (Fig. 10.5) and UV/VIS spectra (Fig. 10.6). These methods can quickly identify some functional groups; for example, unconjugated and conjugated carbonyl moieties (ketone, aldehyde, ester, amide, etc.), OH and NH, aromatic CH, etc. Many heterocycles and aromatic rings show distinctive UV/VIS absorptions. Extensive literature is available for IR and UV correlations, which can help in the preliminary identification of key functional groups and chromophores.

10.1.5 Using NMR Spectroscopy for Structure Elucidation

The next step is to use NMR spectroscopy, an important tool for the structure elucidation of organic molecules (Halabalaki et al., 2014; Breton

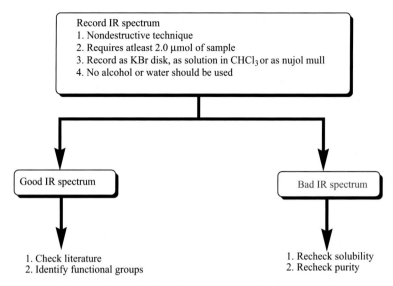

FIGURE 10.5 Determination of functional groups by using infrared spectrophotometry.

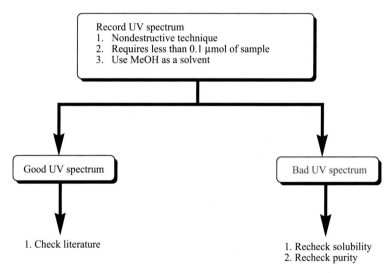

FIGURE 10.6 Determination of unsaturation and conjugated sites (chromophores) by using ultraviolet spectrophotometry.

and Reynolds, 2013). Modern NMR spectrometers are sensitive to organic and inorganic (paramagnetic) impurities. Preparing pure samples for NMR spectroscopic measurements is therefore of paramount importance (Figs 10.7 and 10.8).

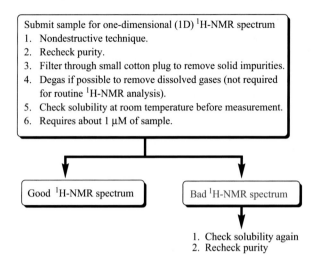

FIGURE 10.7   Quick checks on sample for ^{1}H-NMR spectroscopic measurement.

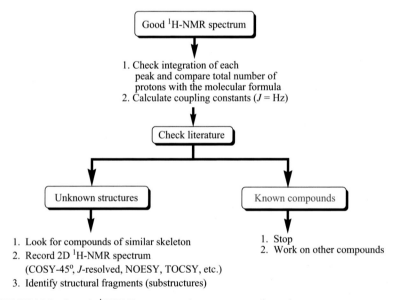

FIGURE 10.8   Steps in ^{1}H-NMR spectroscopic measurement of sample.

Even a cursory look at a one-scan 1D ^{1}H-NMR spectrum can provide useful information about key spectral features. A good quality NMR spectrum can be recorded even with low sample concentrations (2 mM). Very high concentrations (100 mM) are not recommended, even if large quantities are available, as this can introduce artifacts. For example, a one-scan 1D ^{1}H-NMR spectrum of a sample should show mostly well-resolved signals with acceptable integration in the downfield region. Overlapping of signals in the upfield (δ 1–2.5) and mid-spectral regions (δ 3.0–4.5) is not uncommon, as many protons of the sample may resonate in these regions. In contrast, compounds that exhibit complex or broad spectra due to impurities or tautomeric equilibria are typically more difficult to analyze.

For simple unknown compounds, straightforward experiments, such as 1D ^{1}H- and ^{13}C-NMR (Fig. 10.9), COSY (correlation spectroscopy), and nOe/ NOESY spectra are often sufficient. However, the ^{1}H-NMR spectra of complex molecules often contain overlapping signals or higher-order multiplets that may be difficult to interpret. Additionally, good-quality 1D ^{13}C-NMR spectra can be challenging to obtain in a reasonable amount of time with small sample quantities. Compounds with higher molecular weights (750–2,000 Da) often give nOe correlations of weak intensity. In such cases, a different approach that takes advantage of more powerful 2D-NMR techniques (such as TOCSY, HSQC, HMBC, ROESY), is required (Fig. 10.10).

At every stage, a careful literature review is highly recommended. For example, after obtaining the optical rotation and establishing the molecular formula, a literature survey and on-line/off-line spectral library search can quickly help in identifying compounds, which are already reported (Fig. 10.10).

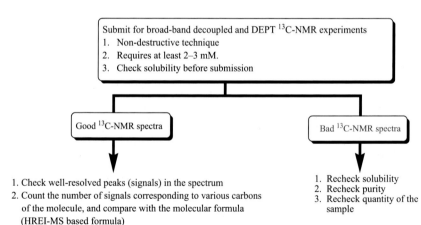

FIGURE 10.9 ^{13}C-NMR spectroscopic measurement of sample.

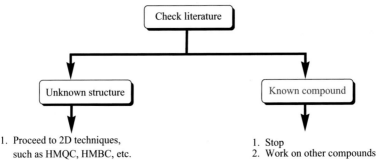

FIGURE 10.10   Various steps in NMR spectroscopic measurement of sample.

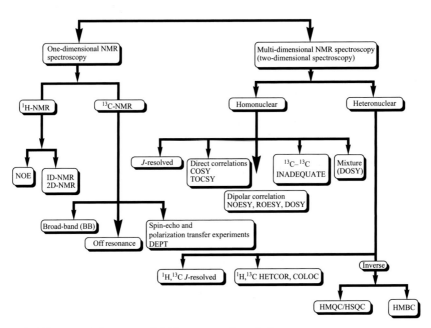

FIGURE 10.11   Various types of NMR spectroscopic methods.

Following is a brief description of various important techniques for NMR spectroscopic analysis (Fig. 10.11):

1. 1D ^{1}H-NMR spectrum provides information about chemical shifts, coupling constants, and integrals.
2. 2D ^{1}H–^{1}H correlations (COSY) identifies J-coupling relationships between protons.

3. 1D ^{13}C-NMR spectrum, such as broad-band (BB) decoupled ^{13}C-NMR spectrum, provides the carbon count and chemical shift values, while multiplicity information can be obtained by recording polarization transfer ^{13}C-NMR experiments, such as DEPT.
4. 2D ^{1}H-^{13}C one-bond heteronuclear techniques, such as HMQC/HSQC provide information about one-bond carbon/proton correlations.
5. 2D ^{1}H-^{13}C long-range heterounclear NMR techniques, i.e., HMBC, provide information about ^{13}C/^{1}H long-range correlations.
6. 2D NOESY, 1D nOe difference spectra or selective 1D NOESY provide stereochemical information from through-space couplings.
7. 2D ^{1}H-^{1}H long-range TOCSY spectrum yields ^{1}H–^{1}H correlations within various spin systems.
8. 2D INADEQUATE spectrum provides information about ^{13}C–^{13}C connectivities.

Once the spectral data has been obtained, the process of interpretation of results, and *putting the jigsaw puzzle together* piece by piece, begins. Comparison of various types of spectra and deducing the structural features from them is a crucial part of this logical process.

10.2 LOGICAL PROTOCOL

The following is a procedure recommended for elucidating the structure of complex organic molecules, based on data from various spectroscopic techniques. It uses a combination of different NMR and other spectroscopic techniques. Computer-based automated or interactive versions of similar approaches have also been devised for structural elucidation of complex compounds, such as *C*omplete *M*olecular *C*onfidence (CMC), CMC-se™ (Bruker), *S*ystematic *E*lucidation of *S*tructure by using *A*rtificial *M*achine *I*ntelligence (SESAMI), X-PERT (Eaton), etc. (Steinbeck, 2004; Blinov et al., 2003; Jaspars, 1999). Many of these packages perform computer-based structure determination, based on available spectroscopic data, including NMR data. Some of them can predict the ^{1}H- and ^{13}C-NMR chemical shifts data for any proposed structure ((ACD, CS Chem Draw Pro (Cambridge Softwares), SpecInfo system (Wiley-VCH), etc.)) (Table 10.2). However, there is no substitute for the hardwork, experience, and intuition of the chemist.

1. The first step in elucidating a structure is to establish the molecular formula by high-resolution mass spectrometry and to calculate the DoU or *D*ouble-*B*ond *E*quivalents (DBEs). An acyclic saturated hydrocarbon has the formula C_nH_{2n+2}. Each double bond or ring represents a degree of unsaturation, while triple bonds represent two degrees of unsaturation. The presence of oxygen or other divalent elements bound by a single bond does not affect the value of DoU (Eq. 10.1).

$$DoU = \frac{2C + 2 + N - X - H}{2}$$

(10.1)

TABLE 10.2 Various Computer-Based Programs Used for Structure Elucidation of Organic Molecules

S. No.	Software	Task
1.	ACD/Labs (Advanced Chemistry Development Company	To predict ^{1}H- and ^{13}C-NMR spectra
2.	CMC-se™ (Bruker)	Perform an automatic analysis of different NMR spectra types and proposes structural solutions
3.	CMC-assist™ (Bruker)	Providing an automated processing and an automated analysis of the NMR data
4.	CASE	A computer-assisted structure elucidation (CASE) program to generate chemical structures by utilizing 1D- and 2D-NMR data
5.	Chemical Concepts	To predict ^{1}H- and ^{13}C-NMR spectra
6.	Sadtler (University of Texas)	To predict ^{1}H- and ^{13}C-NMR spectra
7.	Artificial Neural Networks (ANN)	To predict ^{1}H- and ^{13}C-NMR spectra
8.	PREDICT-IT NMR	To predict ^{1}H- and ^{13}C-NMR spectra
9.	CS ChemDraw Pro (Cambridge Softwares)	To predict ^{1}H- and ^{13}C-NMR spectra
10.	SPECTOOL	To predict ^{1}H- and ^{13}C-NMR spectra
11.	COSMOS (COSMOS software)	To predict ^{1}H- and ^{13}C-NMR spectra
12.	GAUSSIAN (Gaussian, Inc.)	To predict ^{1}H- and ^{13}C-NMR spectra
13.	ACD/PNMR (Advanced Chemistry Development Company)	To predict ^{31}P-NMR spectra
14.	ACD/FNMR (Advanced Chemistry Development Company)	To predict ^{19}F-NMR spectra
15.	X-PERT system (Eaton)	Structure elucidation of organic molecules containing up to 30 skeletal atoms (C, N, O, P, S, B, Si, and the halogens) using a combination of IR, ^{1}H- and ^{13}C-NMR spectral data
16.	SpecInfo system (Wiley-VCH)	To predict ^{1}H- and ^{13}C-NMR spectra and structural interpretation from NMR, IR, and mass spectra
17.	SpecSolv (Wiley-VCH)	Structural interpretation from NMR, IR, and mass spectra

TABLE 10.2 Various Computer-Based Programs Used for Structure Elucidation of Organic Molecules *(cont.)*

S. No.	Software	Task
18.	SESAMI-H, SESAMI-C	Computer-aided structure generator on the basis of 1D- ^{1}H-/^{13}C-NMR and 2D-NMR data
19.	HOUDINI (Side Effects Software)	Computer-aided structure generator on the basis of 1D- ^{1}H-/^{13}C-NMR and 2D-NMR data
20.	COCOA (Pascal program)	Computer-aided structure generator on the basis of 1D- ^{1}H/^{13}C-NMR and 2D-NMR data
21.	CISOC-SES (SGI Indy.)	Computer-aided structure generator on the basis of 1D- ^{1}H-/^{13}C-NMR and 2D-NMR data
22.	NMR-SAM (Spectrum Research, LLC)	Computer-aided structure generator on the basis of 1D- ^{1}H-/^{13}C-NMR and 2D-NMR data
23.	LUCY	Uses 2D-NMR data for computer-aided structure elucidation
24.	StrucEluc	Structure elucidation of organic molecules and biomolecules from 1D- ^{13}C-NMR data

2. Inspect the IR and UV spectra to get preliminary clues about the functional groups and conjugation that might be present in the molecule.

3. A cursory examination of the 1D ^{1}H- and ^{13}C-NMR spectra gives an idea about the ratio of the aromatic to the aliphatic carbons. The ^{1}H-NMR spectrum can also be used to gain additional evidence to confirm the molecular formula (indirectly by counting the protons). The ^{1}H- and ^{13}C-NMR spectra contain valuable information about the type of unsaturation (e.g., the number olefin or aromatic carbons and protons).

4. A good ID ^{1}H-NMR spectrum is always an important starting point for structure elucidation. It makes the deductive process easier if the protons are then labeled with alphabets. Therefore, we first assign each proton appearing in the spectrum an alphabet, starting from the left side of the spectrum. Thus the protons are labeled individually as A, B, C, etc. If a peak or a group of overlapping peaks represents more than one proton, then we need to assign combinations of letters. Thus, four *overlapping* protons can be labeled as a group, such as DEFG. The total number of protons labeled on the spectrum should correspond to the number of protons determined from the high-resolution mass measurement of the molecular ion (HREI–MS). In the case of symmetrical molecules with identical protons, say, in symmetrical halves of the same molecule, a fraction of the actual number of protons present may be observed.

5. Once it has been established through preliminary NMR and mass spectroscopic studies that the substance is new and requires further work, record

the ^{13}C-NMR (BB and DEPT or GASPE), COSY-45°, HMBC (or COLOC, if inverse facilities are not available), HMQC/HSQC (or HETCOR, if inverse facilities are not available), TOCSY (20, 40, 80, and 120 ms) (or delayed COSY/relayed COSY, if TOCSY[1] cannot be performed), and NOE difference spectra.

6. Label each contour *on the diagonal* line of the COSY-45° spectrum with the same letters of the alphabet as used in the 1D ^{1}H-NMR spectrum, and deduce the ^{1}H-^{1}H connectivities from the corresponding cross-peaks.

7. Next, label the HMQC/HSQC spectrum with the same letters at the ^{1}H chemical shifts as those already assigned in the ^{1}H-NMR spectrum. This establishes the one-bond ^{1}H/^{13}C connectivities. Note that if there are two cross-peaks at any particular carbon chemical shift in the HMQC/HSQC spectrum, then these two protons must be the two *geminal* protons attached to that particular carbon. Prepare a table of chemical shifts of the ^{13}C nuclei and their corresponding attached ^{1}H nuclei. By deleting the *geminal coupling interactions* from the COSY spectrum, we arrive at the vicinal ^{1}H/^{1}H connectivities (caution: some weaker long-range couplings, such as "W" coupling interactions, may also appear in the COSY spectrum). Since we already know the direct (one-bond) carbon–hydrogen connectivities of adjacent protonated carbons, *the carbon–carbon connectivities of protonated carbons* can now be deduced. This is the "*pseudo*-INADEQUATE" information that has been derived.

8. Now select one unambiguously identified proton (preferably in an unclustered region of the spectrum) that shows coupling in the COSY spectrum with other protons, and work out the ^{1}H/^{1}H connectivities from this "end of the spin network chain". Repeat this process, starting from some other proton in another coupled spin network in the molecule, in order to arrive at other "spin networks" of protons and their vicinal ^{1}H/^{1}H connectivities (cross-peaks due to geminal couplings have already been identified and deleted from consideration in step "7").

9. Spin networks can also be readily determined from the TOCSY (120 ms) spectrum,[1] in which all the protons in the same spin network (*irrespective of whether they are directly coupled to each other or not*) appear on the same horizontal lines. The relative intensities of cross-peaks within a spin network may provide some information about the relative distances of the coupled protons within that spin network. This can be done by comparing the TOCSY spectra recorded with different mixing delays (say, 20, 40, 80, 120 ms). Delayed COSY or relayed COSY spectra may be recorded if instrumental constraints do not allow measurement of TOCSY spectra, and some information regarding long-range ^{1}H/^{1}H couplings gained from them.

---

[1] TOCSY and HOHAHA are different names applied to the same experiment.

10. Delete the direct connectivities (as determined by the COSY spectra) from the connectivities found in the TOCSY spectra to arrive at the *distant connectivities*. Thus, by combining the COSY and TOCSY data, we can obtain several groups of coupled protons. Hence, by taking into consideration the $^1H/^{13}C$ one-bond connectivities obtained from the HMQC/HSQC spectrum, we can arrive at protonated carbon connectivities. These fragments now need to be joined at their ends either to quaternary carbons or to heteroatoms (e.g., N, O, S).

11. In order to connect these fragments to complete the structure, we can join them together on the basis of long-range $^1H/^{13}C$ connectivities obtained from the HMBC spectrum (or COLOC spectrum). They should be joined together in a way that explains the observed $^2J_{CH}$, $^3J_{CH}$, and $^4J_{CH}$ interactions.

12. Alternatively, the fragments (comprising protonated carbons) thus obtained can be joined together on the basis of nOe difference measurements. NOe difference measurements of protons in different protonated carbon fragments can help us to determine the spatial proximities of protons if the fragments lie close to each other, and hence help in joining the fragments together.

13. If large sample amounts are available and when other methods have failed, the 2D INADEQUATE experiment (or preferably the corresponding inverse experiment) may be used to obtain the C–C connectivity information directly.

The following is a summary of the protocol just described.

1. Obtain a secure molecular formula, and calculate the DoU.
2. Record a 1D ^{1}H-NMR spectrum, and assign each proton a letter of an alphabet.
3. Record the HSQC (or HMQC), and label each proton at the respective chemical shifts with the corresponding letters of the alphabet as in the 1D spectrum.
4. Label the protons along the diagonal line of the COSY-45° spectrum with the same alphabet letters as in the ^{1}H-NMR spectrum. Deduce the vicinal $^1H/^1H$ connectivities from the cross-peaks in the COSY-45° spectrum by deleting all the coupling interactions from geminal protons (geminal protons are readily identified from the HMQC/HSQC or HETCOR spectrum).
5. Identify the various fragments (spin networks) obtained from the TOCSY spectra (20, 40, 60, 120 ms). Alternatively, deduce them from the COSY and HMBC spectra.
6. Using HMBC and nOe difference spectra, connect the fragments obtained using this procedure.

10.3 EXAMPLES

The following are examples of structure elucidation of small organic compounds by using various procedures described above.

10.3.1 2-Propylphenol (1)

A synthetic flavoring agent afforded the molecular formula $C_9H_{12}O$ on the basis of its HREI–MS, m/z 136.0880 (M^+, calculated 136.0888). The EI–MS showed the molecular ion peak at m/z 136 (18.8%), and the base peak at m/z 107 (100%). The DoU was calculated to be 4 by using the following formula:

$$DoU = \frac{2C+2+N-X-H}{2} = \frac{2\times9+2+0-0-12}{2} = 4$$

The base peak at m/z 107 indicated the loss of a CHO unit from the M^+. The IR spectrum exhibited absorptions for OH group at 3440 cm^{-1}, and for C=C at 1589 cm^{-1} (Kohler et al., 1993).

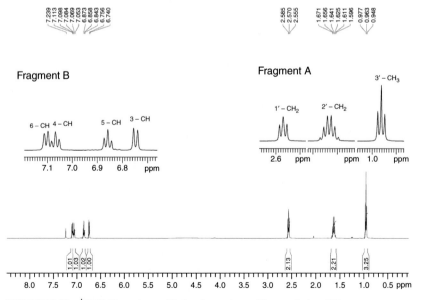

1

The ^{1}H-NMR spectrum (Fig. 10.12) of **1** showed signals in both the aliphatic (δ 0.96–2.57) and aromatic (δ 6.75–7.11) regions. In the aliphatic region, three signals appeared at δ 0.96 (3H), 1.64 (2H) and 2.57 (2H). The upfield 3H triplet

FIGURE 10.12 ^{1}H-NMR spectrum of **1** showing protons of fragments A and B.

at δ 0.96 ($J_{3',2'}$ = 7.0 Hz) indicated a methyl group adjacent to a methylene group (CH_2). The other two 2H resonances (δ 2.57 and 1.64) are consistent with the presence of two mutually coupled methylene groups as indicated by their coupling constants. The downfield methylene protons appeared as a triplet δ 2.57 ($J_{1',2'}$ = 7.5 Hz). This methylene is attached to a CH_2 group, while its downfield shift might be due to its attachment to the aromatic ring at the other end. The other methylene group appeared as a sexet at δ 1.64 ($J_{2',1'/3'}$ = 7.5 Hz). Its splitting pattern indicates that this CH_2 is attached to a CH_3 at one end, and to a CH_2 at the other end. The COSY-45° spectrum helped in the determination of connectivities between the vicinal protons. The methylene protons at δ 2.57 (H$_2$-1′) showed coupling interactions (cross-peak) with the other methylene protons at δ 1.64 (H-2′), which further showed a cross-peak with the methyl protons at δ 0.96 (H$_3$-3′) (Fig. 10.13). This information helped to deduce the aliphatic fragment A (Fig. 10.14).

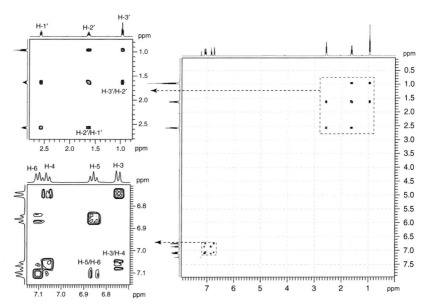

FIGURE 10.13 COSY-45° spectrum of **1** showing correlations between the vicinal protons.

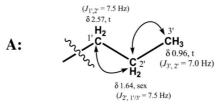

FIGURE 10.14 Fragment A. Double headed arrows indicate COSY interactions between the vicinal protons.

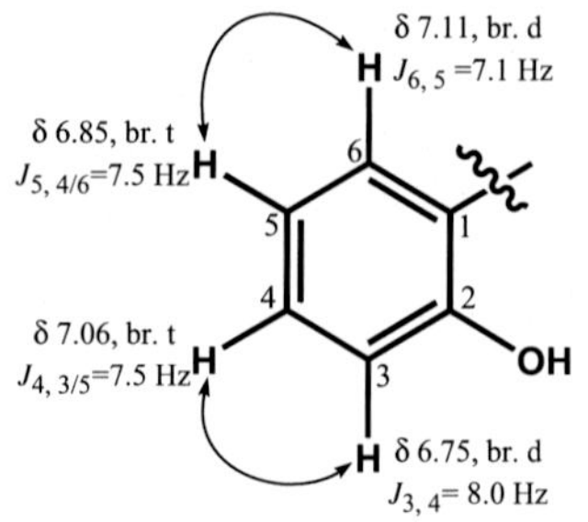

FIGURE 10.15 Fragment B. Double-headed arrows indicate COSY interactions between vicinal protons.

The aromatic region in the ^{1}H-NMR spectrum showed four ^{1}H signals at δ 6.75, 6.85, 7.06, and 7.11, indicating a disubstituted benzene ring. Their mutual couplings in the COSY-45° spectrum further indicated 1,2-disubstitution. The upfield shifts of the aromatic signals at δ 6.85 (br.t, $J_{5,4/6}$ = 7.5 Hz) and 6.75 (br.d, $J_{3,4}$ = 8.0 Hz) can be attributed to the electron-donating effect of an OH, i.e., due to their location *para* and *ortho* to the OH, respectively. The signals at δ 7.06, and 7.11 can be assigned to the protons *meta* to the OH group. The multiplicity or splitting of δ 6.75 and 7.11 signals as broad doublets indicated their coupling with the vicinal (*ortho*) proton. Their signals appeared broad due to minor couplings with protons, *meta* to them. The broad triplets at δ 6.85 and 7.06 indicated aromatic protons, which were coupled to two protons, *ortho* to them. One would expect double doublets for them due to their two coupling partners. However, it is not uncommon in aromatic systems that both *ortho* coupling partners have about the same J value ($J_{5,6} = J_{5,4}$) and the signal appeared as a triplet. In the COSY-45° spectrum, the methine proton resonating at δ 7.11 (H-6) showed coupling with δ 6.85 (H-5). Similarly the proton at δ 7.06 (H-4) showed interaction with δ 6.75 (H-3). This led to the deduction of fragment B (aromatic moiety) (Fig. 10.15). The proton signals in the aliphatic and aromatic regions are shown in the ^{1}H-NMR spectrum (Fig. 10.12).

The BB decoupled ^{13}C-NMR spectrum of **1** showed nine signals (Fig. 10.16). The DEPT-90° and DEPT-135° spectra indicated the presence of four methine (δ 130.2, 127.0, 120.6, and 115.1), two methylene (δ 31.9 and 22.8), and one methyl carbons (δ 14.0). Comparison of DEPT spectra with the BB decoupled ^{13}C-NMR spectrum indicated the presence of two quaternary carbons at δ 153.3 and 128.3. The downfield quaternary carbon (δ 153.3) is probably an aromatic carbon substituted with an OH group. The other quaternary signal at δ 128.3 may belong to the aromatic carbon, to which the aliphatic fragment A is attached.

Direct one-bond ^{1}H-^{13}C connectivities in compound **1** were inferred from the HSQC spectrum, while the concatenated DEPT–HSQC spectrum (Section 8.2.1.1) provided multiplicity information, in addition to the direct ^{1}H/^{13}C connectivities (Fig. 10.17). The methylene carbons appeared as negative peaks (red in the spectrum). The HSQC spectrum indicates that the methylene

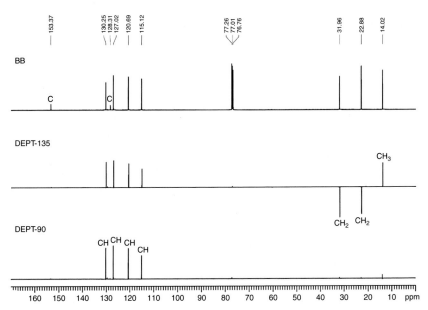

FIGURE 10.16 ^{13}C-NMR spectra of **1** indicating C, CH, CH$_2$, and CH$_3$ in compound **1**. Difference between BB decoupled and DEPT ^{13}C-NMR spectra helps in identifying quaternary carbons (δ 153.3 and 128.3 in this case).

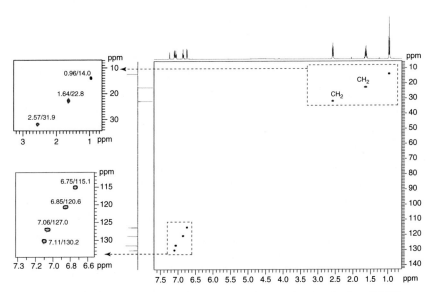

FIGURE 10.17   DEPT–HSQC spectrum of **1** indicating CH$_2$ as red peaks. The direct ^{1}H/^{13}C connectivities of compound **1** are shown on the spectrum.

TABLE 10.3 ^{1}H/^{13}C Connectivities (DEPT–HSQC) for Compound 1

Carbon Chemical Shift δ	Multiplicity	^{1}H/^{13}C Connectivity δ_H (J = Hz)
14.0	CH$_3$	0.96 (t, 7.0)
22.8	CH$_2$	1.64 (sex., 7.5)
31.9	CH$_2$	2.57 (t, 7.5)
115.1	CH	6.75 (br. d, 8.0)
120.6	CH	6.85 (br. t, 7.5)
127.0	CH	7.06 (br. t, 7.5)
130.2	CH	7.11 (br. d, 7.1)

TABLE 10.4 Long-Range ^{1}H/^{13}C Connectivities (from HMBC) for Compound 1 in CDCl$_3$

^{1}H (δ)	$^2J_{CH}$ (δ)	$^3J_{CH}$ (δ)
7.11 (H-6)	–	127.0 (C-4), 153.3 (C-2), 31.9 (C-1′)
7.06 (H-4)	–	130.2 (C-6), 153.3 (C-2)
6.85 (H-5)	–	128.3 (C-1), 115.1 (C-3)
6.75 (H-3)	153.3 (C-2)	128.3 (C-1), 120.6 (C-5)
2.57 (H-1′)	22.8 (C-2′), 128.3 (C-1)	14.0 (C-3′), 153.3 (C-2), 130.2 (C-6)
1.64, (H-2′)	14.0 (C-3′), 31.9 (C-1′)	128.3 (C-1)
0.96, (H-3′)	22.8 (C-2′)	31.9 (C-1′)

protons resonating at δ 2.57 and 1.64 have one-bond connectivities with the carbons at δ 31.9 and 22.8, respectively. The methyl signal at δ 0.96 showed coupling with the carbon signal at δ 14.0. Similarly, the aromatic methine protons appearing at δ 7.11, 7.06, 6.85, and 6.75, exhibited one-bond correlations with the carbon signals at δ 130.2, 127.0, 120.6, and 115.1, respectively (Table 10.3).

Long-range ^{1}H/^{13}C correlations were obtained from the HMBC spectrum (Table 10.4). These correlations were used to connect fragments A and B together to construct the proposed structure. The attachment of the aliphatic side chain with the aromatic ring was deduced from the ^2J correlations of the methylene protons at δ 2.57 (C-1′) with the aromatic carbon C-1 (δ 128.3, cross-peak M), and ^3J correlations with C-2 (δ 153.3, cross-peak O) and C-6 (δ 130.2, cross-peak N) (Fig. 10.18). The splitting pattern of the aromatic proton signals, and their HMBC correlations indicated the presence of an *ortho* disubstituted aromatic ring. The C-1′ methylene protons also showed correlations with the C-2′ methylene carbon (δ 22.8, cross-peak L), and with the C-3′ methyl carbon (δ 14.0, cross-peak K) (Fig. 10.18). Similarly the C-6 aromatic proton (δ 7.11) showed correlations with C-1′ (δ 31.9, cross-peak A), C-2 (δ 153.3, cross-peak C),

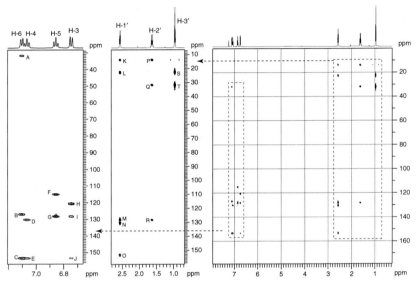

FIGURE 10.18 HMBC spectrum showing long-range ^{1}H/^{13}C correlations in compound **1**.

and C-4 (δ 127.0, cross-peak B). This further established the attachment of the aliphatic moiety, directly to the aromatic ring (C—C linkage) (Fig. 10.19).

On the basis of the above-mentioned observations, the structure of the compound **1** was deduced as 2-propylphenol (Fig. 10.20). The chemical shift assignments, along with the multiplicities are presented in Table 10.5.

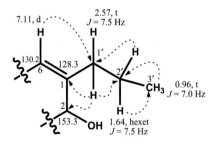

FIGURE 10.19 Key HMBC correlations which help in joining the fragments A and B together.

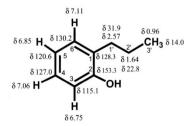

FIGURE 10.20 ^{1}H- and ^{13}C-NMR chemical shift assignments on 2-propylphenol (**1**).

TABLE 10.5 ^{1}H- (500 MHz, CDCl$_3$), and ^{13}C-NMR (125 MHz, CDCl$_3$) Chemical Shift Assignments of Compound 1

Carbon	δ_C	Multiplicity	δ_H
1	128.3	C	–
2	153.3	C	–
3	115.1	CH	6.75 (br. d, 8.0)
4	127.0	CH	7.06 (br. t, 7.5)
5	120.6	CH	6.85 (br. t, 7.5)
6	130.2	CH	7.11 (br. d, 7.1)
1′	31.9	CH$_2$	2.57 (t, 7.5)
2′	22.8	CH$_2$	1.64 (sex., 7.5)
3′	14.0	CH$_3$	0.96 (t, 7.0)

10.3.2 Mangiferin (2)

A xanthone, mangiferin, was isolated from the rhizomes of a medicinal plant *Iris germanica* Linn. of Turkish origin. The FAB–MS ($-$ve) showed the *pseudo* molecular ion peak at m/z 421 [M–H]$^-$. The HRFAB–MS exhibited the [M–H]$^-$ ion at m/z 421.3126, corresponding to the molecular formula C$_{19}$H$_{18}$O$_{11}$ (calculated 421.3235). The molecular formula C$_{19}$H$_{18}$O$_{11}$ indicated eleven degrees of unsaturation (DoU), which was calculated as follows:

$$DoU = \frac{2C + 2 + N - X - H}{2} = \frac{2 \times 19 + 2 + 0 - 0 - 18}{2} = 11$$

The IR spectrum (KBr) of compound **2** displayed absorption bands at 3325 (OH), 1656 (ketone, C=O), and 1565 (aromatic, C=C) cm^{-1}. The UV spectrum displayed λ_{max} at 253, 286 and 329 nm, λ_{min} 224, 273, 299, and 381 nm which is characteristic of a xanthone skeleton.

The ^{1}H-NMR spectrum of the compound showed signals both in the aromatic (δ 6.33–7.41), and in the aliphatic regions (δ 3.40–4.88) (Fig. 10.21). The aromatic region showed only three ^{1}H singlets, resonating at δ 6.35, 6.80, and 7.43. These signals were fully uncoupled, and apparently belonged to two different aromatic rings, as was inferred from the DoU. The upfield appearance of the aromatic proton singlets at δ 6.35 and 6.80 could be attributed to the electron donating effect of OH and/or ether groups, while the downfield proton at δ 7.43 may be due to the electron withdrawing (deshielding) effect of the ketonic carbonyl carbon (Fig. 10.22). The presence of a ketonic carbonyl group was also inferred from the IR spectrum (1656 cm^{-1}, C=O). The COSY-45° spectrum did not show any coupling between these aromatic protons, which further supported their presence on two different aromatic rings. Based on these observations, fragments A and B were initially deduced (Fig. 10.22).

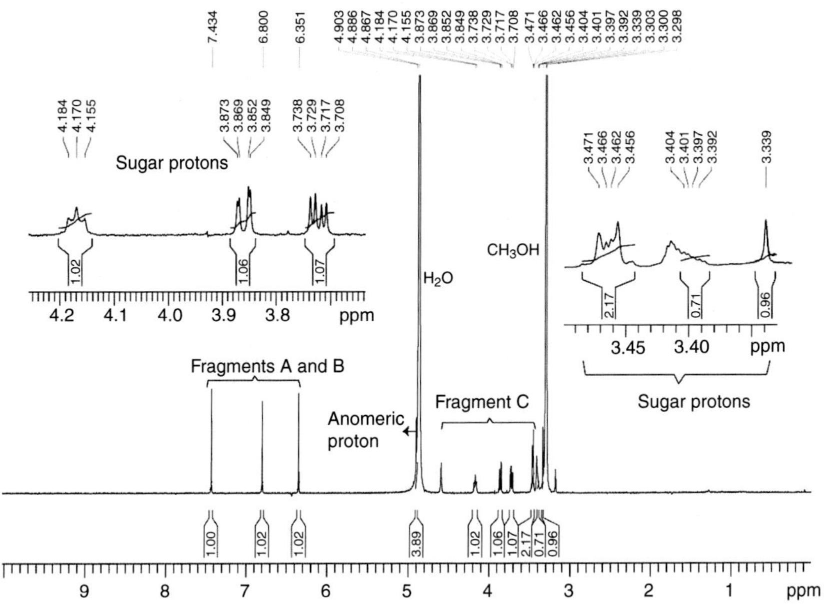

FIGURE 10.21 ^{1}H-NMR spectrum of compound **2** (in CD$_3$OD) showing signals for fragments A, B and C. The expansion of the fragment C regions, i.e., 3.33–3.47 and 3.70–4.18 shows the sugar protons, while the anomeric proton at δ 4.88 is hidden in the broad moisture peak at δ 4.88.

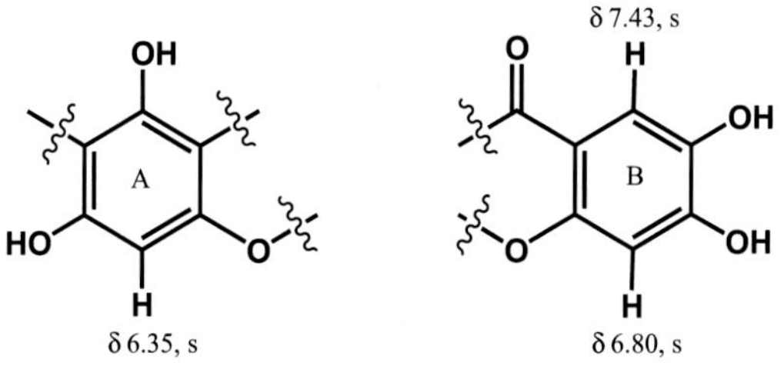

A

B

FIGURE 10.22   Fragments A and B.

The aliphatic proton signals between δ 3.40 and 4.88 indicated the presence of a sugar moiety, which was further inferred from the presence of a characteristic anomeric proton doublet at δ 4.88 ($J_{1',2'}$ = 9.8 Hz). The large coupling constant of the anomeric proton signal ($J_{1',2'}$ = 9.8 Hz) indicated that the sugar is attached through a β-C-glycosidic linkage to the aglycone part. Since the anomeric proton resonance was hidden in the broad peak of water, therefore the ^{1}H-NMR spectrum was again recorded in DMSO-d_6 (Fig. 10.23) which helped to determine the coupling constant (J) of the anomeric proton. The lesson learned is that when some expected signals are not visible due to overlapping, recording

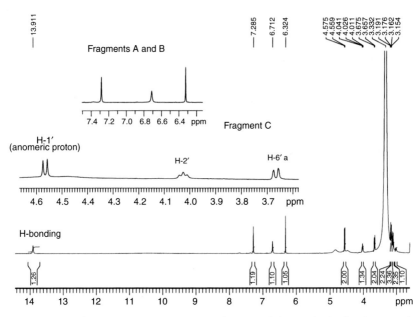

FIGURE 10.23 ¹H-NMR spectrum of compound **2** (in DMSO-d_6) clearly showing the anomeric proton signal (δ 4.57), and the downfield OH group (δ 13.9) which is hydrogen bonded to the ketonic carbonyl.

of the ¹H-NMR spectrum in a different solvent might solve the problem. This is due to solvent-induced chemical shift changes.

The remaining signals in the sugar moiety were assigned with the help of the COSY-45° spectrum (Fig. 10.24). The anomeric proton (H-1′) at δ 4.88 showed vicinal coupling with H-2′ (δ 4.15) (cross-peak A), which in turn showed coupling with H-3′ (δ 3.44) (cross-peak B). Since the H-3′ and H-4′ appeared as multiplets at the same chemical shift i.e., δ 3.44, therefore their vicinal coupling was not visible. The interaction between H-4′ (δ 3.44) and H-5′ (δ 3.40) was, however, visible as cross-peak E. The C-6′ methylene protons of the sugar moiety (δ 3.85, 3.71) showed geminal couplings with each other (cross-peak C), and a vicinal coupling with H-5′ (δ 3.40) (cross-peak D). These observations helped to deduce fragment C as the sugar moiety (Fig. 10.25).

The TOCSY spectrum of **2** (Fig. 10.26) showed one-spin system of seven protons which resonated in the region of δ 3.40 − 4.88. This further supported the presence of a sugar moiety in the molecule. The cross-peaks representing correlations between all the protons of this spin systems are shown in the TOCSY spectrum. The TOCSY spectrum showed no correlation between the three aromatic protons, further supporting the presence of two aromatic rings.

The BB decoupled ¹³C-NMR data of compound **2** showed resonances for nineteen carbons, which can be resolved through DEPT-90° and DEPT-135° experiments (Fig. 10.27). The comparison of BB decoupled with DEPT-90° and

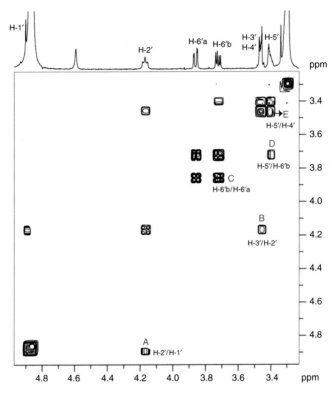

FIGURE 10.24   COSY-45° spectrum of compound **2** showing couplings between geminal and vicinial protons.

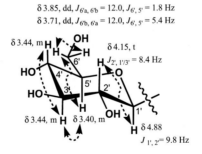

FIGURE 10.25   Fragment C.

135° ^{13}C-NMR spectra indicated the presence of 1 methylene, 8 methine and 10 quaternary carbons. The DEPT spectrum showed five methine carbon signals at δ 71.7, 72.5, 75.3, 80.1 and 82.6, and one methylene carbon at δ 62.8 which further supported the presence of a glucose moiety in molecule. These chemical shifts are characteristic of a glucose unit. The remaining three downfield methine carbons at δ 94.8, 103.4, and 108.9 indicated the presence of aromatic rings. The upfield shift

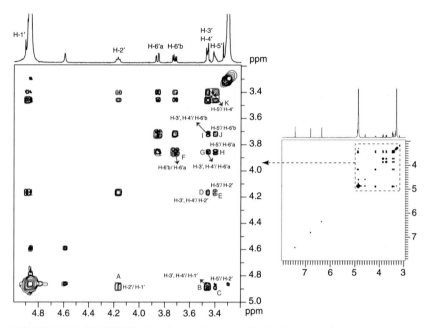

FIGURE 10.26 TOCSY (60 ms) spectrum of compound **2** showing correlations between protons of one spin system of the sugar moiety.

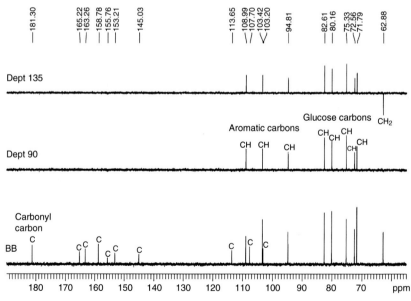

FIGURE 10.27 ^{13}C-NMR spectrum of compound **2** showing C, CH, and CH$_2$ signals. No methyl signal was observed.

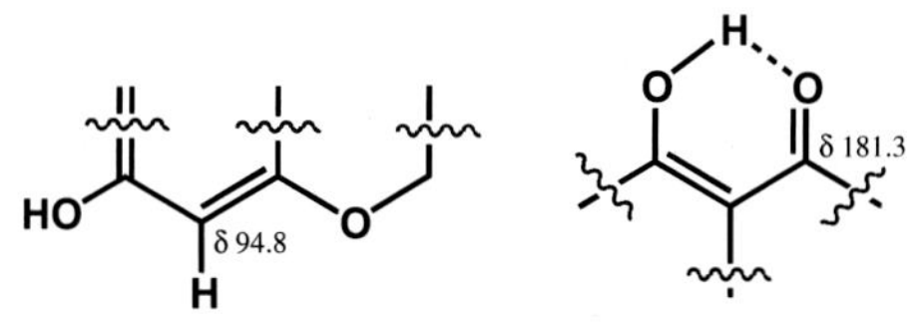

FIGURE 10.28   Fragments D and E.

of the aromatic methine carbon at δ 94.8 further indicated that the vicinal (*ortho*) carbons have oxygen substituents (fragment D). The most downfield signal in the ^{13}C-NMR spectrum at δ 181.3 was due to a ketonic carbonyl carbon. Its chemical shift indicated its chelation with a hydroxyl group (fragment E) (Fig. 10.28). The remaining quaternary carbons resonated at δ 165.2, 163.2, 158.7, 155.7, 153.2, 145.0, 113.6, 107.7, and 103.2. The downfield shift of the quaternary carbons was due to their attachment with the electronegative OH substituents.

The one-bond ^{1}H-^{13}C connectivities (Table 10.6) were inferred from the HSQC spectrum. The aromatic protons at δ 6.35, 6.80 and 7.43 exhibited their

TABLE 10.6 ^{1}H/^{13}C Connectivities (HSQC) for Compound 2

Carbon Chemical Shift δ	^{1}H/^{13}C Connectivity δ
62.8	3.71, 3.85
71.7	3.44
72.5	4.15
75.3	4.88
80.1	3.44
82.6	3.40
94.8	6.35 (s)
103.2	–
103.4	6.80 (s)
107.7	–
108.9	7.43 (s)
113.6	–
145.0	–
153.2	–
155.7	–
158.7	–
163.2	–
165.2	–
181.3	–

Fragment A Fragment B Fragment C Fragment D

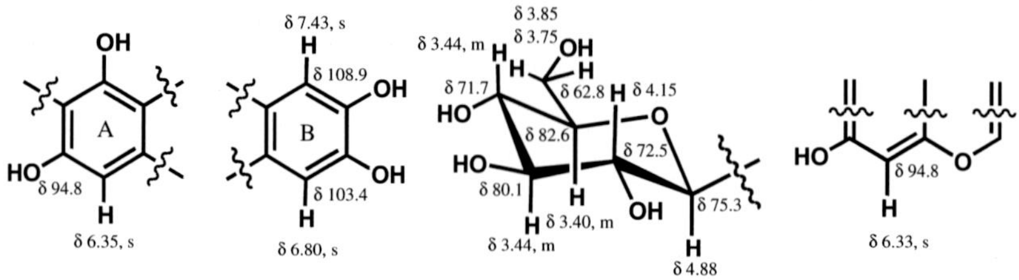

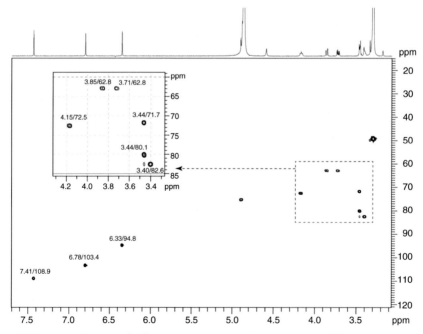

FIGURE 10.29 One-bond $^1H/^{13}C$ connectivities in fragments A, B, C, and D.

direct heteronuclear correlations with the carbons resonating at δ 94.8, 103.4 and 108.9, respectively (Fig. 10.29). The anomeric H-1′ proton doublet (δ 4.88), showed connectivity with the carbon at δ 75.3. The upfield shift of the anomeric carbon further supported a *C*-glycosidic linkage. The remaining methine protons of the sugar moiety, resonating at δ 4.15 (1H), 3.44 (2H) and 3.40 (1H), showed direct correlations with the respective carbons at δ 72.5, 80.1 and 71.7, and 82.6. The presence of a methylene carbon is supported by the correlations of the two protons at δ 3.85 and 3.71 with the same carbon which appeared at δ 62.8, as shown in Fig. 10.30.

The cross-peaks in the HMBC spectrum can be used to join different fragments together and to construct the final structure of compound **2** (Fig. 10.31).

FIGURE 10.30 The direct $^1H/^{13}C$ connectivities of compound **2** in on the HSQC spectrum.

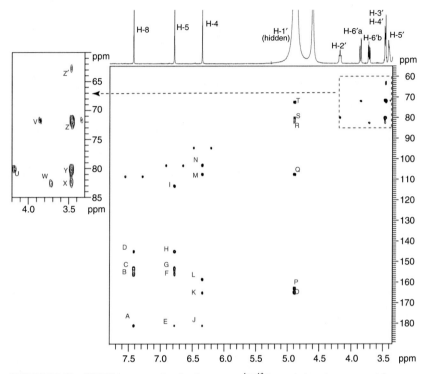

FIGURE 10.31 HMBC spectrum showing long-range ^{1}H/^{13}C correlations in compound **2**.

The proton at δ 6.35 (fragment A) showed ^2J correlations with the carbons that resonated at δ 165.2 (cross-peak K) and 158.7 (cross-peak L) (fragment D). In this way, fragment A can be joined to fragment D. The same proton (δ 6.35) showed ^3J correlations with the carbon atoms at δ 107.7 (cross-peak M) and 103.2 (cross-peak N) and ^4J correlation with the carbonyl carbon at δ 181.3 (cross-peak J) (Table 10.7). Since the carbonyl group is chelated with an OH group, therefore the chemical shift of both the carbonyl and the hydroxyl-bearing carbons appeared in the downfield region of the ^{13}C-NMR spectrum, i.e., δ 181.3 and 163.2, respectively, as compared to the resonances of such carbon atoms in the unchelated compounds. In this way fragments A, D, and E could be joined together to construct a larger fragment F (Fig. 10.32).

The downfield aromatic proton resonating at δ 7.43 (ring B) showed ^2J and ^3J correlations with the quaternary carbons at δ 153.2 (^2J) (cross-peak C), 155.7 (^3J) (cross-peak B), and 145.0 (^3J) (cross-peak D) in the HMBC spectrum. Their chemical shifts indicated that these are the oxygen-bearing quaternary carbons. The proton at δ 6.80 also showed ^2J correlations with same carbons at δ 155.7 (cross-peak F), and 145.0 (cross-peak H), which further supported that these protons are located on the same ring. Both these protons (δ 7.43 and 6.80) also showed correlations with the carbonyl carbon (δ 181.3) (cross-peaks A and E, respectively). This helped to combine fragments B and F together (Fig. 10.33).

TABLE 10.7 Long-Range ¹H/¹³C Connectivities (From HMBC) for Compound 2 in CD₃OD

¹H (δ)	²J_{CH} (δ)	³J_{CH} (δ)	⁴J_{CH} (δ)
6.35 (H-4)	165.2 (C-3) 158.7 (C-4a)	103.2 (C-9a)	181.3 (C-9)
6.80 (H-5)	155.7 (C-5a) 145.0 (C-6)	113.6 (C-8a)	181.3 (C-9)
7.43 (H-8)	153.2 (C-7)	155.7 (C-5a) 181.3 (C-9) 145.0 (C-6)	–
4.88 (H-1′)	72.5 (C-2′) 107.7 (C-2)	80.1 (C-3′) 82.6 (C-5′) 163.2 (C-1) 165.2 (C-3)	–
3.71, 3.85 (H-6′)	82.6 (C-5′)	71.7 (C-4′)	–

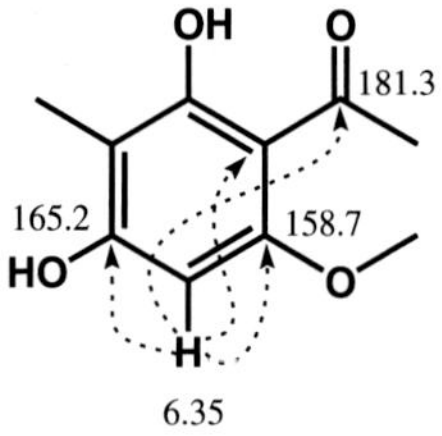

FIGURE 10.32 Key HMBC correlations, which helped in joining fragments A, D and E into a new fragment F.

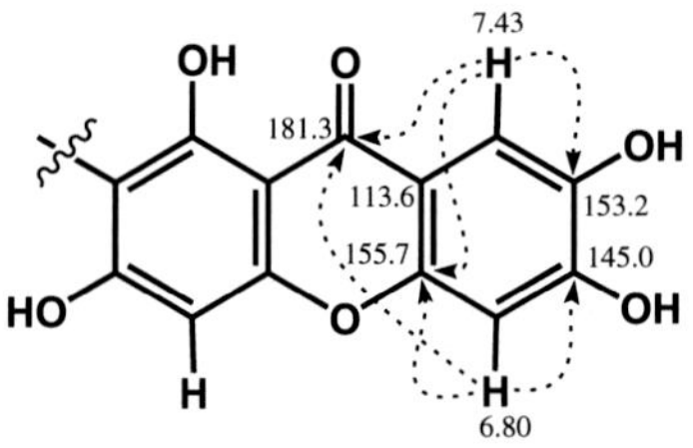

FIGURE 10.33 Key HMBC correlations between fragments B and F.

Finally we wanted to find out the point of attachment of the glucose moiety with the aromatic (xanthone) moiety. The anomeric proton at δ 4.88 showed ²J correlation with C-2 (δ 107.7) (cross-peak Q), and ³J correlations with C-1 (δ 163.2) (cross-peak P) and C-3 (δ 165.2) (cross-peak O). This showed that the glucose moiety is attached to C-2 of xanthone (Fig. 10.34). The above data helped to deduce the structure **2** for the compound (mangiferin) (Wu et al., 2010). The complete chemical shift assignments are presented in Table 10.8.

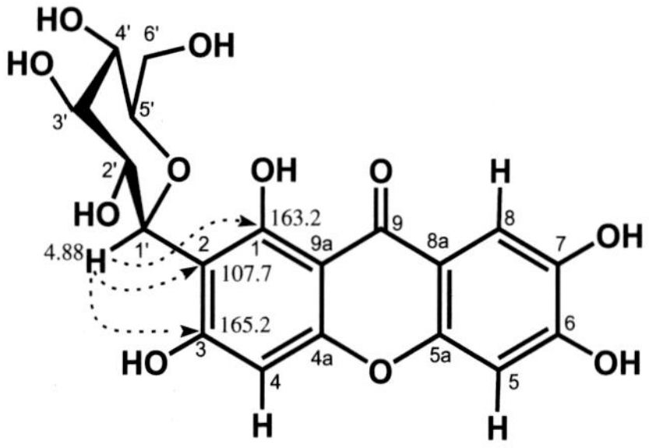

$C_{19}H_{18}O_{11}$
$m/z = 422$

FIGURE 10.34 Key HMBC correlations of the anomeric proton with the aglycone (xanthone) indicates their connectivity.

TABLE 10.8 ^{1}H- (500 MHz, CD$_3$OD), and ^{13}C-NMR (125 MHz, CD$_3$OD) Chemical Shift Assignments of Compound 2

Carbon	δ_C	Multiplicity	δ_H
1	163.2	C	–
2	107.7	C	–
3	165.2	C	–
4	94.8	CH	6.35 (s)
4a	158.7	C	–
5	103.4	CH	6.80 (s)
5a	155.7	C	–
6	145.0	C	–
7	153.2	C	–
8	108.9	CH	7.43 (s)
8a	113.6	C	–
9	181.3	C	–
9a	103.2	C	–
1'	75.3	CH	4.88
2'	72.5	CH	4.15
3'	80.1	CH	3.44
4'	71.7	CH	3.44
5'	82.6	CH	3.40
6'	62.8	CH$_2$	3.71, 3.85

10.3.3 (R)-(−)-Carvone (3)

A substance **3** was isolated from the essential oil of *Mentha spicata* Linn. (spearmint). HREI–MS of compound **3** showed the molecular ion at *m/z* 150.1046 (calcd, 150.1045), corresponding to the molecular formula $C_{10}H_{14}O$ which established four degrees of unsaturation (DoU). The UV absorbance at λ_{max} 231 and 243 nm suggested the presence of an α,β-unsaturated ketone moiety in compound **3**. The IR absorption at 1742 cm^{-1} showed the presence of an enone carbonyl group. The presence of an α,β-unsaturated ketone was inferred from the UV spectrum (231 and 243 nm).

The ^{1}H-NMR spectrum showed signals resonated between δ 1.72 and 6.72 (Fig. 10.35). The most downfield proton resonating at δ 6.72 (dd, $J_{vic(3,4a)}$ = 1.5 Hz, $J_{vic(3,4b)}$ = 3.5 Hz) was assigned to the γ proton of the conjugated olefin. The splitting pattern indicated that this proton is coupled to two vicinal protons. The COSY-45° spectrum also supported the vicinal couplings between the protons at δ 6.72 and 2.42 (cross-peak A), and allylic coupling of δ 6.72 with the methyl signal at δ 1.75 (cross-peak B) (Fig. 10.36). These observations led to fragment A (Fig. 10.37).

Other downfield ^{1}H-NMR signals appeared at δ 4.77 (1H, s) and 4.72 (1H, s) indicating the presence of methylene olefinic protons. Since these protons appeared as only slightly broad singlets, therefore the vicinal carbon might be a quaternary carbon. In the COSY-45° spectrum, geminal coupling between these two protons was also observed (cross-peak C), along with their allylic coupling

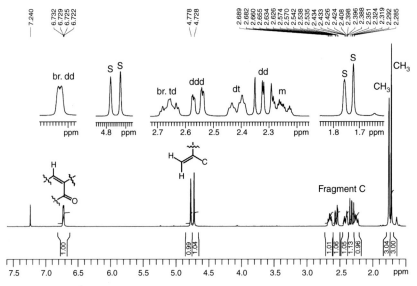

FIGURE 10.35 ^{1}H-NMR spectrum of compound **3** (in $CDCl_3$) showing proton signals of fragments A, B and C.

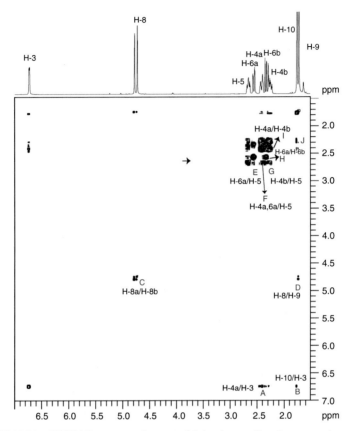

FIGURE 10.36   COSY-45° spectrum of compound **3** showing couplings between various geminal and vicinal protons.

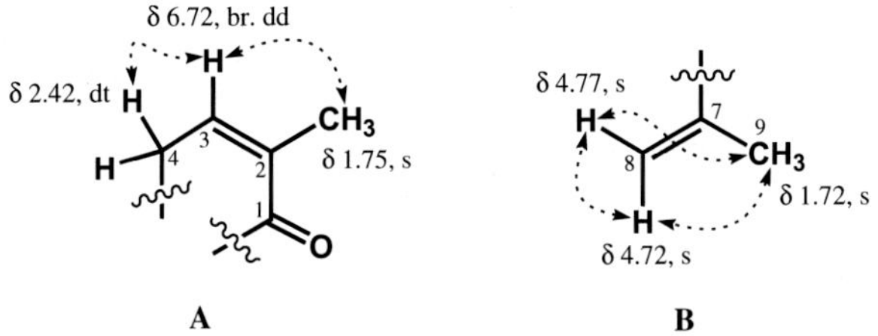

δ 6.72, br. dd

δ 2.42, dt

δ 1.75, s

A

δ 4.77, s

δ 1.72, s

δ 4.72, s

B

FIGURE 10.37   Vicinal and allylic couplings of fragments A and B in COSY-45° spectrum.

with the methyl protons at δ 1.72 (cross-peak D). Based on these observations, fragment B was constructed (Fig. 10.37).

Five upfield proton signals appeared at δ 2.65 (br. td, $J_{5,6b/4a}$ = 14.5 Hz, J = 3.5 Hz), 2.54 (ddd, $J_{gem(6a,6b)}$ = 16.1 Hz, $J_{vic(6a,5)}$ = 4.0 Hz, J_W = 2.0 Hz), 2.42 (dt, $J_{gem(4a,4b)}$ = 18.0 Hz, $J_{vic(4a,5)}$ = 5.0 Hz, $J_{vic(4a,3)}$ = 1.5 Hz), 2.31

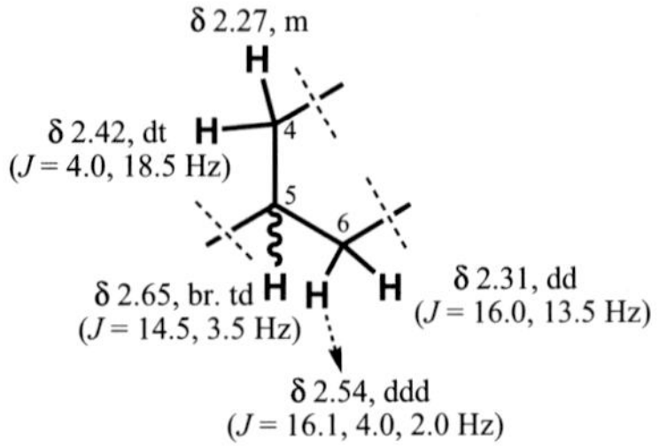

δ 2.27, m

δ 2.42, dt (J = 4.0, 18.5 Hz)

δ 2.65, br. td (J = 14.5, 3.5 Hz)

δ 2.31, dd (J = 16.0, 13.5 Hz)

δ 2.54, ddd (J = 16.1, 4.0, 2.0 Hz)

C

FIGURE 10.38   Fragment C.

(dd, $J_{gem(6b,6a)}$ = 16.0 Hz, $J_{vic(6b,5)}$ = 13.5 Hz), and 2.27 (m). All these signals had 1H integration, and interesting splitting patterns. This indicated the presence of an asymmetric center in the molecule. The 1H signal at δ 2.65 appeared as a broad doublet of triplet ($J_{5,6b/4a}$= 14.5 Hz, $J_{5,6}$ = 3.5 Hz), indicating its couplings with four protons on its neighboring carbons. The larger coupling constant $J_{5,6b/4a}$ = 14.5 Hz showed the *trans* disposition between these protons, while the smaller coupling constant $J_{5,6}$ = 3.5 indicated a dihedral angle θ ~ 60°, according to Karplus equation. This further supported that these protons are part of a ring system. The signals at δ 2.54, 2.42, 2.31, and 2.27 were assigned to the methylene protons vicinal to the methine proton at δ 2.65, on the basis of their coupling constants and cross-peaks in the COSY-45° spectrum (cross-peaks E, F, and G). Based on these observations, fragment C was constructed (Fig. 10.38).

In the ^{1}H-NMR spectrum, we also observed two 3H signals, resonating at δ 1.75 and 1.74, indicating the presence of two methyl groups. The downfield shifts of these methyl protons suggested their attachment on an olefinic moiety.

The ^{13}C-NMR spectrum of compound **3** further helped to deduce the structure. A comparison of the DEPT-90° with the DEPT-135° spectra showed signals for two methine carbons, which resonated at δ 144.6 and 42.4 (Fig. 10.39). The downfield methine signal at δ 144.6 was assigned to the γ carbon of the α, β-unsaturated ketone moiety (fragment A) (Fig. 10.40). The negative phase signals in the DEPT-135° spectrum belong to the three methylene carbons (δ 110.4, 43.1, and 31.1). The downfield shift of the methylene carbon at δ 110.4 was due to its olefinic nature $\left(H_2C=C\!\!\begin{smallmatrix}\diagup\\\diagdown\end{smallmatrix}\right)$ (fragment B). The remaining signals in the DEPT-135° spectrum indicated the presence of two methyl groups (δ 20.5 and 15.6). A comparison of the DEPT spectra with the BB decoupled ^{13}C-NMR spectrum indicated three quaternary carbons (δ 199.8, 146.6, and 135.3). The most downfield quaternary carbon at δ 199.8 can be assigned to the α, β-unsaturated ketonic carbonyl carbon (fragment A).

The DEPT–HSQC spectrum of **3** was very useful, not only in determining the one-bond ^{1}H/^{13}C connectivities, but also in assigning the multiplicity to various carbon signals. The three methylene signals (δ 110.4, 43.1, and 31.1) appeared

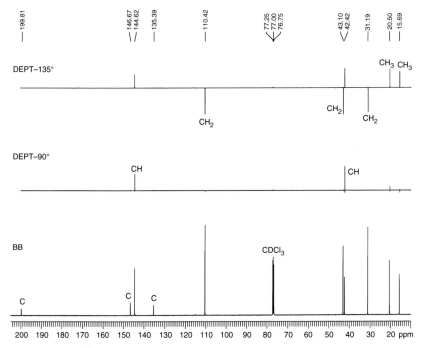

FIGURE 10.39 ^{13}C-NMR spectra of compound **3**, showing C, CH, and CH$_2$ signals.

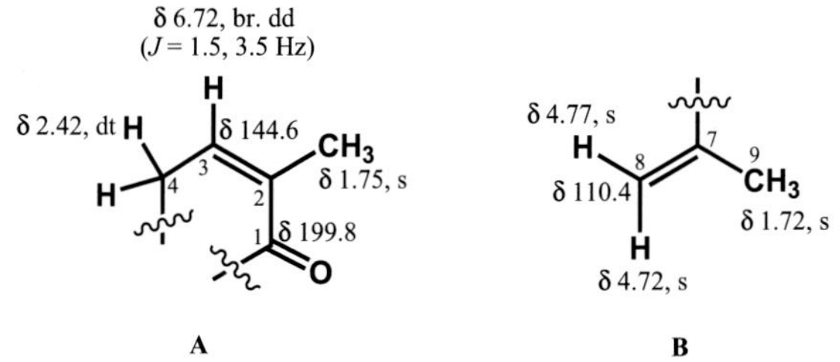

δ 6.72, br. dd
(*J* = 1.5, 3.5 Hz)

δ 2.42, dt **H**

H

δ 144.6

CH$_3$
δ 1.75, s

δ 199.8

O

A

δ 4.77, s

H
δ 110.4

CH$_3$
δ 1.72, s

H
δ 4.72, s

B

FIGURE 10.40 Fragments A and B with ^{13}C-NMR chemical shifts.

as negative peaks (red color) in the DEPT–HSQC spectrum (Fig. 10.41). The most downfield methine proton signal at δ 6.72 showed connectivity with the carbon atom at δ 144.6, while the other methine protons at δ 2.64 showed connectivity with the carbon atom at δ 42.4. The remaining two methyl groups at δ 1.75 and 1.72 were found to be attached to the carbons at δ 15.6 and 20.5, respectively. With these observations, we could assign the carbon chemical shift to fragments A–C (Fig. 10.42). The HSQC data of compound **3** is presented in Table 10.9.

Concatenated HSQC–TOCSY experiments (Section 8.2.2.1) were used to deduce the ^{1}H-/^{13}C based ^{1}H-/^{1}H couplings in a spin system (Fig. 10.43). The

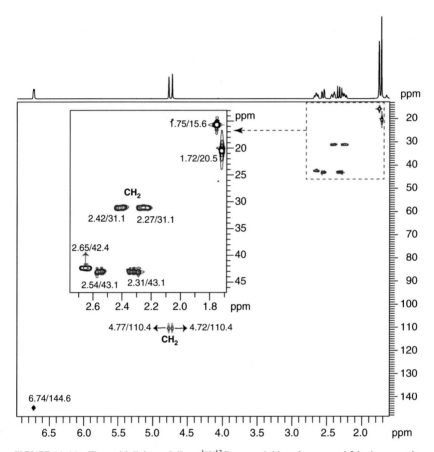

FIGURE 10.41 The multiplicity and direct ¹H/¹³C connectivities of compound **3** is shown on the DEPT–HSQC spectrum.

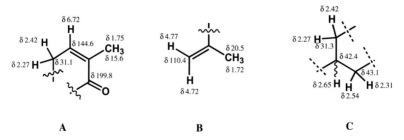

FIGURE 10.42 One-bond ¹H/¹³C connectivities in fragments A, B, and C.

TABLE 10.9 ^{1}H/^{13}C Connectivities and Multiplicities (HSQC) for Compound 3

Carbon Chemical Shift δ	Multiplicity	^{1}H/^{13}C Connectivity δ
15.6	CH$_3$	1.75, s
20	CH$_3$	1.72
31.1	CH$_2$	2.42, 2.27 m
42.4	CH	2.65
43.1	CH$_2$	2.31, 2.54
110.4	CH$_2$	4.72, 4.77 s
135.3	C	–
144.6	CH	6.72
146.6	C	–
199.8	C	–

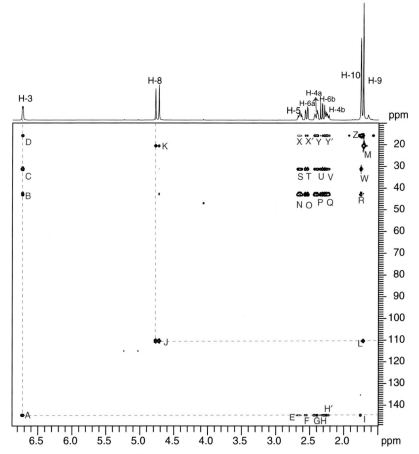

FIGURE 10.43 HSQC–TOCSY spectrum of compound **3** indicating one spin system F_1 plane and carbon connectivity in F_2 plane.

proton at δ 6.72 showed correlations with the carbons at δ 15.6 (cross-peak D), 31.1 (cross-peak C), and 42.4 (cross-peak B). The same proton at δ 6.72 showed TOCSY correlations with the protons at δ 2.64 (cross-peak E), 2.54 (cross-peak F), 2.42 (cross-peak G), 2.31 (cross-peak H), 2.27 (cross-peak H′), and 1.75 (cross-peak I). This data showed that all these protons are part of the same spin system. Similarly, the olefinic protons at δ 4.77 and 4.72 exhibited correlations with the carbon at δ 20.5 (cross-peak K), and TOCSY correlations with the proton at δ 1.74 (cross-peak L). These observations showed that there are two distinct spin systems in the molecule.

The HMBC experiments were used to join different structural fragments together and thus deduce the carbon skeleton through proton–carbon multiple bond correlations (Fig. 10.44). The most downfield olefinic proton, resonating at δ 6.72, showed ^2J correlations with the carbon signal at δ 31.1 (cross-peak B′), and ^3J correlation with the carbon signals at δ 15.6 (cross-peak C), 42.4

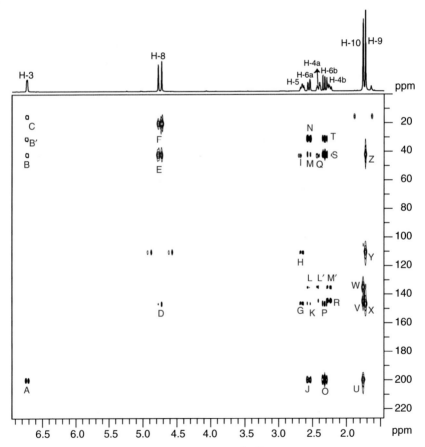

FIGURE 10.44 HMBC spectrum showing long-range ^{1}H/^{13}C correlations of compound **3**.

TABLE 10.10 Long-Range $^1H/^{13}C$ Correlation (From HMBC) for Compound 3 in CDCl$_3$

^1H (δ)	$^2J_{CH}$ (δ)	$^3J_{CH}$ (δ)
6.72 (H-3)	135.3 (C-2)	15.6 (C-10)
		199.8 (C-1)
		42.4 (C-5)
2.27 (H-4a)	144.6 (C-3)	135.3 (C-2)
2.42 (H-4b)	144.6 (C-3)	135.3 (C-2)
	42.4 (C-5)	
2.65 (H-5)	31.1 (C-4)	110.4 (C-8)
	43.1 (C-6)	
2.54(H-6a)	42.4 (C-5)	31.1 (C-4)
		146.6 (C-7)
2.31 (H-6b)	199.8 (C-1)	31.1 (C-4)
	42.4 (C-5)	146.6 (C-7)
4.72 (H-8a)	146.6 (C-7)	42.4 (C-5)
		20 (C-9)
4.77 (H-8b)	146.6 (C-7)	42.4 (C-5)
		20 (C-9)
1.72 (H-9)	146.6 (C-7)	42.4 (C-5)
		110.4 (C-8)
1.75 (H-10)	135.3 (C-2)	199.8 (C-1)
		144.6 (C-3)

(cross-peak B), and 199.8 (cross-peak A) (Table 10.10). These interactions helped to join fragments A and C together to form a new fragment D (Fig. 10.45).

The structure of fragment D is further supported by the long-range ^2J, and ^3J correlations of the methyl protons at δ 1.75 with the carbon atoms at δ 135.3 (C-2) (cross-peak W), 199.8 (C-1) (cross-peak U), and 144.6 (C-3) (cross-peak V), respectively. Similarly, H-4 (δ 2.42, and 2.27) showed correlations with C-3 (δ 144.6) (cross-peak R), and C-2 (δ 135.5) (cross-peak M′), while H-6 (δ 2.54,

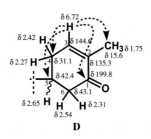

FIGURE 10.45   Key HMBC correlations that help to join fragments A and C into a new fragment D.

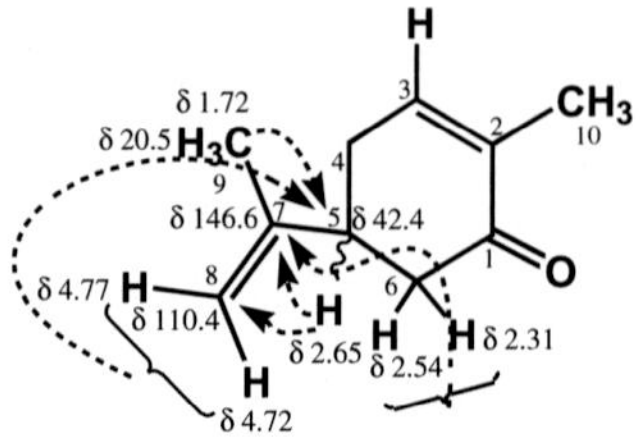

FIGURE 10.46 Key HMBC correlations between fragments B and D.

and 2.31) showed ^2J correlations with C-1 (δ 199.8) (cross-peaks J and O), and C-5 (δ 42.4) (cross-peaks M and Q). Finally, fragment D was combined with the fragment B with the help of HMBC correlations (Fig. 10.46). The methyl protons (δ 1.72) and methylene protons (δ 4.77 and 4.72) showed ^3J correlations with C-5 (δ 42.4) (cross-peaks Z and E, respectively). Similarly H-6 showed ^3J correlations with the quaternary carbon (δ 146.6) of fragment B (cross-peak R). Thus the final structure of the compound was deduced as carvone (**3**). Complete assignments of compound **3** are presented in Table 10.11.

The stereochemistry at the C-5 position was deduced as R by comparing the specific rotation values, i.e., observed $[\alpha]_D^{25}$ value (-69.3) *versus* the reported value (-61).

This close comparison indicated that the compound in hand has the same stereochemistry as that reported for $(R)(-)$-carvone. The NOESY spectrum further supported the through space (spatial) couplings of H-5 (δ 2.65) with H-4b (δ 2.27), H-6b (δ 2.31), H-8 (δ 4.72), and H-9 (δ 1.72) (Figs 10.47 and 10.48). Thus the final structure of compound **3** was deduced as $R(-)$-carvone (Elmastas, et al., 2006).

TABLE 10.11 ^{1}H- (500 MHz, CDCl$_3$), and ^{13}C-NMR (125 MHz, CDCl$_3$) Chemical Shift Assignments of Compound 3

Carbon	δ_C	Multiplicity	δ_H
1	199.8	C	–
2	135.3	C	–
3	144.6	CH	6.72 (s)
4	31.1	CH$_2$	2.27 (m), 2.42
5	42.4	CH	2.65
6	43.1	CH$_2$	2.31, 2.54
7	146.6	C	–
8	110.4	CH$_2$	4.72, 4.77 (s)
9	20	CH$_3$	1.72
10	15.6	CH$_3$	1.75 (s)

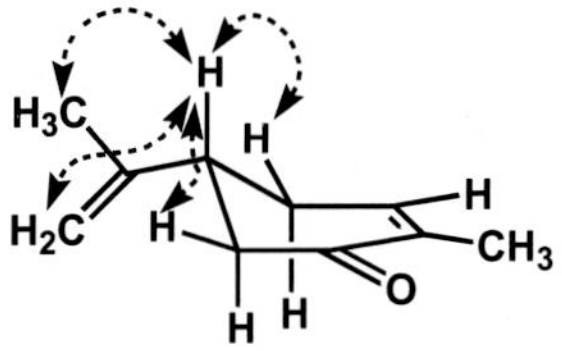

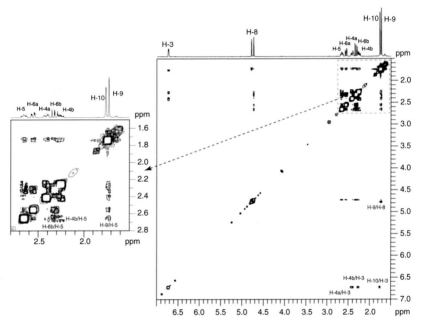

FIGURE 10.47 Through-space coupling (NOESY) interactions of compound **3**.

FIGURE 10.48 NOESY spectrum of compound **3** showing through space nOe couplings.

10.3.4 Eriodictyol (4)

Eriodictyol was isolated from a medicinal plant *Lyonia ovalifolia* Wall. Drude. The HREI–MS of compound **4** showed the molecular ion peak (M$^+$) at *m/z* 288.0419 (calculated for $C_{15}H_{12}O_6$, 288.0426). This indicated ten degrees of unsaturation (DoU). The UV spectrum displayed absorptions at 332 and 288 nm. A comparison of the UV absorptions with the literature indicated the presence of a flavanone skeleton. The IR spectrum showed absorption bands at 3423, 1655, and 1630–1598 cm^{-1} for hydroxyl, conjugated carbonyl and C=C groups, respectively.

The ^{1}H-NMR spectrum of compound **4** showed signals both in the aromatic (δ 6.90–5.86), and in the aliphatic regions (δ 5.28–2.67) (Fig. 10.49). The two upfield ^{1}H-NMR signals appearing as doublets at δ 5.88 ($J_{8,6} \approx$ 2.0 Hz), and 5.86 ($J_{6,8} \approx$ 2.0 Hz) can be assigned to protons of an aromatic ring based on their

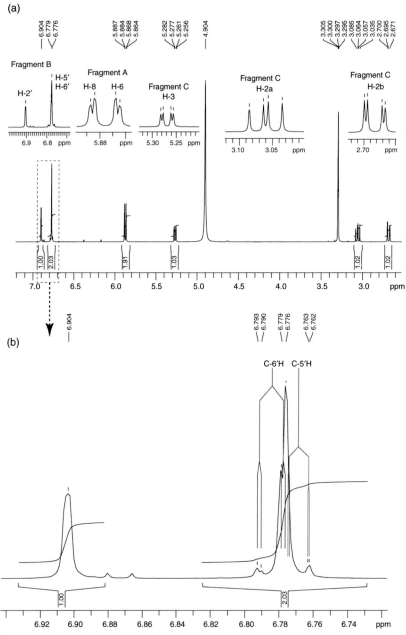

FIGURE 10.49   (a) ¹H-NMR spectrum (CD₃OD) of **4** showing protons of fragments A, B and C. (b) Expansion of the dotted parts showing a broad signlet at δ 6.90 and two overlapping signal, i.e., a double doublet at δ 6.78 showing *ortho* coupling with C-5′H ($J_{6',5'}$ = 7.0 Hz) and *meta* coupling with C-2′H ($J_{6',2'}$ = 1.5 Hz). The other close double doublet resonated at δ 6.77 with *ortho* and *para* couplings ($J_{5',6'}$ = 7.0 Hz, $J_{5',2'}$ = 0.5 Hz).

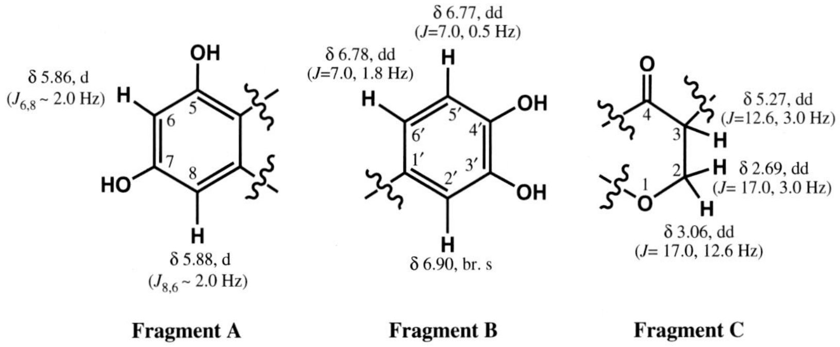

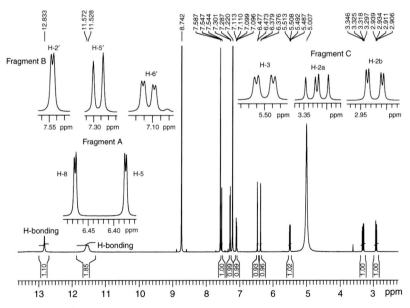

δ 6.77, dd
(J=7.0, 0.5 Hz)

δ 5.86, d
(J₆,₈ ~ 2.0 Hz)

Fragment A   **Fragment B**   **Fragment C**

FIGURE 10.50   Fragments A, B and C.

coupling constants, which showed a *meta* coupling between them. The upfield shifts of these signals indicated the presence of hydroxyl or ether groups at their *ortho* positions. From this data we could construct fragment A (Fig. 10.50).

The aromatic region of the ^{1}H-NMR spectrum also showed three downfield signals. This includes a broad singlet at δ 6.90 (br. s), and two overlapping signals at δ 6.78 (dd, $J_{6',5'}$ = 7.0 Hz, $J_{6',2'}$ = 1.8 Hz), and 6.77 (dd, $J_{5',6'}$ = 7.0 Hz, $J_{5',2'}$ = 0.5 Hz). These signals showed the presence of a tri-substituted aromatic ring. Since the splitting of these protons was not clear, therefore we recorded the ^{1}H-NMR spectrum in C_5D_5N (Fig. 10.51) to obtain a better resolution spectrum. The ^{1}H-NMR spectrum in pyridine-d_6 clearly showed an ABX spin system. The

FIGURE 10.51 ^{1}H-NMR spectrum (C_5D_5N) of **4** showing protons of fragments A, B, and C.

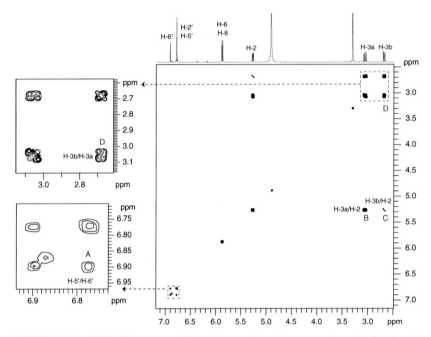

FIGURE 10.52   COSY-45° spectrum of **4** showing correlations between various vicinal and geminal protons.

protons of the ABX spin system appeared at δ 7.54 (d, $J_{2',6'}$ = 1.8 Hz), 7.30 (d, $J_{5',6'}$ = 8.4 Hz), and 7.11 (dd, $J_{6',5'}$ = 8.4 Hz, $J_{6',2'}$ = 1.8 Hz) (the downfield shifts of the protons were due to the effect of the solvent C_5D_5N).

The COSY-45° spectrum indicated *meta* couplings between the protons at δ 6.90 and 6.78 (cross-peak A) (Fig. 10.52). These observations led us to construct fragment B (Fig. 10.50).

The aliphatic region showed the presence of three double doublets at δ 5.27, 3.06 and 2.69. The downfield shift of the signal at δ 5.27 ($J_{2,3a}$ = 12.6 Hz, $J_{2,3b}$ = 3.0 Hz) indicated the presence of the electron withdrawing carbonyl group. The coupling constant (J) of the remaining two doublets at δ 3.06 (J_{gem} ≈ 17.0 Hz, J_{vic} = 12.6 Hz), and 2.69 (J_{gem} ≈ 17.0 Hz, J_{vic} = 3.0 Hz) showed their geminal disposition. Both methylene protons were vicinally coupled with the methine proton at δ 5.27. Further the COSY-45° spectrum indicated the vicinal couplings between protons at δ 5.27 and 3.06 (cross-peak B), and between the protons at δ 5.27 and 2.69 (cross-peak C). Cross-peak D is due to the geminal coupling between methylene δ 3.06 and 2.69. These observations led us to construct fragment C (Fig. 10.50).

The BB decoupled ^{13}C-NMR spectrum of **4** showed a total of 15 resonances (Fig. 10.53). A comparison between DEPT-90° with DEPT-135° spectra indicated the presence of one methylene carbon (negative phase signal in DEPT-135°),

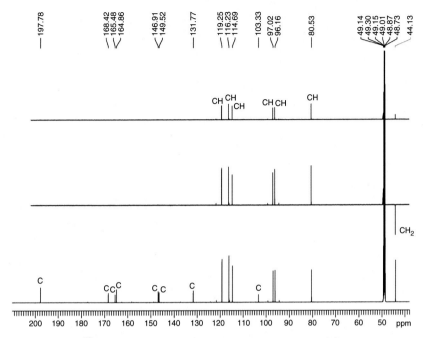

FIGURE 10.53 ^{13}C-NMR spectra indicating C, CH, and CH$_2$, in compound **4**.

and six methine carbons. No methyl carbon signal was detected. The remaining signals in the BB decoupled spectrum indicated eight quaternary carbons. The two upfield methine carbon resonances at δ 96.1 and 97.0 were due to the aromatic carbons with electron donating (OH) or ether groups in their vicinity (*ortho* positions). Therefore they can be assigned to fragment A.

The signal at δ 197.7 was due to a carbonyl carbon. Its downfield shift might be due to chelation of the carbonyl with the OH in fragment A. With this observation fragments A and C were combined to make a larger fragment D (Fig. 10.54).

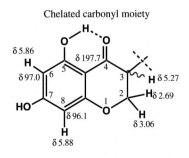

Fragment D

FIGURE 10.54 Fragment D was constructed by combining fragments A and C together.

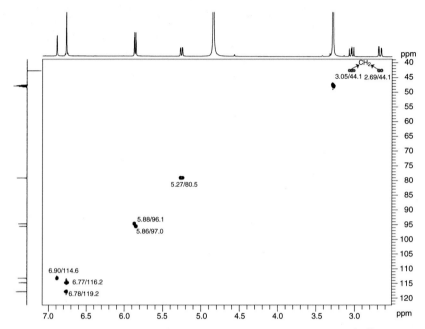

FIGURE 10.55 DEPT–HSQC spectrum indicating CH_2 as red peaks. The direct $^1H/^{13}C$ connectivities of compound **4** are shown on the spectrum.

As discussed in Section 8.2.1.1, the concatenated NMR technique DEPT–HSQC not only provides direct one-bond $^1H/^{13}C$ connectivities, but due to the addition of the DEPT pulse sequence in the experiment, the methylene carbons appear as negative peaks, thereby helping in the identification of the CH_2 groups easily.

The DEPT–HSQC spectrum of compound **4** showed two negative signals (red color) at δ 3.06, and 2.69. Both these protons showed connectivity with the methylene carbon at δ 44.1. The C-3 methine proton at δ 5.27 showed connectivity with the carbon that resonated at δ 80.5. The downfield shift of this aliphatic carbon was due to the electron withdrawing effect of the ketonic carbonyl group at C-4. The remaining aromatic methine protons at δ 6.90, 6.78, 6.77, 5.88, and 5.86 showed connectivities with the carbons at δ 114.6, 119.9, 116.2, 96.1, and 97.0, respectively, as shown in Fig. 10.55 and Table 10.12. This helped in the complete $^1H/^{13}C$ assignment of fragments B and D (Fig. 10.56).

The concatenated HSQC–TOCSY spectrum is similar to the TOCSY spectrum, but additionally it gives dispersion in ^{13}C dimension as well. Often the overlap in ^1H-NMR spectrum can be easily resolved in the HSQC–TOCSY spectrum since there is much more dispersion in the F_1 dimension. In the HSQC–TOCSY spectrum, the total correlation allows one to unravel the entire spin system systematically.

TABLE 10.12 DEPT–HSQC Correlations in Compound 4

Proton Chemical Shift δ	Multiplicity	Carbon Chemical Shift δ
6.90	CH	114.6
6.78	CH	119.2
6.77	CH	116.2
5.88	CH	96.1
5.86	CH	97.0
5.27	CH	80.5
3.06, 2.69	CH$_3$	44.1

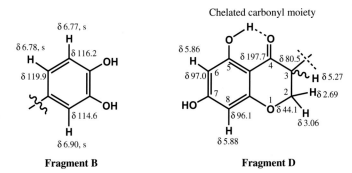

Fragment B   **Fragment D**

FIGURE 10.56   Fragments B and D with ^{13}C chemical shifts.

Here in the HSQC–TOCSY spectrum, the connectivity of H-3 (δ 5.27/80.5, cross-peak H) appeared with C-2 (δ 44.1) in the ^{13}C- dimension (cross-peak I) (Fig. 10.57). In the ^{1}H- dimension it showed couplings with the C-2 protons at δ 3.06, and 2.69 as indicated by cross-peaks J, and L. The cross-peaks A-C and D-F showed connectivities only in the ^{13}C dimension since there is no spin system existing in fragment B.

The HMBC spectrum of **4** helped to combine fragments B and D together based on long-range ^{1}H/^{13}C correlations. H-2 (5.27) showed ^2J correlations with 131.7 (C-1′, cross-peak M), ^3J correlations with δ 119.9 (C-2′, cross-peak N) and 114.6 (C-6′, cross-peak O) (Figs 10.58 and 10.59). This indicated the attachment of fragment B with C-2 of the fragment D. This was further deduced by ^3J correlations of H-2′ (δ 6.776, cross-peak H) and H-6′ (δ 6.90, cross-peak D) with C-2 (δ 80.5). Similarly H-6 (δ 5.86) showed HMBC correlations with C-5 (δ 165.4, cross-peak J), C-7 (δ 168.4), and C-10 (δ 103.3), while H-8 (δ 5.89) showed HMBC correlations with C-7 (δ 168.4), and C-9 (δ 164.8). This further supported the presence of "OH" groups at C-5 and C-7 (Fig. 10.59,

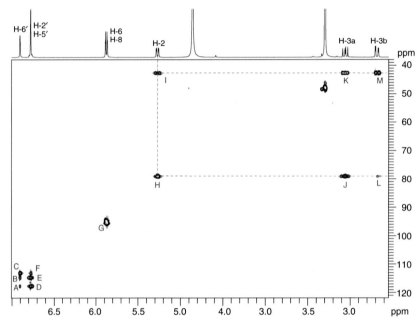

FIGURE 10.57 HSQC–TOCSY spectrum (concatenated NMR experiment) of compound **4** indicating one spin system F_1 plane and carbon connectivity in F_2 plane.

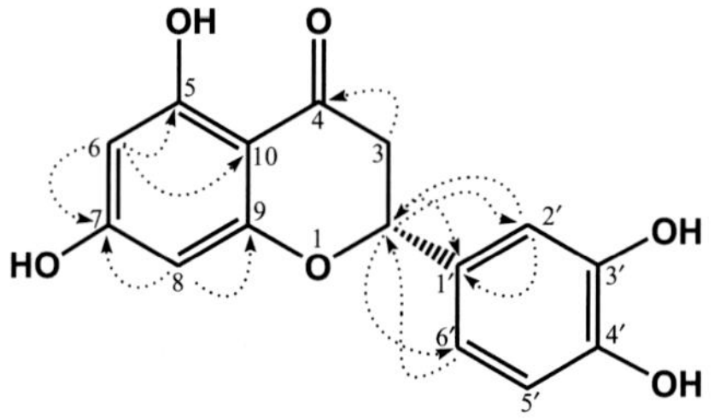

FIGURE 10.58 Key HMBC correlations in compound **4**.

Table 10.13). The protons at δ 3.06 (H-3α), and 2.69 (H-3β) showed HMBC correlations with the ketonic carbonyl carbon at δ 197.7 (C-4) (cross-peaks P and S, respectively), which further supported the presence of a flavanone skeleton in compound **4**. The complete chemical shift assignments are presented in Table 10.14

A literature survey revealed that the enantiomer of eriodictyol with "*R*" configuration has specific rotation with positive sign (+3.5), while its enantiomer with "*S*" configuration, has specific rotation with negative sign (−3.5). In our case, the specific rotation was found to be of negative value (−4.2). Therefore the configuration at C-2 was deduced as "*S*". On the basis of the above-cited

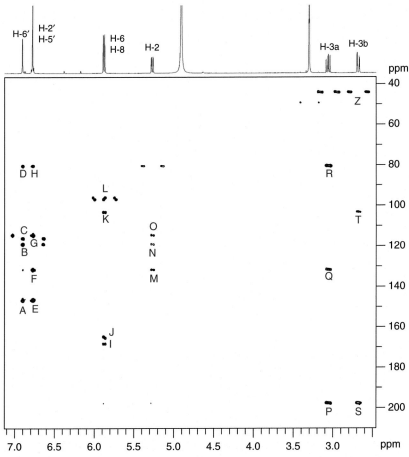

FIGURE 10.59 HMBC spectrum showing long-range ¹H/¹³C correlations in compound **4**.

spectroscopic data, the structure of eriodictyol was unambiguously identified as (*S*)-2-(3,4-dihydroxyphenyl)-5,7-dihydroxychroman-4-one (**4**).

10.3.5 3α-Hydroxylupanine (5)

A lupine alkaloid was isolated by us from *Leontice leontopetalum* L. (Taleb et al., 1991). The HREI–MS of compound **5** showed the M⁺ at 264.18378, in agreement with the molecular composition $C_{15}H_{24}N_2O_2$ (calculated 264.18588) with five degrees of unsaturation. Each proton (or sets of close-lying protons) of 3α-hydroxylupanine (**5**) were first assigned letters of alphabets (or set of letters, if more than one proton overlapped in an area), on the ¹H-NMR spectrum (Fig. 10.60). Thus, the farthest downfield proton is assigned letter A, and protons upfield to proton A are designated B, C, D, etc. In the case of overlapping

TABLE 10.13 Long-Range ^{1}H/^{13}C Correlations (From HMBC) for Compound 4 in CDCl$_3$

δ_H	$^2J_{CH}$	$^3J_{CH}$
6.90 (H-2′)	116.2 (C-5′)	119.9 (C-2′)
		80.5 (C-2)
		146.9 (C-4′)
6.78 (H-6′)	–	146.5 (C-3′)
		131.7 (C-1′)
6.77 (H-5′)	146.5 (C-3′)	114.6 (C-6′)
	131.7 (C-1′)	80.5 (C-2)
5.88 (H-8)	164.8 (C-9)	103.3 (C-10)
	168.4 (C-7)	
5.86 (H-6)	168.4 (C-7)	103.3 (C-10)
	165.4 (C-5)	
5.27 (H-2)	131.7 (C-1′)	119.9 (C-2′)
		114.6 (C-6′)
3.06 (H-3a)	197.7 (C-4)	131.7 (3J)
	80.5 (C-2)	
2.69 (H-3b)	197.7 (C-4)	103.3 (C-10)

TABLE 10.14 The ^{1}H- (600 MHz, CD$_3$OD), and ^{13}C-NMR (150 MHz, CD$_3$OD) Chemical Shift Assignments of Compound 4

Carbon No.	δ_C	Multiplicity	δ_H (J = Hz)
2	80.5	CH	5.27 (dd, $J_{2,3\alpha}$ = 12.6, $J_{2,3\beta}$ = 3.0)
3	44.1	CH$_2$	3.06 (dd, $J_{3\alpha,3\beta}$ = 17.0, $J_{3\alpha,2}$ = 12.6)
			2.69 (dd, $J_{3\beta,3\alpha}$ = 17.0, $J_{3\beta,2}$ = 3.0)
4	197.7	C	–
5	165.4	C	–
6	96.1	CH	5.86 (1H, $J_{6,8}$ = 2.0)
7	168.4	C	–
8	97.0	CH	5.88 (1H, $J_{8,6}$ = 2.0)
9	164.8	C	–
10	103.3	C	–
1′	131.7	C	–
2′	114.6	CH	6.90 (s)
3′	146.5	C	–
4′	146.9	C	–
5′	116.2	CH	6.77 (dd, J = 7.0, 0.5 Hz)
6′	119.2	CH	6.78 (dd, J = 7.0, 1.8 Hz)

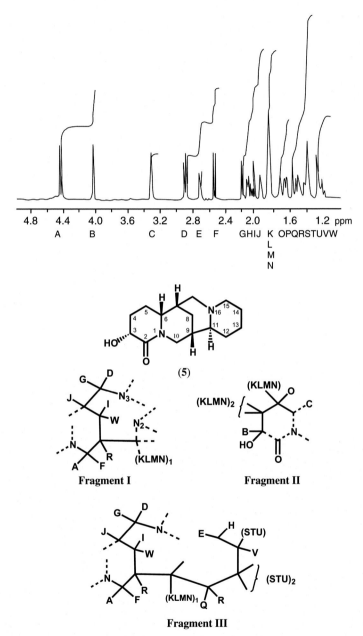

FIGURE 10.60 ¹H-NMR spectrum and substructures of 3α-hydroxylupanine, an alkaloid isolated from *Leontice leontopetalum*. The fragments, I, II, and III are labeled with the alphabetic assignment of every proton or a set of alphabets for close-lying protons.

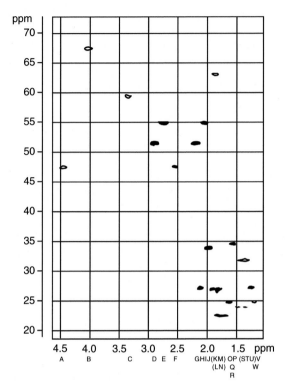

FIGURE 10.61   Hetero-COSY spectrum of 3α-hydroxylupanine (5).

multiplets, for example, at δ 1.86 (integration 4H), a group of letters (KLMN) was assigned. When all the protons in the ¹H-NMR spectrum had been labeled, we examined the hetero-COSY spectrum (Fig. 10.61), in order to determine which protons were attached directly to which carbon atoms by observing the one-bond ¹H/¹³C interactions.

The hetero-COSY results were then tabulated by writing the letters and ¹³C-NMR values together, with the data for CH, CH₂, and CH₃ groups being placed in *separate* columns (Table 10.15).

Table 10.15 therefore provides us the direct connectivities between the protons and the carbons to which they are attached in one glance. For instance the carbon which resonates at δ 47.23 (C-10) is attached to the protons A (δ 4.52) and F (δ 2.53), whereas the carbon resonating at δ 33.66 (C-9) is coupled to proton J (δ 1.96). Similarly, the carbon at δ 51.31(C-17) shows cross-peaks with protons D (δ 2.89) and G (δ 2.26), whereas the carbon at δ 34.27 (C-9) is coupled to proton R (δ 1.58).

Once all the direct C/H connectivity relationships had been determined, we examined the COSY-45° spectrum (Fig. 10.62). Since all the protons of

TABLE 10.15 Correlations of Methylene and Methine Protons With Attached Carbons

CH Protons (δ of CH)	CH$_2$ Protons (δ of CH)
B (67.5)	A, F (47.2)
C (59.2)	D, G (51.3)
	Q, P (51.3)
J (33.6)	STU-2H (23.8)
R (34.3)	E, H (54.8)
	STU-1H, W (24.5)
KLMN-1H (63.2)	K, V (26.9)
	KLMN-2H (36.7)
	KLMN-1H,O (22.4)

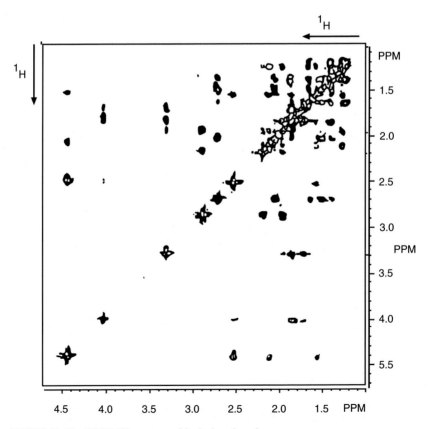

FIGURE 10.62   COSY-45° spectrum of 3α-hydroxylupanine.

methylene groups in 3α-hydroxylupanine (**5**) were already identified earlier from the hetero-COSY spectrum, the corresponding *geminal* coupling interactions could readily be determined. By deleting such geminal coupling interactions, the COSY-45° spectrum can be simplified greatly, and it allows the vicinal coupling interactions to be seen with greater clarity, since what is of interest at this stage is *to deduce C—C connectivity information of protonated carbon atoms from the vicinal proton coupling interactions.* Since we had already established the $^1H/^{13}C$ one-bond connectivities from the HETCOR spectrum (Table 10.15), a combination of these two sets of data told us which protonated carbons are bound to one another.

The COSY-45° spectrum (Fig. 10.62) of the compound showed that one of the protons of a CH_2 group (with ^{13}C chemical shift of δ 47.22)—i.e., proton A—was coupled to proton R (δ 1.58). The latter was attached to the methine carbon having a chemical shift of δ 34.27, indicating that the CH_2 carbon (δc 47.22) was adjacent to a CH carbon (with $δ_C$ 34.27). Similarly, proton R showed cross-peaks with protons W (δ 1.28) and I (δ 2.12) as well as with one of the protons present in the 4H multiplet (KLMN) at δ 1.88, leading to fragment I. Other important cross-peaks were shown by proton J (δ 1.96) with proton D (δ 2.89), G (δ 2.26), I (δ 2.12), and W (δ 1.28), confirming the substructure of fragment I (Fig. 10.60).

Similarly, proton B (δ 4.02) was coupled to one of the protons in the 4H multiplet labeled KLMN (δ 1.85), whereas proton C (δ 3.31) was coupled to proton O (δ 1.72) as well as to one of the protons present in the 4H multiplet labeled KLMN (δ 1.85). The various coupling interactions of 3α hydroxylupanine (**5**) obtained by considering both hetero-COSY and COSY-45° data led to deducing the two fragments II and III (Fig. 10.60).

The spin systems in 3α-hydroxylupanine were further investigated by applying the TOCSY experiment (Fig. 10.63) recorded, with a mixing times of up to 100 ms. The various coupling connectivities present in the TOCSY spectra not only confirmed their assignments but also helped to join fragments I and II together (fragment III). The spectrum obtained, with a mixing time of 20 ms, resembled closely the COSY-45° spectrum, showing mainly direct connectivities, while with longer mixing intervals the magnetization was seen to spread to more distant protons within individual spin systems. This spreading of the magnetization with the increase of the mixing delay (20, 40, 60, and 100 ms) can be seen in the projections of the TOCSY spectra at δ 4.52 and 3.39 (Fig. 10.64). These considerations led to structure **5** for 3α-hydroxylupanine.

10.3.6 (+)-Buxalongifolamidine (6)

A steroidal alkaloid **6** was isolated from the leaves of *Buxus longifolia* Boiss. of Turkish origin (Atta-ur-Rahman et al., 1993). The HREI–MS of compound **6** showed the molecular ion at *m/z* 578.3533 corresponding to the molecular

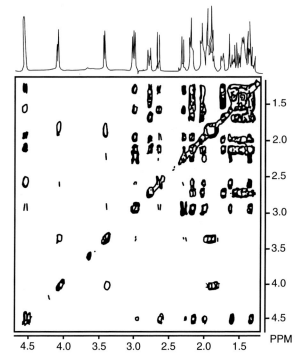

FIGURE 10.63 HOHAHA spectrum of 3α-hydroxylupanine (5) recorded at 100 ms.

formula $C_{35}H_{50}N_2O_5$ representing twelve degrees of unsaturation (DoU) in the molecule. The compound showed a peak at m/z 563, resulting from the loss of a methyl group from the M$^+$ ion.

(a)   (b)

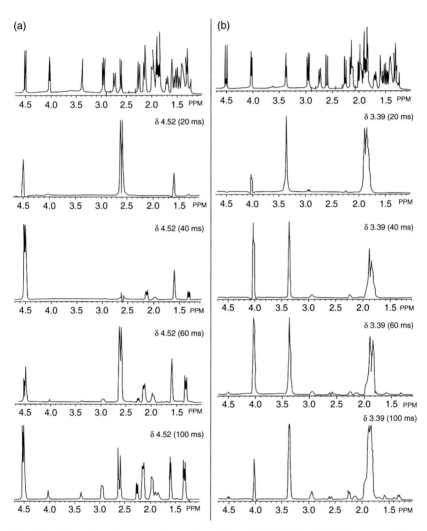

FIGURE 10.64 The slice plots taken at (a) δ 4.52 and (b) δ 3.39 of TOCSY spectra of 3α-hydroxylupanine (**5**).

The ¹H-NMR spectrum (Fig. 10.65) of **6** revealed the presence of three tertiary methyl groups at δ 0.74, 0.93 and 1.16, while the secondary methyl group resonated as a doublet at δ 1.25 ($J_{20,21}$ = 6.6 Hz).

These observations favor a cycloartenol-type triterpenoidal skeleton, as commonly present in *Buxus* alkaloids. An AB doublet resonating at δ 3.65 and 3.70 ($J_{31\alpha,31\beta}$ = 8.7 Hz) was assigned to the C-31 hydroxymethylenic protons. A doublet at δ 5.85 ($J_{1,2}$ = 10.3 Hz) due to the C-1 vinylic proton showed coupling with the C-2 vinylic proton resonating at δ 5.87 as a doublet of a doublet ($J_{1,2}$ = 10.3 Hz, $J_{2,3}$ = 3.3 Hz). The C-2 vinylic proton, on the other hand,

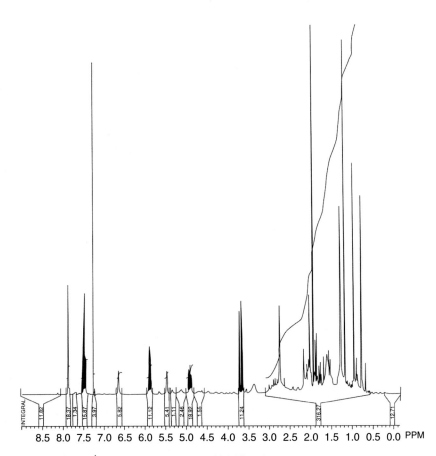

FIGURE 10.65 ¹H-NMR spectra of buxalongifolaidine (**6**).

showed COSY cross-peaks with the C-3 methine proton. The C-3 proton resonated as a doublet of double doublets at δ 4.80 ($J_{3,NH}$ = 10.0 Hz, $J_{3,2}$ = 3.3 Hz,
$J_{3,5}$ = 1.7 H$_z$), showing cross-peaks with the C-2 vinylic, benzamidic NH, and W
coupling with the C-5 methine proton in the COSY-45° spectrum (Fig. 10.66).
These observations suggested the presence of fragment I, including ring A of
the skeleton.

Fragment I

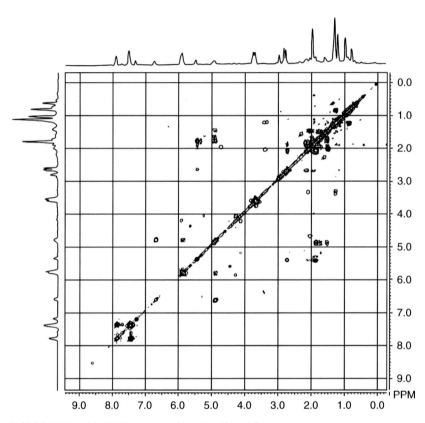

FIGURE 10.66   COSY 45° spectrum of buxalongifolamidine.

The C-19 methylene protons appeared as a broad singlet at δ 2.72 and showed weak allylic interactions with the vinylic C-11 proton that appeared at δ 5.45 as a multiplet which in turn showed vicinal coupling with the C-12 methylene protons. This indicated the presence of a trisubstituted olefinic bond which when tailored into ring C of the skeleton gives rise to fragment II.

A multiplet centered at δ 4.90 was assigned to the C-16 proton, which exhibited COSY interactions with the C-15 methylene and C-17 methine protons. Since an acetoxy group was inferred from the IR absorptions and the presence of a singlet at δ 1.90 in the ID ^{1}H-NMR spectrum, C-16 seemed to be a plausible site of its substitution in ring D of the skeleton (fragment III).

Fragment II

Fragment III

TABLE 10.16 ^{1}H/^{13}C-NMR Connectivities (From HMQC) for Compound 6

Carbons	Chemical Shift (δ)	Multiplicity (DEPT)	^{1}H/^{13}C Connectivity δ (J = Hz)
1	132.5	CH	5.85 (d, J = 10.3)
2	131.7	CH	5.87 (dd, J = 10.3, 3.3)
3	56.1	CH	4.80 (ddd, J = 10.0, 3.3, 1.7)
4	45.1	C	–
5	53.6	CH	2.68 (m)
6	29.7	CH$_2$	1.25 (m)
7	34.3	CH$_2$	1.81 (m)
8	41.9	CH	2.00 (m)
9	133.6	C	–
10	79.5	C	–
11	124.0	CH	5.45 (m)
12	37.3	CH$_2$	1.85 (m)
13	46.2	C	–
14	47.7	C	–
15	43.8	CH$_2$	1.50 (m)
16	75.1	CH	4.90 (m)
17	56.5	CH	2.05 (m)
18	16.8	CH$_3$	0.93 (s)
19	44.3	CH$_2$	2.72 (s)
20	65.4	CH	3.40 (m)
21	11.7	CH$_3$	1.25 (d, J = 6.6)
30	19.9	CH$_3$	1.16 (s)
31	78.6	CH$_2$	3.65 (d, J = 8.7) 3.70 (d, J = 8.7)
32	17.3	CH$_3$	0.74 (s)
N$_b$-CH$_3$	35.8	CH$_3$	2.03 (s)
N$_b$-CH$_3$	44.1	CH$_3$	2.03 (s)
1'	135.3	C	–
2'	127.4	CH	7.85 (m)
3'	128.5	CH	7.48 (m)
4'	130.3	CH	7.50 (m)
5'	128.5	CH	7.48 (m)
6'	127.4	CH	7.85 (m)
Ph-CO-N	167.5	C	–
CO-C̲H$_3$	21.2	CH$_3$	1.90 (s)
C̲O-CH$_3$	171.2	C	–

Other major peaks in the ^{1}H-NMR spectrum were a six-proton singlet at δ 2.03 assigned to the N(CH$_3$)$_2$ protons and groups of three- and two-proton multiplets centered at δ 7.50 and 7.85 due to the aromatic protons of the benzamidic moiety substituted on C-3 of ring A.

The ^{13}C-NMR spectra (CDCl$_3$) (BB and DEPT) of compound **6** showed resonances for all 35 carbons. A notable feature of the ^{13}C-NMR BB decoupled spectrum was the appearance of a downfield signal for a quaternary carbon at δ 79.5, which was assigned to C-10, bearing a hydroxy group. The chemical shifts for various carbons are in the Table 10.16. One-bond correlations between ^{1}H- and ^{13}C-nuclei were established by inverse ^{1}H-detected HMQC experiment (Fig. 10.67); the results are presented in Table 10.16.

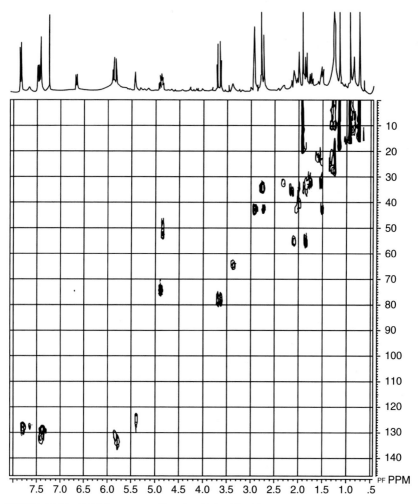

FIGURE 10.67   HMQC spectrum of buxalongifolamidine (**6**).

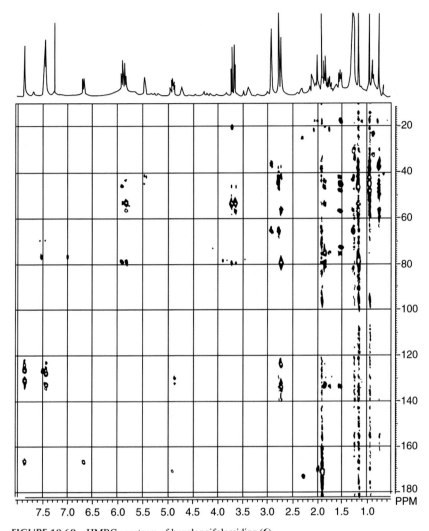

FIGURE 10.68   HMBC spectrum of buxalongifolamidine (**6**).

The ^{1}H/^{13}C long-range coupling information obtained from the HMBC experiment (Fig. 10.68) allowed the various fragments to be connected together. The proton at δ 3.65 (C-31αH) showed $^2J_{CH}$ interaction with C-4 (δ 45.1) and $^3J_{CH}$ interactions with C-3 (δ 56.1) and C-5 (δ 53.6). The C-31β proton (δ 3.70) showed $^2J_{CH}$ interaction with C-4, $^3J_{CH}$ interactions with C-3 (δ 56.1) and C-30 (δ 19.9), and $^4J_{CH}$ interaction with C-10 (δ 79.5). The protons at δ 2.72 (C-19) showed $^2J_{CH}$ couplings with C-9 (δ 133.6) and C-10 (δ 79.5). The $^3J_{CH}$ connectivities between C-5 (δ 53.6) and C-19 (δ 2.72) protons were also observed.

TABLE 10.17 Long Range ^{1}H/^{13}C Connectivities (From HMBC) for Buxalongifolamidine (5) in CDCl$_3$

^{1}H (δ)	$^2J_{CH}$ (δ)	$^3J_{CH}$ (δ)	$^4J_{CH}$ (δ)
7.85 (H-2′, 6′)	128.5 (C-3′)	167.5 (amide C=O)	
5.87 (H-2)		79.5 (C-10)	
		45.1 (C-4)	
5.85 (H-1)	79.5 (C-10)	53.6 (C-5)	
		44.3 (C-19)	
4.90 (H-16)	171.2 (C-CH$_3$)	46.2 (C-13)	
	56.5 (C-17)		
4.80 (H-3)	131.7 (C-2)		79.5 (C-10)
3.70 (H-31 βH)	45.1 (C-4)	56.1 (C-3)	79.5 (C-10)
		19.9 (C-30)	
3.65 (H-31 αH)	45.1 (C-4)	56.1 (C-3)	
		53.6 (C-5)	
2.72 (H-19)	133.6 (C-9)	41.9 (C-8)	
	79.5 (C-10)	53.6 (C-5)	
1.81 (H-7)	41.9 (C-8)	17.3 (C-32)	
		133.6 (C-9)	
1.50 (H-15)	75.1 (C-16)	56.6 (C-17)	
	47.7 (C-14)	41.9 (C-8)	
1.25 (H-21)	65.4 (C-20)	56.6 (C-17)	
1.16 (H-30)	78.6 (C-31)	56.1 (C-3)	
	45.1 (C-4)	53.6 (C-5)	

The C-2 protons (δ 5.87) showed $^3J_{CH}$ couplings with C-10 (δ 79.5) and C-4 (δ 45.1). The proton at δ 1.16 (C-30) showed $^3J_{CH}$ connectivities with C-3 (δ 56.1). Other connectivities are shown in Table 10.17. The TOCSY spectrum (100 ms) (Fig. 10.69) of buxalongifolamidine further confirmed the ^{1}H-^{1}H connectivities within each spin-system. On the basis of these studies, structure **6** was assigned to (+)-buxalongifolamidine.

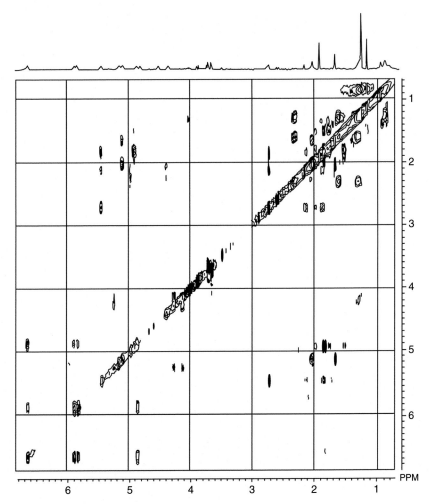

FIGURE 10.69 HOHAHA spectrum (100) of buxalongifolamidine (**6**).

REFERENCES

Atta-ur-Rahman, Noor-e-ain, F., Ali, R.A., Choudhary, M.I., Pervin, A., Turkoz, S., Sener, B., 1993. Two steroidal alkaloids from *Buxus longifolia*. Phytochemistry 32 (4), 1059–1063.

Blinov, K.A., Carlson, D., Elyashberg, M.E., Martin, G.E., Martirosian, E.R., Molodtsov, S., Williams, A.J., 2003. Computer-assisted structure elucidation of natural products with limited 2D NMR data: application of the StrucEluc system. Magn. Reson. Chem. 41 (5), 359–372.

Breton, R.C., Reynolds, W.F., 2013. Using NMR to identify and characterize natural products. Nat. Prod. Rep. 30 (4), 501–524.

Elmastaş, M., Dermirtas, I., Isildak, O., Aboul-Enein, H.Y., 2006. Antioxidant activity of *S*-carvone isolated from spearmint (*Mentha spicata* L. Fam. Lamiaceae). J. Liq. Chrom. Rel. Technol. 29 (10), 1465–1475.

Halabalaki, M., Vougogiannopoulou, K., Mikros, E., Skaltsounis, A.L., 2014. Recent advances and new strategies in the NMR-based identification of natural products. Curr. Opin. Biol. 25, 1–7.

Jaspars, M., 1999. Computer assisted structure elucidation of natural products using two-dimensional NMR spectroscopy. Nat. Prod. Rep. 16 (2), 241–248.

Kohler, H.P., Van der Maarel, M.J., Kohler-Staub, D., 1993. Selection of *Pseudomonas* sp. strain HBP1 Prp for metabolism of 2-propylphenol and elucidation of the degradative pathway. Appl. Environ. Microbiol. 59 (3), 860–866.

Steinbeck, C., 2004. Recent developments in automated structure elucidation of natural products. Nat. Prod. Rep. 21 (4), 512–518.

Taleb, H.Al-Tel., Sabri, S.S., Zarga, M.H.A., Pervin, A., Shah, Z., Atta-ur-Rahman, Rycroft, D.S., 1991. Alpine alkaloid from *Leontice leontopetalum*. Phytochemistry 30 (7), 2393–2395.

Wu, Z., Wei, G., Lian, G., Yu, B., 2010. Synthesis of mangiferin, isomangiferin, and homomangiferin. J. Org. Chem 75 (16), 5725–5728.

Glossary

Absolute-value-mode spectrum The spectrum is produced by recording the square root of the sum of the squares of the real (R) and imaginary (I) parts of the spectrum ($R^2 + I^2$).

Absorption-mode spectrum The spectrum in which the peaks appear with Lorentzian line shapes. NMR spectra are normally displayed in absolute-value mode.

Acquisition time The time taken to digitize the free induction decay. In 2D NMR experiments, this takes place during time t_2 when the FID is "acquired."

Aliasing False peaks produced that originate outside the spectral window, when the digitization rate is less than twice the Nyquist frequency.

Analog-to-digital converter The computer hardware that converts the voltage from the detector into a binary number, usually 12 or 16 bits long.

Antiphase The phases opposite to that of an individual line in a multiplet.

Apodization Applying weighting functions for enhancing sensitivity or resolution.

Attached proton test (APT) A one-dimensional ^{13}C NMR experiment that is used for distinguishing quaternary carbon (C) and CH_2 signals from CH and CH_3 signals. In the APT ^{13}C-NMR spectrum, the methine (CH) and methyl (CH_3) signals appear in positive phase, whereas the quaternary (C) and methylene (CH_2) signals appear in negative phase.

Artifacts False peaks, noise, or other unwanted, signals in the NMR spectrum.

Band-pass filter A filter used for selecting the desired coherence order, p.

Carrier frequency The transmitter frequency that consists of high-frequency pulses.

Chemical exchange Chemical exchange is a process in which a nucleus is exchanging with other nuclei, the exchange occurring between two or more centers inter- and/or intramolecularly. These centers generally have different electronic environments.

Chemical shift The difference between the nuclear precession frequency and the carrier frequency.

Coherence A condition under which nuclei precess with a given phase relation and can exchange spin states via transitions between two eigenstates. Coherence may be zero-quantum, single-quantum, double-quantum, etc., depending on the ΔM_z of the transition corresponding to the coherence. Only single-quantum coherence can be detected directly.

Coherence order, p The difference in the magnetic quantum number, M_z of the two energy levels connected by the same coherence.

Coherence pathway selection Choosing a corresponding phase cycle so that undesired resonances are suppressed and only the desired magnetization is observed.

Coherence transfer The transfer of coherence between two different types of nuclei.

Composite pulse A composite "sandwich" of pulses that replaces a single pulse; employed to compensate for B_1 field inhomogeneities, phase errors, or offset effects.

Concatenated NMR Concatenated hybrid NMR techniques are based on linking two NMR pulse sequences in a series. DEPT-HSQC, TOCSY-HSQC, and NOESY-HSQC are some common concatenated NMR techniques.

Solving Problems with NMR Spectroscopy. http://dx.doi.org/10.1016/B978-0-12-411589-7.00016-4

Constant-time experiment An NMR experiment in which the total duration of t_1 is kept constant.

Continuous wave A method of recording an NMR spectrum in which the field B1 is applied continuously and either the magnitude of B_0 or the radiofrequency is varied so that the nuclei are brought successively into resonance.

Contour plot The most widely applied method for recording 2D spectra. The peaks appear as concentric circular lines (*contours*), with the two axes of the 2D spectrum representing chemical shifts or coupling constants.

Convolution The random noise in a spectrum can be smoothed out by the application of convolution functions. The NMR spectrum is sampled at regular frequency intervals, and each ordinate in the resulting new spectrum is multiplied by the corresponding ordinate on the convoluting function. The sum of the products is normalized and plotted.

Convolution difference To emphasize the narrow peaks in the spectrum of a macromolecule and to suppress the broad resonances, we can apply a severe broadening function that is calculated to affect the narrow lines much more than the broad lines. This causes the narrow peaks to broaden, and the resulting spectrum is subtracted from the original spectrum to give the convolution difference spectrum, in which each narrow line appears on shallow depressions.

Correlation time (τ_c) A parameter related to the mean time during which a molecule maintains its spatial geometry between successive rotations. For internuclear vectors, τ_c is approximately equal to the average time for it to rotate through an angle of 1 radian.

COSY (^{1}H–^{1}H correlation spectroscopy) An important 2D experiment that allows us to identify the protons coupled to one another.

Cross-peaks The off-diagonal peaks in a 2D experiment that appear at the coordinates of the correlated nuclei.

Cross-relaxation The mutual intermolecular or intramolecular relaxation of magnetically equivalent nuclei, for example, through dipolar relaxation. This forms the basis of nOe experiments.

Cryogenically cooled NMR probes These are specially designed NMR probes in which the coil and/or built-in signal preamplifier are cooled with a stream of helium gas at ∼20 K or with liquid nitrogen. This increases the sensitivity of the NMR signal and reduces the level of the thermal noise. The increase in the signal-to-noise ratio is about fourfold greater than that obtained with a conventional room temperature probe.

CYCLOPS A four-step phase cycle that corrects *dc* imbalance in the two channels of a quadrature detector system.

Dead time A very short delay introduced before the start of acquisition that allows the transmitter gate to close and the receiver gate to open.

Density matrix A description of the state of nuclei in quantum-mechanical terms.

DEPT (distortionless enhancement by polarization transfer) A one-dimensional ^{13}C-NMR experiment commonly used for spectral editing that allows us to distinguish between CH_3, CH_2, CH, and quaternary carbons.

Detectable magnetization The magnetization precessing in the $x'y'$-plane induces a signal in the receiver coil that is detected. Only single-quantum coherence is directly detectable.

Detection period The FID is acquired in the last segment of the pulse sequence. This is the detection period, and, in 2D experiments, this is the t_2 domain.

Deuterium lock Signals of deuterated solvents (such as $CDCl_3$, CD_3OD, DMSO-d_6, etc.) are now used in modern NMR spectroscopy as a lock to offset the effect of the natural drift of the applied magnetic field B_0. This is called a "deuterium lock." Currently, NMR spectrometers are designed to constantly monitor the resonance frequency of the deuterium signal of the solvent and make necessary adjustment to the magnetic field B_0. This keeps the resonance frequency constant. Deuterium signals are also used to accurately define 0 ppm, as the difference between the resonant frequency of the deuterium lock signal and TMS (0 ppm) is well known.

Diagonal peaks Cross-sections of peaks that appear on or near a diagonal line in a 2D spectrum. The projection produces the 1D spectrum. They give no shift-correlation information.

Diamagnetism Substances that have no unpaired electrons are diamagnetic. When placed in an applied magnetic field, their induced magnetic fields oppose the applied magnetic field.

Digital resolution The distance, in hertz, between successive data points; it determines the extent of peak definition by the data points.

Digitization The process of converting voltage by an analog-to-digital converter (ADC) to a corresponding binary number for digital processing; used to represent free induction decay in a digital form.

Digitization rate The rate that represents the number of analog-to-digital conversions per second during data acquisition; must be twice the highest signal frequency.

Dipolar coupling The direct through-space coupling interaction between two nuclei. It is responsible for nOe and relaxation and represents the main mechanism for spin-lattice relaxation of spin ½ nuclei in diamagnetic liquids.

Dispersion mode A Lorentzian line shape that arises from a phase-sensitive detector (which is 90° out of phase with one that gives a pure-absorption mode line). Dispersion-mode signals are dipolar in shape and produce long tails. They are not readily integrable, and we need to avoid them in a 20 spectrum.

DOSY An experiment which separates the ^{1}H-NMR signals of different species in a mixture according to their diffusion coefficients. The DOSY technique provides a way to separately record the ^{1}H-NMR spectra of different compounds in a mixture, based on their diffusion coefficients in the solution state. This technique does not require any particular sample preparation or chromatographic separation for mixture analysis.

Double-quantum coherence Coherence between states that are separated by magnetic quantum numbers of ±2. This coherence cannot be detected directly, but must be converted into single-quantum coherence before detection.

Dynamic nuclear polarization (DNP)-NMR A technique for the improvement of inherently low sensitivity of certain nuclei in NMR spectroscopy. DNP-NMR is a hyper-polarization method that achieves significantly larger nuclear polarizations than the common thermal-equilibrium polarization. This involves the transfer of electronic polarization from a highly polarized small molecule (a DNP agent) to the coupled nuclei in solution state.

DNP-MAS NMR This is the application of dynamic nuclear polarization (DNP) technique in solid-state NMR spectroscopy by using magic angle spinning (MAS) probe. This provides significant enhancement to the S/N ratio in solid-state NMR experiments by transferring the large electron spin polarization to the coupled nuclear spins.

Dynamic range Range determined by the number of bits in the output of the analog-to-digital converter; it represents the ratio between the largest and smallest signals observable in the digitized spectrum.

Earth field NMR (EF-NMR) An NMR technique that uses a special NMR spectrometer that employs the globally available earth's homogeneous magnetic field for the detection of nuclear spins in a sample. This allows the use of large quantities of samples without fear of artifact signals arising due to inhomogeneous magnetic field B_0. Currently, prepolarization methods (polarization by a relatively crude electromagnet with a magnetic field about 350 times larger than the earth's magnetic field) are used for the initial magnetization. This greatly improves the sensitivity of the EF-NMR signals.

Editing Obtaining a given set of subspectra to supply some desired information, for example, the multiplicity of signals.

Equilibration delay A time period introduced between successive scans to allow the nuclei to relax to their respective equilibrium states (i.e., to become realigned along the z-axis).

Ernst angle The angle by which the nuclear magnetization vector must be tipped to give the best signal-to-noise ratio for a given combination of spin-lattice relaxation time and the time between successive pulses.

Evolution period The delay time in the second segment of a pulse sequence, during which the nuclei precess under the influence of chemical shift and/ or spin coupling; it incremented successively in 2D experiments, thereby producing the t_1 domain.

F_1-axis The axis of a 2D spectrum resulting from the Fourier transformation of the t_1 domain signal.

F_2-axis The axis of a 20 spectrum resulting from the Fourier transformation of the transposed digitized FIDs (t_2 domain).

Fast Fourier transform (FFT) A procedure for carrying out Fourier transformation at high speed, with a minimum of storage space being used.

Field frequency lock The magnitude of the field B_0 is stabilized by locking onto the fixed frequency of a resonance in the solution, usually ^2H of the solvent.

Field gradients In an inhomogeneous magnetic field, the magnetization will become subdivided into a number of smaller vectors, each of which will precess at a slightly differing frequency. These smaller vectors spread out (dephase) during the evolution period, thereby reducing the net transverse magnetization of the sample. Such field gradients broaden the NMR signals and lead to a more rapid disappearance of the FID.

First-order multiplet First-order multiplets are those in which the difference in chemical shifts of the coupled protons, in hertz (Hz), divided by the coupling constant between them is about 8 or more ($\Delta v/J > 8$). These multiplets have a symmetrical disposition of the lines about their midpoints. The distance between the two outermost peaks of a first-order multiplet is the sum of each of the coupling constants.

Flip angle The angle by which a vector is rotated by a pulse.

Fourier transformation A mathematical operation by which the FIDs are converted from time-domain data into the equivalent frequency-domain spectrum.

Free induction decay A decay time-domain beat pattern obtained when the nuclear spin system is subjected to a radiofrequency pulse and then allowed to precess in the absence of *Rf* fields.

Frequency spectrum A plot of signal amplitude versus frequency, produced by the Fourier transformation of a time-domain signal.

Gated decoupling The decoupler is "gated" during certain pulse NMR experiments, so spin decoupling occurs only when the decoupler is switched on and not when it is switched off; used to eliminate either ^1H–^{13}C spin coupling or nuclear Overhauser effect in a 1D ^{13}C-NMR spectrum, and employed as a standard technique in many other ^1H-NMR experiments, such as APT and *J*-resolved.

Heteronuclear correlation spectroscopy (HETCOR) Shift-correlation spectroscopy in which the chemical shifts of different types of nuclei (e.g., ^1H and ^{13}C) are correlated through their mutual spin-coupling effects. These correlations may be over one bond or over several bonds.

Heteronuclear multiple bond correlation (HMBC) A two-dimensional inverse NMR experiment which correlates the chemical shifts of ^1H with hetero-atoms (e.g., $^1H/^{13}C$ or $^1H/^{15}N$), separated from each other with two or more chemical bonds. This is a ^1H-detected long-range heteronuclear technique.

Heteronuclear multiple quantum coherence (HMQC) A two-dimensional inverse experiment used to correlate directly bonded carbon–proton nuclei. This provides similar results as the HSQC experiments, but detects heteronuclear multiple quantum coherences.

Heteronuclear single quantum coherence (HSQC) An important inverse NMR technique which correlates the hetero-atom (such as ^{15}N or ^{13}C) with the directly attached proton (1H). The HSQC spectrum is a ^1H-detetced two-dimensional plot where every cross-peak represents a direct correlation between the proton and the attached heteroatom (^{15}N, ^{13}C, etc.).

High-pass filter A filter used to select a coherence order $\geq p$.

INADEQUATE Correlation spectroscopy in which the directly bonded ^{13}C nuclei are identified.

Increment A short, fixed time interval by which time interval t_1 is repeatedly increased in a 2D NMR experiment.

INEPT (insensitive nuclei enhanced by polarization transfer) Polarization transfer pulse sequence used to record the NMR spectra of insensitive nuclei, for example, ^{13}C, with sensitivity enhancement; it may be used for spectral editing.

In-phase A term applied to a multiplet structure when the individual lines of the multiplet have the same phase.

Inverse experiments Heteronuclear shift-correlation experiments in which magnetization of the less sensitive heteronucleus (e.g., ^{13}C) is detected through the more sensitive magnetization (e.g., ^1H).

Inverse polarization transfer These NMR techniques were developed to study insensitive nuclei (such as ^{13}C or ^{15}N) through their effects on a high-sensitivity nucleus (typically ^1H). The techniques include HMQC, HSQC, and HMBC experiments.

Inversion recovery A pulse sequence used to determine spin-lattice relaxation times.

J-scaling A method for preserving a constant scaling factor for all ^{13}C multiplets.

J-spectroscopy A 2D experiment in which the chemical shifts are plotted along one axis (F_2-axis), whereas the coupling constants are plotted along the other axis (F_1-axis). This is an excellent procedure for uncoding overlapping multiplets.

Laboratory frame The Cartesian coordinates (x, y, and z) are stationary with respect to the observer, in contrast to the rotating frame, in which they rotate at the spectrometer frequency.

Larmor frequency The nuclear precession frequency about the direction of B_0. Its magnitude is given by $\gamma B_0/2\pi$.

Longitudinal magnetization This is the magnetization directed along the z-axis parallel to the B_0 field.

Longitudinal operators Orthogonal operators, designated as I_z.

Lorentzian line shape The normal line shape of an NMR peak that can be displayed either in absorption or in dispersion mode.

Low-pass filter A filter used for selecting coherence orders $\geq p$.

Magnetic field gradients See field gradients.

Magnetic quantum number (m_I) The number that defines the energy of a nucleus in an applied field B_0. It is quantized, that is, has values of $-I$, $-I$, $+1$,..., $+I$. In the case of ^1H or ^{13}C nuclei, the spin quantum number is $\pm\frac{1}{2}$, corresponding to the alignment of the z-component of the nuclear magnetic moment with or against the applied magnetic field, B_0.

Magnetic resonance imaging (MRI) A technique, based on magnetic field gradients, that is widely used as an aid in clinical diagnosis. MRI has proved to be a promising tool for tumor detection.

Magnetization vector (M) The resultant of individual magnetic moment vectors for an ensemble of nuclei in a magnetic field.

Magnetogyric ratio (γ) The intrinsic property of a nucleus that determines the energy of the nucleus for specific values of the applied field, B_0. It can be either positive (e.g., ^{1}H, ^{13}C) or negative (e.g., ^{15}N, ^{29}Si).

Matched filter The multiplication of the free induction decay with a sensitivity enhancement function that matches exactly the decay of the raw signal. This results in enhancement of resolution, but broadens the Lorentzian line by a factor of 2 and a Gaussian line by a factor of 2.5.

Mixing period The third time period in 2D experiments, such as NOESY, during which mixing of coherences can occur between correlated nuclei.

Modulation The variation in amplitude and/or phase of an oscillatory signal by another function, for example, modulation of the nuclear precession frequency of one nucleus by the nuclear precession frequency of a correlated nucleus in COSY spectra.

Multiple-pulse experiment An experiment in which more than one pulse is applied to the nuclei.

Multiple-quantum coherence When two or more pulses are applied, they can act in cascade and excite multiple-quantum coherence, that is, coherence between states whose total magnetic quantum numbers differ by other than ±1 (e.g., 0, ±2, ±3). Such coherence is not directly detectable but must be converted into single-quantum coherence to be detected.

Multiple receiver technology The latest NMR spectrometers can incorporate multiple receivers for parallel acquisition of signals from two or more nuclei simultaneously. This major technological development in NMR spectroscopy has shortened measurement times and given birth to new types of NMR experiments.

Nonselective pulse A pulse with wide frequency bandwidth that excites all nuclei of a particular type indiscriminately.

Nuclear magnetic moment (μ) The consequence when a nucleus has both a charge and an angular momentum. The magnitude of the observed magnetic moment is given by $m_I = h\gamma/2\pi$ (where m_I = magnetic quantum number and γ = magnetogyric ratio).

Nuclear Overhauser effect (nOe) The change in intensity in the signal of one nucleus when another nucleus lying spatially close to it is irradiated, with the two nuclei relaxing each other via the dipolar mechanism.

Nuclear Overhauser effect correlation spectroscopy (NOESY) A 2D method for studying nuclear Overhauser effects.

Nyquist frequency The highest frequency that can be characterized by sampling at a given rate; it is equal to one-half the rate of digitization.

Off-resonance decoupling Applying an unmodulated radiofrequency field B_2 of moderate intensity in the proton spectrum just outside the spectral window of the carbon spectrum shrinks the multiplet structure of the carbon signals. Some multiplet structure is, however, still left behind, allowing the CH_3 (quartets), CH_2 (triplets), CH (methines), and C (quaternary carbons) to be distinguished.

Paramagnetism Magnetic behavior of nuclei that contain unpaired electrons. When such substances are placed in a magnetic field, the induced magnetic field is parallel to the applied magnetic field.

Phasing A process of phase correction that is carried out by a linear combination of the real and imaginary sections of a 1D spectrum to produce signals with pure absorption-mode peak shapes.

Phase cycling As employed in modern NMR experiments, repeating the pulse sequence with all the other parameters being kept constant and only the phases of the pulse(s) and the phase-sensitive detector reference being changed. The FIDs are acquired and coadded. The procedure is used to eliminate undesired coherences or artifact signals, or to produce certain desired effects (e.g., multiple-quantum filtration).

Phase-sensitive data acquisition NMR data are acquired in this manner so that peaks are recorded with pure absorption-mode or pure dispersion-mode line shapes.

Phase-sensitive detector A detector whose output is the product of the input signal and a periodic reference signal, so that it depends on the amplitude of the input signal and its phase relative to the reference.

Polarization The results of an excess of nuclear magnetic moment vectors (μ_z) aligned with or opposed to the applied magnetic field, B_0. The magnitude of the polarization effect is determined by the magnetogyric ratio, the applied field B_0, and the temperature.

Polarization transfer Application of certain pulse sequences causes the transfer of the greater-equilibrium polarization from protons to a coupled nucleus, for example, ^{13}C, which has a smaller magnetogyric ratio.

Polarization transfer in reverse Same as inverse polarization transfer.

Precession A characteristic rotation of the nuclear magnetic moments about the axis of the applied magnetic field B_0 at the Larmor frequencies.

Preparation period The first segment of the pulse sequence, consisting of an equilibration delay. It is followed by one or more pulses applied at the beginning of the subsequent evolution period.

Presaturation Selective irradiation of a nucleus prior to application of a nonselective pulse causes it to be saturated, so its resonance is suppressed. This technique may be employed for suppressing solvent signals.

Product operator A set of orthogonal operators that can be connected together by a product. The product operator approach can be used to analyze pulse sequences by applying simple mathematical rules to weakly coupled spin systems.

Projection The one-dimensional spectrum produced when one of the two axes of a 2D spectrum is collapsed and the resulting "projection spectrum" recorded.

Pulse field gradients (PFG) In this technique, the homogeneous magnetic field B_0 is linearly disturbed in a systematic and predictable manner. The PFG technique is extensively employed to enhance the quality and sensitivity of the NMR experiments. The use of PFG to perform signal selection in high-resolution NMR spectroscopy has been widely recognized. It is routinely used in multidimensional NMR experiments, such as HSQC, NOESY, ROESY, etc.

Pulse sequence A series of *Rf* pulses, with intervening delays, followed by detection of the resulting transverse magnetization.

Pulse width The time duration for which a pulse is applied; it determines the extent to which the magnetization vector is bent.

Quadrature detection A method for detecting NMR signals that employs two phase-sensitive detectors. One detector measures the x-component of the magnetization; the other detector measures the y-component. The reference phases are offset by 90°.

Quadrature images Any imbalances between the two channels of a quadrature detection system cause ghost peaks, which appear as symmetrically located artifact peaks on opposite sides of the spectrometer frequency. They can be eliminated by an appropriate phase-cycling procedure, for example, CYCLOPS.

Real and imaginary parts Two equal blocks of frequency that result from Fourier transformation of the FIDs.

ROESY An NMR experiment that provides nOe information. It is particularly useful for molecules of intermediate molecular mass (1000–3000 amu).

Rotating frame The frame is considered to be rotating at the carrier frequency, v_0. This allows a simplified picture of the precessing nuclei to be obtained.

Saturation If the Rf field is applied continuously, or if the pulse repetition rate is too high, then a partial or complete equalization of the populations of the energy levels of an ensemble of nuclei can occur and a state of saturation is reached.

Scalar coupling Coupling interaction between nuclei transmitted through chemical bonds (vicinal and geminal couplings).

Spin-echo Fourier transform (SEFT) ^{13}C-NMR experiments which are used for the assignment of resonances on the basis of the number of attached protons (multiplicity assignments of methyl, methylene, methine, and quaternary carbons). Gated broad-band proton decoupling is employed to modulate the phase and intensity of the ^{13}C signals.

Selective polarization transfer A polarization transfer experiment in which only one signal is enhanced selectively.

Shaped Pulses Long duration pulses, involving shaped Rf oscillation (maximum in the center), are used for selective excitation and nonselective inversion in many of NMR experiments.

Shimming A process for adjusting the homogeneity of the static magnetic field B_0. Various coils are used to adjust the homogeneity of a magnetic field by changing the current flowing through (called "electrical current shims"). The Shimming process is routinely carried out prior to any NMR experiment to eliminate the magnetic field inhomogeneities.

Signal averaging A procedure for improving the signal-to-noise ratio involving repeating an experiment n times so that the signal-to-noise ratio increases by $n^{1/2}$.

Signal-to-noise ratio Ratio of signal intensity to the root-mean-square noise level.

Sine-bell An apodization function employed for enhancing resolution in 2D spectra displayed in the absolute-value mode. It has the shape of the first half-cycle of a sine function.

Single-quantum coherence Coherence between states whose total quantum numbers differ by ±1. The only type of coherence that can be observed directly.

Solid-state MAS NMR Solid-state magic-angle spinning (MAS) is an NMR technique often used to record NMR experiments in a solid state. By spinning the sample at the magic angle (ca. 54.74°) with a frequency of 1–100 kHz, with respect to the direction of the magnetic field, the normally broad lines become narrower, increasing the resolution of the NMR spectrum. Because of the use of MAS, solid-state NMR spectroscopy has become a routinely used technique.

Spin-echo The refocusing of vectors in the xy-plane caused by a (τ-180°-τ) pulse sequence produces a spin-echo signal. It is used to remove field inhomogeneity effects or chemical shift precession effects.

Spin-lattice relaxation time Time constant for reestablishing equilibrium, involving return of the magnetization vector to its equilibrium position along the z-axis.

Spin-locking If a continuous B_1 field is applied along the y'-axis immediately after a 90°_x pulse, then all the magnetization vectors become spin-locked along the y'-axis, and negligible free precession occurs.

Stacked plot The 2D NMR spectrum is presented as a series of 1D spectra parallel to the F_1-axis stacked together at a number of F_2 values (or vice versa), one below the other.

STD-NMR The saturation transfer difference (STD) NMR is a simple and robust technique used for the study of interactions between a large molecule (receptor) and a small- to medium-sized molecule (ligand) in the solution state. It is based on the nuclear Overhauser effect (nOe), and the quantitative measurement of enhancement in signal intensity of the interacting specific protons of the ligand.

Symmetrization A procedure for removing artifact signals that appear nonsymmetrically disposed on either side of the diagonal.

t_1 **noise** Noise peaks parallel to the F_1-axis that arise due to instrumental factors, such as random variations of pulse angles, phases, B_0 inhomogeneity, or temperature. It is not removed by signal averaging.

T_2 **Measurement** This is transverse or spin-spin relaxation time T_2. It is measured by inversion recovery methods. It represents the de-coherence of transverse nuclear spin magnetization M_{xy}.

Tip angle (α) The angle by which the magnetization vector is rotated by the applied pulse.

Transient Time-domain signal (FID) acquired in an FT experiment.

Transmitter Coil of wire and accompanying electronics from which Rf energy is applied to the NMR sample.

Transverse magnetization Magnetization existing in the $x'y'$-plane.

Transverse operators Designated as I_x or I_y in the product operator description of NMR spectroscopy.

Zero-filling A procedure used to improve the digital resolution of the transformed spectrum (e.g., in the t_1 domain of a 2D spectrum) by adding zeros to the FID so that the size of the data set is adjusted to a power of 2.

Zero-quantum coherence The coherence between states with the same quantum number. It is not observable directly.

Subject Index

A

Absorption-mode spectrum, 304
Absorption spectroscopy, 128
ABX spin system, 473
Acetone methyl protons
 magnetization vector of, 173
Acquisition time, 46
Active coupling, 271
Aliphatic proton signals, 453
AMX spin system, 271
 schematic drawing of, 271
Analog-to-digital converter (ADC), 51, 57, 103
 resolution, sensitivity via increase in, 60
Angular Larmor frequency, 304
Angular velocity, 36, 43
Antiphase *z*-magnetization, 149
Apodization, 67
 functions, 205, 207
Artifact signals, 133, 302
Attached proton test (APT), 140–147
 disadvantage of, 145
 pulse sequence for, 142
 signal intensities in, 144
 spectrum in, 176
AX spin system, 84
 energy levels of, 84, 268

B

Back-transfer peaks, 389
Bench-top NMR spectrometers, 415–417
Biofluid analysis, 104
Bio-NMR spectroscopists, 23
Boltzmann constant, 426
Boltzmann distribution excess, 6, 7, 26, 37, 227
 population, 177
Boltzmann equation, 6
Boltzmann equilibrium
 populations at, 231
Boltzmann polarization, 124
Bovine phospholipase, phase-sensitive
 absorption-mode NOESY spectrum
 of, 305
Broad-band (BB) decoupled ^{13}C-NMR
 spectrum, 146, 441

Broad-band decoupling, 165, 286
Broad-band multinuclear probes, 11
Brownian motion, 426
Bruker instruments, 31, 200
Bruker spectrospin
 liquid nitrogen cooled cryogenic prodigy
 probe by, 420
Bulk magnetization vector, 36
(+)-Buxalongifolamidine, 484–493
 ^{13}C-NMR spectra, 490
 COSY-45° spectrum, 486, 488
 ^{1}H/^{13}C-NMR connectivities, 489
 long range, 492
 HMBC spectrum of, 491
 HMQC spectrum of, 490
 ^{1}H-NMR spectrum, 486, 487
 HOHAHA spectrum of, 493
 HREI–MS of, 484
 from leaves of *Buxus longifolia* Boiss, 484
 TOCSY spectrum, 491
Buxatenone, NOESY spectrum of, 343, 376
Buxus longifolia Boiss
 (+)-buxalongifolamidine from, 484

C

Capacitors, 15
Carbon–carbon connectivities, 382, 383
Carbon-detected 2D HETCOR experiment,
 314
Carbon–proton coupling interactions, 352
Carbon resonances, 373
Carbonyl carbon, 459
Carr–Purcell–Meiboom–Gill (CPMG) spin
 echo pulse sequence, 139, 141
(R)-(–)-Carvone, 462–471
 ^{13}C-NMR chemical shifts, 465
 ^{13}C-NMR spectrum of, 464, 465
 COSY-45° spectrum of, 463
 DEPT–HSQC spectrum of, 464
 ^{1}H- and ^{13}C-NMR chemical shift
 assignment, 470
 ^{1}H/^{13}C connectivities of, 467
 HMBC correlations, 469, 470
 ^{1}H-NMR spectrum of, 462

U

V

Printed in the United States
By Bookmasters